Forecasting Methods for Renewable Power Generation

Scrivener Publishing
100 Cummings Center, Suite 541J
Beverly, MA 01915-6106

Publishers at Scrivener
Martin Scrivener (martin@scrivenerpublishing.com)
Phillip Carmical (pcarmical@scrivenerpublishing.com)

Forecasting Methods for Renewable Power Generation

Edited by

Jai Govind Singh

Rupendra Kumar Pachauri

and

Sasidharan Sreedharan

Scrivener Publishing

WILEY

This edition first published 2025 by John Wiley & Sons, Inc., 111 River Street, Hoboken, NJ 07030, USA and Scrivener Publishing LLC, 100 Cummings Center, Suite 541J, Beverly, MA 01915, USA

For more information about Scrivener publications please visit www.scrivenerpublishing.com.

Wiley Global Headquarters
111 River Street, Hoboken, NJ 07030, USA

For details of our global editorial offices, customer services, and more information about Wiley products visit us at www.wiley.com.

Library of Congress Cataloging-in-Publication Data

ISBN 9781394249435

Front cover images supplied by Pixabay.com
Cover design by Russell Richardson

Set in size of 11pt and Minion Pro by Manila Typesetting Company, Makati, Philippines

Printed in the USA

10 9 8 7 6 5 4 3 2 1

Contents

Preface

As we stride toward a sustainable future, the integration of renewable power sources into our energy mix becomes increasingly pivotal. However, the intermittent nature of these resources poses unique challenges for power system operators, planners, and policymakers alike.

Forecasting Methods for Renewable Power Generation emerges as a comprehensive guide to navigating these challenges. In this edited book, we delve deep into the realm of forecasting techniques tailored specifically for renewable energy, demand, and electricity price. This book serves as a platform for experts from diverse backgrounds, including academia, industry, and research institutions, to share their insights, methodologies, and practical experiences in forecasting renewable power generation, demand patterns, and electricity prices. By integrating theoretical foundations with real-world applications, it offers the readers a holistic understanding of the complexities involved in predicting variables crucial for the efficient operation and planning of modern power systems.

The chapters in this volume cover a wide spectrum of topics, ranging from statistical methods to cutting-edge machine learning algorithms and hybrid models. Each chapter not only elucidates the underlying principles but also provides extensive guidance on implementing these methods using various software tools and platforms. Furthermore, this book explores the interdisciplinary nature of forecasting in the context of renewable energy, drawing connections between meteorology, economics, and engineering. By fostering collaboration and knowledge exchange across disciplines, it aims to accelerate the development and adoption of innovative forecasting solutions that can drive the transition toward a more sustainable and resilient energy landscape.

Whether you are a researcher, practitioner, policymaker, or student, ***Forecasting Methods for Renewable Power Generation*** offers valuable insights and practical strategies to address the forecasting methods in renewable energy integration. We hope that this book serves as a catalyst for further advancements in the field.

Editors
Dr. Jai Govind Singh
Dr. Rupendra Kumar Pachauri
Dr. Sasidharan Sreedharan

1

Solar Power Forecasting Using Hybrid Deep Learning Networks Combined with Variational Mode Decomposition

Krishna Prakash Natarajan and Jai Govind Singh*

Department of Energy, Environment and Climate Change, School of Environment Resources and Development, Asian Institute of Technology, Pathum Thani, Thailand

Abstract
Solar power forecasting is beneficial for better operation of electrical systems to satisfy the rising energy demand using renewable energy. However, the uncertainty of solar irradiance degrades the prediction accuracy and model generalization ability. Therefore, hybrid deep learning methods with decomposition techniques are proposed here to ease the impact caused by the nonlinearity and nonstationarity of solar radiation. Firstly, the historical time-series solar power data is decomposed into band-limited intrinsic mode functions using variational mode decomposition. Secondly, the band-limited signals and the multivariate meteorological and time features are independently predicted using the hybrid network; finally, the aggregation results in the solar power prediction. The outcomes are compared with other benchmarked models.

Keywords: Solar power forecasting, hybrid deep learning networks, variational mode decomposition, recurrent neural networks (RNN), long short-term memory (LSTM) networks

1.1 Introduction

Solar power forecasting (SPF) is a rapidly evolving field driven by the growing importance of solar energy in the transition to sustainable and

**Corresponding author*: jgsingh@ait.ac.th

Jai Govind Singh, Rupendra Kumar Pachauri and Sasidharan Sreedharan (eds.) Forecasting Methods for Renewable Power Generation, (1–20)

renewable power sources. It helps grid operators manage the integration of variable solar generation into the grid while ensuring grid stability and reliability. Precise solar power forecasts allow operators to proactively respond to output variations and balance supply and demand in real time, minimizing grid disruptions and ensuring smooth operation [1]. Furthermore, forecasting solar power generation is crucial for facilitating renewable energy integration, energy trading, providing stakeholders with information to facilitate informed decision-making, improving system reliability, and unlocking the full potential of solar energy as a clean and abundant resource [2, 3].

There are several types of SPF methods, each with its characteristics, advantages, and limitations. Based on the time horizon, forecasting methods are divided into three types: long-term, medium-term, and short-term power forecasting [4]. Short-term forecasting predicts solar power output over a horizon ranging from a few minutes to several hours ahead. It is essential for operational planning and real-time control of solar power plants, allowing grid operators to optimize energy scheduling and manage grid stability. Medium-term forecasting predicts solar power output over a horizon of several hours to several days ahead. It helps renewable energy developers, grid operators, and market participants make informed decisions about resource allocation, energy trading, and investment planning. Long-term forecasting predicts solar power output over days to weeks or even months ahead. It supports strategic decision-making, such as capacity planning, infrastructure development, and policy formulation, by providing insights into long-term trends and variability in solar power generation [3, 5]. Deterministic forecasting aims to estimate future solar power output based on historical time-series data and meteorological forecasts [6].

The forecasting models are divided into physical, statistical, and deep learning [7]. Physical models of solar irradiance forecasting leverage knowledge of atmospheric physics, radiative transfer processes, and solar geometry to simulate solar radiation under different weather conditions. They account for atmospheric scattering, cloud cover, and terrain effects, offering valuable insights into solar irradiance variability. However, physical models require detailed input parameters and may be computationally intensive [8].

Early research in SPF predominantly relied on statistical methods, such as time-series analysis, autoregressive integrated moving averages, and exponential smoothing techniques. While effective for short-term forecasting, statistical methods may struggle to capture nonlinear and complex relationships in solar irradiance data [9, 10]. Machine learning (ML) techniques, particularly artificial neural networks (ANNs), have gained popularity in solar power and irradiance forecasting owing to their ability to model complex, nonlinear relationships [11]. ANNs, including feedforward and recurrent neural networks (RNNs), have been extensively used to predict solar power based on meteorological data, such as temperature, humidity, and cloud cover. Random forests, K-nearest neighbor, support vector machines, and gradient boosting algorithms are also employed for solar power or irradiance forecasting [12–15]. ML approaches often outperform traditional statistical methods, especially in capturing nonlinear dependencies and improving forecast accuracy [16].

Hybrid forecasting approaches aim to overcome the limitations of individual methods and improve forecast accuracy across different time horizons. In recent years, with the emergence of deep learning, RNN [17], long short-term memory networks (LSTMs) [18], and convolutional neural networks [19–21], have become the essential units of hybrid models in renewable energy forecasting. Furthermore, integrated hybrid models may leverage the distinct characteristics of many models to predict future data patterns based on historical data. However, relying solely on hybrid models will not yield all the useful information in the dataset. As a result, integrated hybrid models frequently combine with data decomposition to strengthen their anti-interference capacity. Signal decomposition, therefore, becomes a critical component of data processing. Wavelet decomposition [22], empirical mode decomposition [23], and variational mode decomposition (VMD) [24] are commonly used decomposition methods. Combining these methods with deep learning models achieves better accuracy and performance.

1.2 Methodology

This section presents the models and theories used in the proposed forecasting framework.

1.2.1 Variational Mode Decomposition

VMD [25] decomposes a signal into a specified number of band-limited intrinsic mode functions (IMFs) or modes that represent the underlying oscillatory components of the signal. VMD decomposes a real-valued input signal f into discrete sub-signals or modes, u_k, with specific sparsity properties. Each mode k is considered compact around a central pulsation ω_k, which will be determined alongside the decomposition. The goal is to determine a specific set of IMFs u_k and associated center frequencies

ω_k that yield the minimum value for the constrained variational problem, defined by

$$\min_{\{u_k\},\{\omega_k\}} \left\{ \Sigma_k \| \partial_t [(\delta(t) + \frac{j}{\pi t}) * u_k(t)] e^{-j\omega_k t} \|_2^2 \right\} \quad (1.1)$$
$$\text{subject to } \Sigma_k u_k = f$$

where $\{u_k\}=\{u_1,\ldots, u_k\}$ and $\{\omega_k\}=\{\omega_1,\ldots, \omega_k\}$ are the set of all modes and their center frequencies, respectively, δ is Dirac function, $\|.\|_2$ represents the L2 distance, and * represents the convolution operation.

The constraint variation problem is transformed into a nonconstraint variation by using the second penalty component α and the Lagrangian multiplication operator $\lambda(t)$.

$$\mathcal{L}(\{u_k\},\{\omega_k\},\lambda) = \alpha \Sigma_k \| \partial_t [(\delta(t) + \frac{j}{\pi t}) * u_k(t)] e^{-j\omega_k t} \|_2^2 + \| f(t) - \Sigma_k u_k(t) \|_2^2 + \langle \lambda(t), f(t) - \Sigma_k u_k(t) \rangle \quad (1.2)$$

The "saddle point" of the expanded Lagrange formula is found by updating u_k^{n+1}, ω_k^{n+1}, and λ_k^{n+1} using the alternate direction method of multipliers (ADMM). In order to update the mode $u_k(t)$, the following sub-optimization problem is considered at the n-th iteration.

$$u_k^{n+1} \leftarrow \underset{u_k}{\arg\min} \, \mathcal{L}(\{u_{i<k}^{n+1}\},\{u_{i\geq k}^{n}\}\{\omega_i^n\},\lambda^n) \quad (1.3)$$

This is resolved in the spectral domain in VMD, leading to an update in the frequency domain.

$$\hat{u}_k^{n+1}(\omega) = \frac{\hat{f}(\omega) - \sum_{i\neq k} \hat{u}_i(\omega) + \frac{\hat{\lambda}(\omega)}{2}}{1 + 2\alpha(\omega - \omega_k)^2} \quad (1.4)$$

In the next step, the center frequency ω_k is updated by solving the following sub-optimization problem iteratively

$$\omega_k^{n+1} \leftarrow \underset{\omega_k}{\arg\min}\, \mathcal{L}(\{u_i^{n+1}\},\{\omega_{i<k}^{n+1}\}\{\omega_{i\geq k}^{n}\},\lambda^n) \tag{1.5}$$

to get an update for ω_k^{n+1} in the dual frequency domain.

$$\omega_k^{n+1} = \frac{\int_0^\infty \omega \mid \hat{u}_k(\omega) \mid^2 d\omega}{\int_0^\infty \mid \hat{u}_k(\omega) \mid^2 d\omega} \tag{1.6}$$

The new frequency is estimated as the centroid of the power spectrum of the associated modes. Overall, the VMD updates all of the sub-signals constantly via the frequency domain and subsequently converts them into the time domain using the inverse Fourier transform.

1.2.2 Long Short-Term Memory

The RNN effectively models time-series data. Unlike traditional feed-forward neural networks, the output is feedback to the input, acting as dynamic memory to process the input sequence. RNN comprises an input layer, an output layer, a recurrent layer, a series of weight matrices, and activation functions. The weights are shared between hidden units across each time step, and the network produces the same outcome by performing the same task on all layers as shown in Figure 1.1. For each time step t, the activation function and the predicted output $\hat{y}$ are expressed as follows:

$$h_1 = g(W_x X + b_h) \tag{1.7}$$

$$h_t = g(W_x X + W_h h_{t-1} + b_h) \tag{1.8}$$

$$\hat{y}_t = softmax(W_y h_t) = \frac{e^{W_y h_t}}{\Sigma_1^j e^{W_y h_{tj}}} \tag{1.9}$$

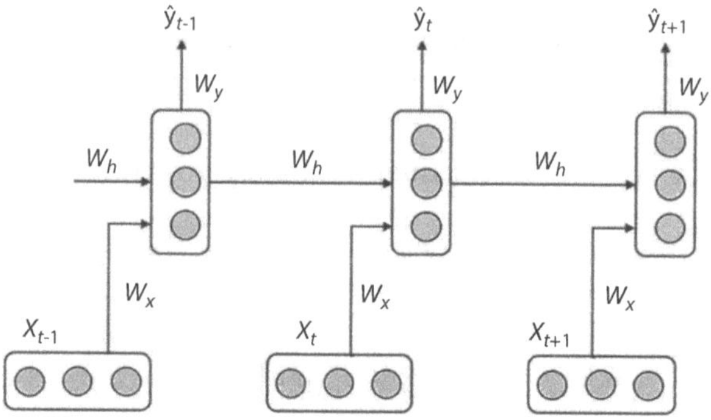

Figure 1.1 Traditional RNN.

where X is an input vector, h_t is a hidden state vector at time t, g is the logistic function, $\mathfrak{b}$ are bias vectors, and W_x, W_h, and W_y are weight matrices shared across time steps. The model sets the weight matrices during training to minimize an overall loss function. Errors must be propagated over many steps to compute gradients concerning inputs. The contribution of gradient values diminishes gradually as the computation propagates to previous time steps, thus decreasing the capability to learn long sequences. A powerful variant of the RNN, the LSTM, is used to model time-series forecasting to overcome the long-term dependency and vanishing/exploding gradient problems.

Employing LSTM units in RNN layers will allow errors to propagate throughout the entire network. The LSTM introduced in [26] is designed to store information for a long period. LSTM addresses gradient issues by introducing an additional cell state C in the hidden layer(s).

The structure of an LSTM network in Figure 1.2 consists of cells that form the memory blocks, and are recurrently connected comprising of gates, namely, forget gates, input gates, and output gates. For each gate in the network, the input weights W and bias $\mathfrak{b}$ are all initialized as a column matrix, respectively, as below.

$$W = \begin{bmatrix} W_f \\ W_i \\ W_C \\ W_O \end{bmatrix}, \beta = \begin{bmatrix} \beta_f \\ \beta_i \\ \beta_C \\ \beta_O \end{bmatrix} \tag{1.10}$$

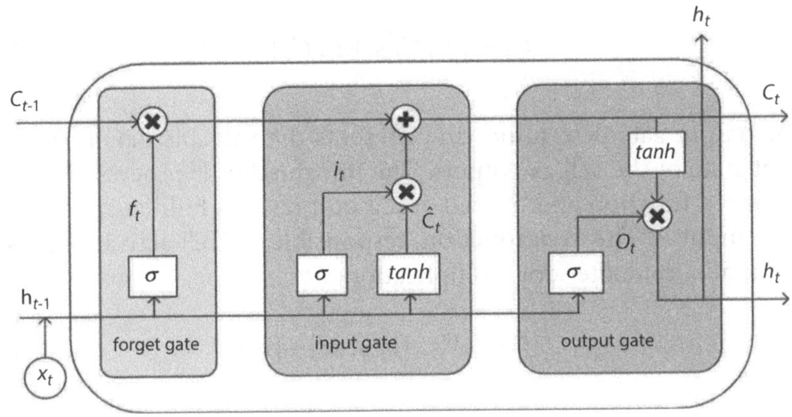

Figure 1.2 Structure of an LSTM cell.

The forget gate eliminates unnecessary or trivial information by multiplying the hidden layer's previous state h_{t-1} and the current input x_t with the weight matrices and by adding a bias. It uses a sigmoid function, producing values between 0 and 1 for each cell status. An output of 0 for a particular cell state means that information is forgotten and outputting 1 means that information is retained for future use.

$$f_t = \sigma(W_f \,.\, [h_{t-1}, x_t] + \beta_f) \tag{1.11}$$

An input gate adds useful information to a cell state controlled by a sigmoid function acting as a filter alike the filter gate. The cell state is updated using the *tanh* function. The cell state can be updated by adding a *tanh*-activated vector containing all possible values of previous hidden state and current input. The *tanh*-generated vectors are multiplied with regulated values. The following represent the cell state equations:

$$i_t = \sigma(W_i \,.\, [h_{t-1}, x_t] + \beta_i) \tag{1.12}$$

$$\hat{C}_t = tanh(W_C \,.\, [h_{t-1}, x_t] + \beta_C) \tag{1.13}$$

$$C_t = (f_t . C_{t-1} + i_t . \hat{C}_t) \tag{1.14}$$

The output gate determines and presents the valuable details from the current state of the cell as outputs. The information is processed through the sigmoid function and filtered based on previous hidden state h_{t-1} and current input x_t. The *tanh* function, responsible for cell activation, results in the final output after dot multiplication.

$$O_t = \sigma(W_O . [h_{t-1}, x_t] + \beta_O) \tag{1.15}$$

$$h_t = O_t . tanh(C_t) \tag{1.16}$$

Given an input sequence at time t, $X_t = [x_1, x_2, \ldots x_t]$, the final prediction of an LSTM layer is a vector of all the outputs, represented by $\hat{y}_t = [h_1, h_2, \ldots h_t]$.

1.2.3 Gated Recurrent Units

GRUs combine the input and forget ports from LSTMs into one update gate. The update port retains previous state information it receives. On the other hand, the reset port decides whether the current state should be combined with the information earlier. The functions are outlined below:

$$z_t = \sigma(W_z . [h_{t-1}, x_t] + \beta_z) \tag{1.17}$$

$$s_t = \sigma(W_r . [h_{t-1}, x_t] + \beta_r) \tag{1.18}$$

$$\tilde{h}_t = \tanh(s_t W . [h_{t-1}, x_t] + \beta) \tag{1.19}$$

$$h_t = (1 - z_t)\tilde{h}_t + z_t h_{t-1} \tag{1.20}$$

where W_z, W_r, and W denote the weight matrices; σ denotes the logistic sigmoid function; β_z, β_r, and β are the bias; z_t, s_t, h_t denotes the update gate, reset gate, and hidden layer, respectively.

GRU demonstrates the potential to capture long-term dependencies from massive sequential data without excluding information from the earlier part of the data sequences. GRU is used to capture the temporal dependencies in the data.

1.3 Proposed Methodology for Solar Power Forecasting

Hybrid deep learning models enhance time-series forecasting. Deep learning approaches are employed for enhancing forecasting precision and resilience. Utilizing LSTM and GRU models, the proposed hybrid system aims to capture diverse solar power patterns and correlations within time-series data. The LSTM and GRU models are well adapted to the vanishing gradient problem and the long-term dependencies. Therefore, hybrid techniques synergistically use the merits of different architectures to enhance prediction accuracy. To do this, the proposed approach combines the predictions of these models.

A hybrid deep learning approach using LSTM-GRU models for day-ahead SPF is implemented in this study. The original time-series solar power data is pre-processed to make the data suitable for VMD decomposition. Neural networks require data to be normalized within the range [0 1]. In this framework, the solar power prediction is divided into two stages. The decomposition of the dataset into n subseries called IMFs using VMD is the first stage. The n number of IMFs is expressed as IMF_1, IMF_2,…IMF_{n-1}, IMF_n. The IMFs are separated into relatively high-frequency (h) and low-frequency (k) subseries. The lower-order IMFs are high-frequency useful information, and the higher-order IMFs are the low frequency subseries. An h number of independent LSTM models is built and trained for the high-frequency subseries and a k-independent GRU model for the low-frequency subseries. LSTM modules produce h predictions and GRU modules produce k prediction values. These h+k prediction results from each component are aggregated to obtain the final prediction results. The framework of the proposed prediction model is shown in Figure 1.3.

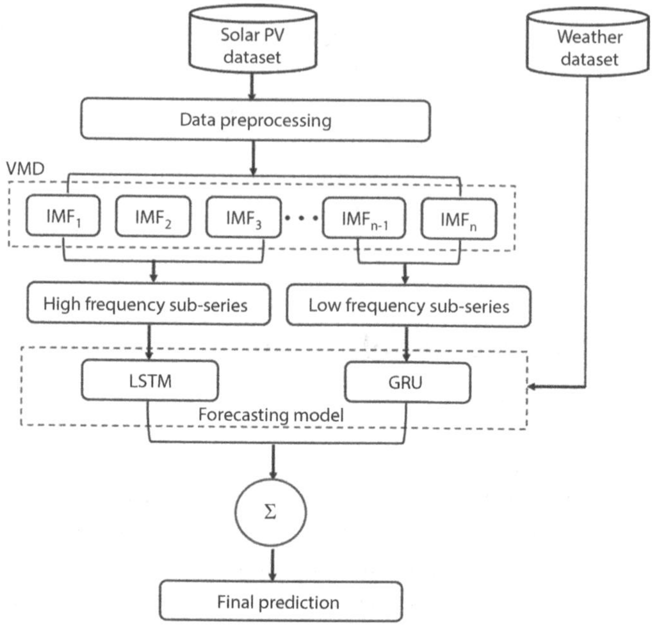

Figure 1.3 The framework of the proposed prediction method.

1.4 Experimental Results and Discussion

1.4.1 Solar PV Dataset

This study is based on historical solar power generation data collected from a 4-kW plant from January 2019 to December 2022, with a resolution of 5 min. The features of the dataset include the status of the day (day or night), day temperature (°C), dc power from panels to inverter (W), solar radiation (W/m^2), and panel temperature (°C). Figure 1.4 shows the variability of the power generation in typical days of April 2022. There is volatility in solar power generation in these three successive days. The hourly distribution of PV power production reaches its maximum at midday and falls to zero at night. One-hour data is derived from the real-time 5-min interval dataset as the average value for corresponding time intervals.

Solar power generation depends on many factors, mainly solar irradiation. Figure 1.5 (a and b). shows the mean solar power generated by hours for summer and winter, which indicates that the generation is low for winter. Figure 1.6 shows the mean solar radiation by month. Solar radiation is particularly low in January, February, and December, resulting in reduced power generation during these winter months.

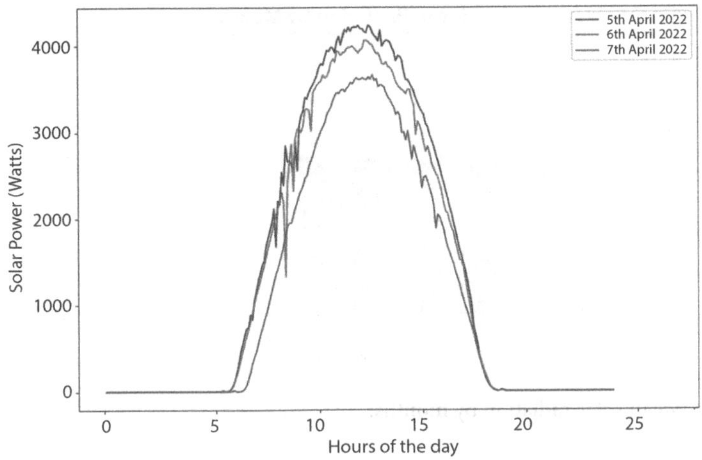

Figure 1.4 Solar power generation on typical days.

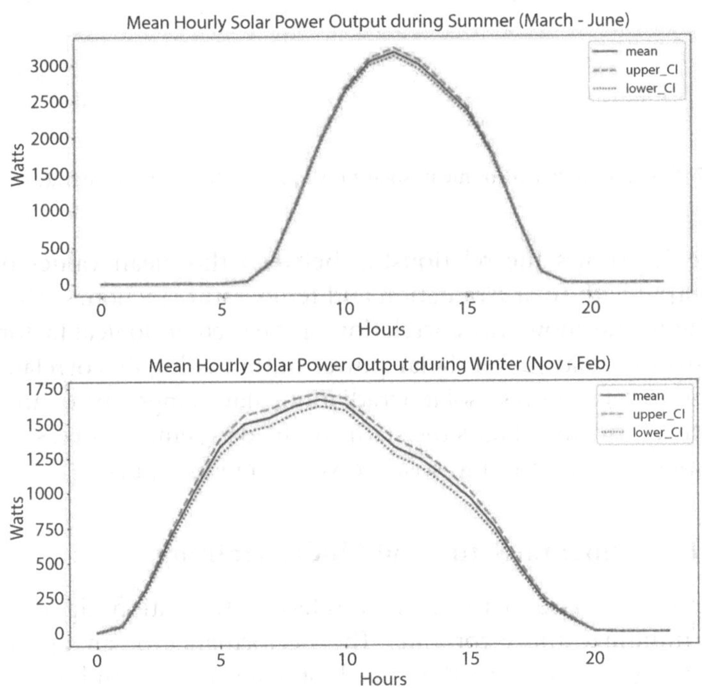

Figure 1.5 Mean solar power generated by hour for (a) summer and (b) winter.

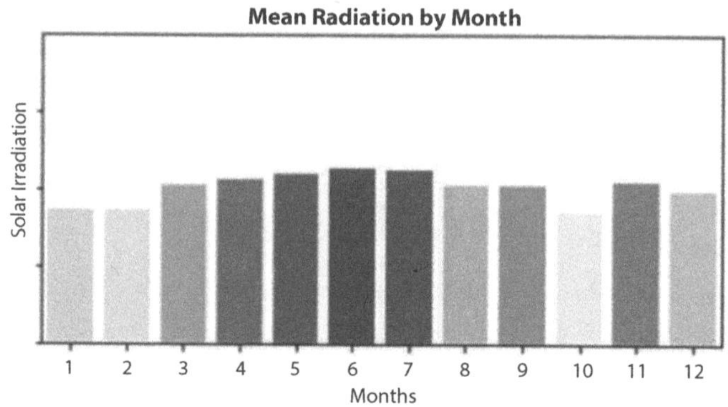

Figure 1.6 Mean solar radiation by months.

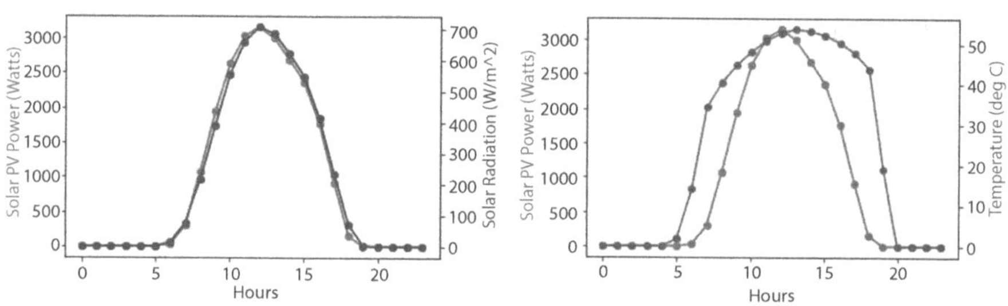

Figure 1.7 Correlation between mean solar power, irradiation, and temperature.

Figure 1.7 shows the relationship between the mean values of solar power output with solar irradiation and temperature by hours. The scatter plot in Figure 1.8 shows the correlation of the meteorological factors plotted against the solar power. Solar power output is highly correlated with three important features: solar irradiation, day temperature, and panel temperature. The model uses the status of the day, temperature, solar radiation, panel temperature, and solar power output as inputs.

1.4.2 Experimental Setup and Model Training

In the training phase of recurrent models, optimization algorithms are used to minimize the error rate. The performance of an optimizer is generally characterized by the speed of convergence and the model's capabilities. Adaptive moment estimation (Adam) is the gradient-based optimizer for model instantiation and training. The activation function used here for determining the final response of the network is sigmoid,

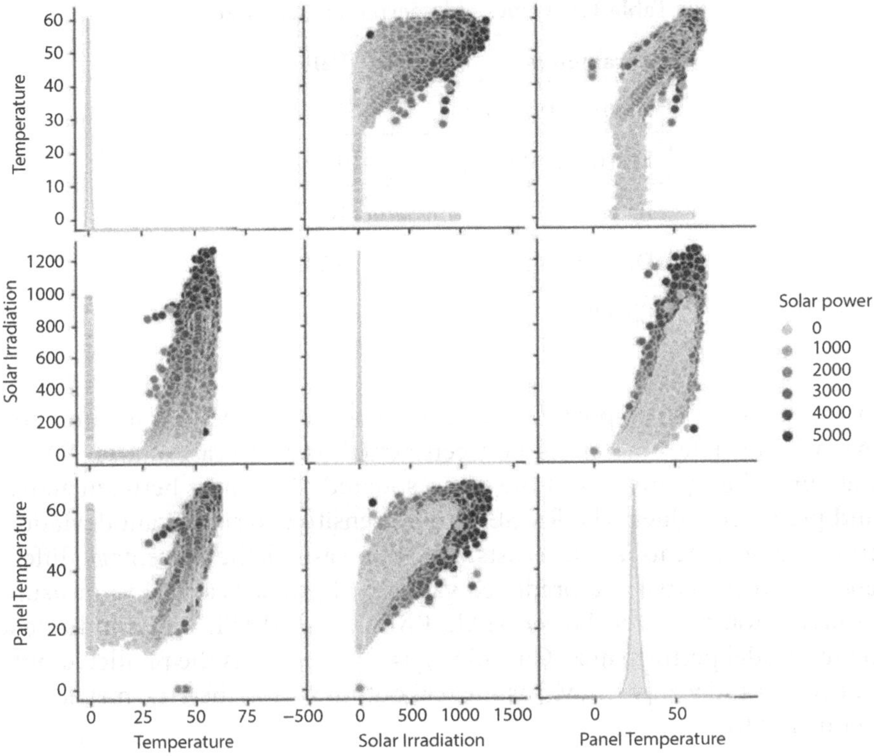

Figure 1.8 Correlation of the features plotted against the solar power.

which rescales the data to values in the interval [0, 1]. In order to avoid overfitting and enhance the prediction ability, a stochastic regularization process called dropout is applied to the models. The value of the dropout rate used here is 0.2. The MSE serves as the loss function for the network. The models are produced in Python using the open-source Keras package with TensorFlow's backend. Table 1.1 shows the model parameters, which include the number of neurons, Adam's learning rate, batch size, and number of learning epochs, determined by grid search hyperparameter tuning. The original dataset is divided into training, validation, and test sets of 80%, 10%, and 10%, respectively. Each batch in the training process validates the model. After the model has been trained, its performance has been evaluated using the test data.

Evaluation metrics such as mean absolute error (MAE), root mean square error (RMSE), coefficient of determination (R^2), and mean absolute percentage error (MAPE) are used to measure the accuracy and

Table 1.1 Values of hyperparameters used.

Parameters	**Value**
Neurons per layer	48
Learning rate	0.001
Batch size	16
Dropout rate	0.2
Epochs	100

effectiveness of solar power prediction models. The MAE measures the average absolute differences between actual and forecast values. RMSE measures the square root of the mean squared differences between actual and predicted values. The RMSE is more sensitive to significant deviation from measurements and forecasts. MAPE measures the percentage difference between actual and predicted values and provides a relative measure of prediction accuracy. Lower MAE, RMSE, and MAPE values indicate a better model performance. Consider $\hat{y}_i = [\hat{y}_1,\ldots,\hat{y}_N]$ as the predicted output vector and $y_i = [y_1,\ldots,y_N]$ as the real output vector, then the metrics are defined as follows:

$$MAE = \frac{1}{N}\Sigma_{i=1}^{N}|y_i - \hat{y}_i| \tag{1.21}$$

$$RMSE = \sqrt{\frac{1}{N}\Sigma_{i=1}^{N}(\hat{y}_i - y_i)^2} \tag{1.22}$$

$$MAPE = \frac{1}{N}\sum_{i=1}^{N}\frac{|y_i - \hat{y}_i|}{y_i}x100 \tag{1.23}$$

R^2 is a measure that provides information about the best fit of a model. The numerator and denominator denote the sum of the residuals squared and the sum of the distance from the mean of all squared data, respectively.

It is expressed as a percentage and ranges from 0 to 1, with higher values signifying a stronger positive linear relationship.

$$R^2 = 1 - \frac{\Sigma(y_i - \hat{y}_i)^2}{\Sigma(y_i - \overline{y})^2} \tag{1.24}$$

1.4.3 Experimental Results

The performance of the proposed model is evaluated using RNN, LSTM, and GRU deep learning models and, finally, the proposed hybrid LSTM-GRU model. The prediction is compared with and without the VMD method of decomposition. The exogenous variables include the day's status, temperature, solar radiation, and panel temperature. These variables and historical solar power data are used as inputs for prediction. The inputs are then normalized in the range of [0, 1] using the min-max scaling technique. The VMD is carried out on the scaled solar power data and is decomposed into different IMFs. The number of decompositions is 9, and the IMFs are shown in Figure 1.9. It is clear that IMF6 to IMF9 captures low-frequency signals, whereas the useful information of high frequency is with IMF1 to IMF5.

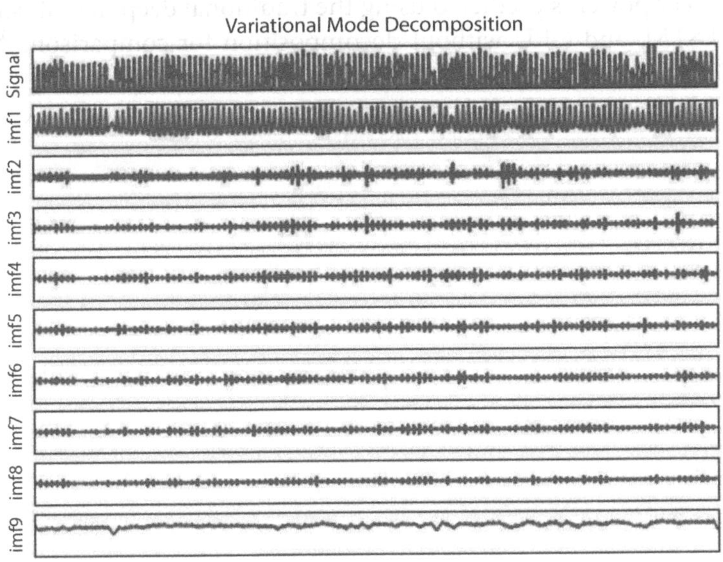

Figure 1.9 Solar power dataset decomposed with VMD.

After the decomposition, the low-frequency and high-frequency sub-series are combined with the exogenous variables to form a multivariate dataset. The feature vector for the subsequent time step is formed using the sliding window technique and is delivered as input to the forecasting model. The results of the proposed methodology are compared to deep recurrent networks RNN, LSTM, and GRU for day-ahead SPF using the metrics defined in Section 1.4.2.

The solar power prediction for a typical day using different models is shown in Figure 1.10. The performance metrics of the proposed hybrid model with VMD decomposition, as shown in Table 1.2, have the minimum values for MAE, RMSE, and MAPE and maximum value for R^2. The value of MAE is 0.292 W, RMSE is 1.054 W, MAPE is 7.07%, and R^2 is 99.7%. This proves the effectiveness of combining LSTM and GRU, which helps improve prediction accuracy. The scatter plot of predicted values fitted using the proposed model versus actual values is shown in Figure 1.11, which is almost a straight line illustrating the level of performance accuracy. The deviation of the predicted values from the true values is almost minimal.

The LSTM and GRU models achieve similar R^2 of 99.6%, and the MAPE values are 8.3% and 7.89%, respectively. The RNN model is ranked last with 9.93% MAPE, MAE of 0.735 W, RMSE of 1.211 W, and 98.9% R^2. Figure 1.12 illustrates that the proposed model can predict PV power generation even under a cloudy day where solar power is highly fluctuating.

The solar power is predicted using the traditional deep neural networks RNN, LSTM, and GRU without decomposition for comparison. Among

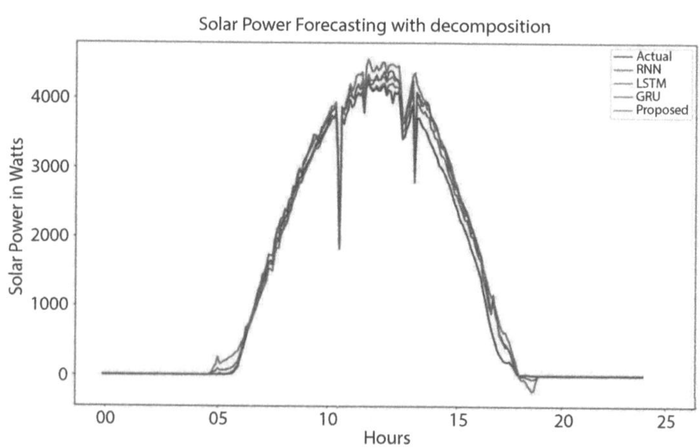

Figure 1.10 Solar power prediction for a typical day.

these models, GRU performs better with smaller prediction errors. The proposed approach improves MAPE values by 16% to 25% compared to other single reference models without VMD. This demonstrates that the VMD decomposition improves the forecasting accuracy of the deep-learning networks.

To summarize, among the models RNN, LSTM, and GRU, both LSTM and GRU show slightly similar values of MAE and RMSE. This shows

Table 1.2 Performance metrics of the proposed hybrid model with VMD decomposition.

	Model	MAE (W)	RMSE (W)	MAPE (%)	R^2
Without VMD	RNN	0.939	1.632	11.52	0.982
	LSTM	0.824	1.487	10.23	0.987
	GRU	0.728	1.203	9.88	0.989
With VMD	RNN	0.735	1.211	9.93	0.989
	LSTM	0.609	1.164	8.30	0.996
	GRU	0.576	1.136	7.89	0.996
	Proposed	0.292	1.054	7.07	0.997

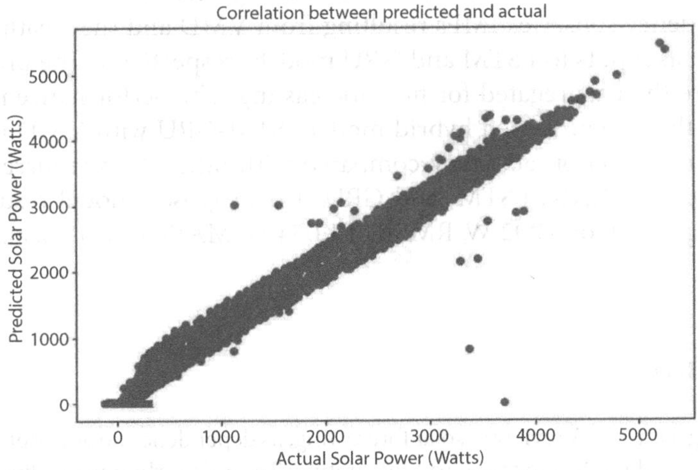

Figure 1.11 Scatter plot of solar power prediction versus actual values.

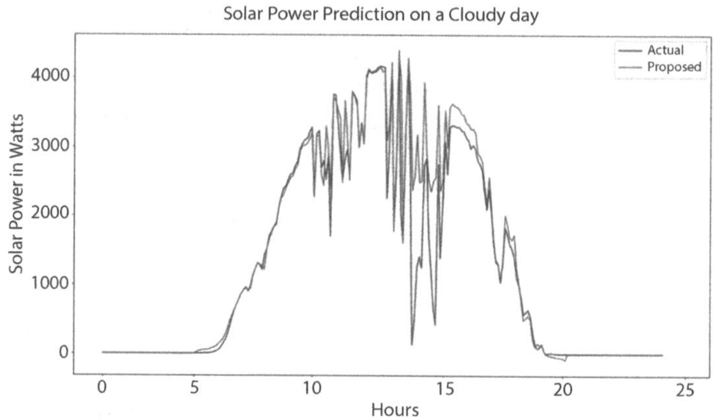

Figure 1.12 Solar power prediction for a typical cloudy day.

they are better at extracting the temporal dependencies and even patterns inherent in solar power data, making forecasting models more effective. The advantages of these two models are combined in the proposed forecasting model.

1.5 Conclusion

This study presents a hybrid deep learning framework combined with VMD for a day-ahead SPF. LSTM and GRU models are used in the framework to predict the solar power output. The high-frequency subseries and low-frequency subseries IMFs resulting from VMD and the weather data are given as inputs to LSTM and GRU models, respectively. The predicted values are then aggregated for final forecasting. The performance metrics indicate that the proposed hybrid model LSTM-GRU with VMD is satisfactory in prediction accuracy compared with other deep recurrent networks, such as RNN, LSTM, and GRU. The proposed model resulted in an average MAE of 0.292 W, RMSE of 1.054 W, MAPE of 7.07%, and R^2 of 99.7%.

References

1. Yang, D., *et al.*, A review of solar forecasting, its dependence on atmospheric sciences and implications for grid integration: Towards carbon neutrality. *Renew. Sustain. Energy Rev.*, 161, 112348, 2022, doi: 10.1016/j.rser.2022.112348.

2. Ye, H., Yang, B., Han Y and Chen, N., State-of-The-Art Solar Energy Forecasting Approaches: Critical Potentials and Challenges. *Front. Energy Res.*, 10, 875790, 2022, doi: 10.3389/fenrg.2022.875790.
3. Bazionis, I.K., *et al.*, A taxonomy of short-term solar power forecasting: Classifications focused on climatic conditions and input data. *IET Renew. Power Gener.*, 17, 2411–2432, 2023, doi: 10.1049/rpg2.12736.
4. Ma, J. and Ma, X., A review of forecasting algorithms and energy management strategies for microgrids. *Syst. Sci. Control Eng.*, 6, 1, 237–248, 2018, doi: 10.1080/21642583.2018.1480979.
5. Ahmed, R., Sreeram, V., Mishra, Y., Arif, M.D., A review and evaluation of the state-of-the-art in PV solar power forecasting: Techniques and optimization. *Renew. Sustain. Energy Rev.*, 124, 109792, 2020, doi: 10.1016/j.rser.2020.109792.
6. Iheanetu, K.J., Solar Photovoltaic Power Forecasting: A Review. *Sustainability*, 14, 17005, 2022, doi: 10.3390/su142417005.
7. Rahimi, N., *et al.*, A Comprehensive Review on Ensemble Solar Power Forecasting Algorithms. *J. Electr. Eng. Technol.*, 18, 2, 719–733, 2023, doi: 10.1007/s42835-023-01378-2.
8. Sobri, S., Koohi-Kamali, S., Rahim, N.A., Solar photovoltaic generation forecasting methods: A review. *Energy Convers. Manage.*, 156, 459–497, 2018, doi: 10.1016/j.enconman.2017.11.019.
9. Dou, Y., Tan, S., Xie, D., Comparison of machine learning and statistical methods in the field of renewable energy power generation forecasting: a mini review. *Front. Energy Res.*, 11, 1218603, 2023, doi: 10.3389/fenrg.2023.1218603.
10. Chodakowska, E., Nazarko, J., Nazarko, Ł., Rabayah, H.S., Abendeh, R.M., Alawneh, R., ARIMA Models in Solar Radiation Forecasting in Different Geographic Locations. *Energies*, 16, 13, 5029, 2023, doi: 10.3390/en16135029.
11. Mas'ud, A.A., Comparison of three machine learning models for the prediction of hourly PV output power in Saudi Arabia. *Ain Shams Eng. J.*, 13, 4, 101648, 2022, doi: 10.1016/j.asej.2021.11.017.
12. Gaboitaolelwe, J., *et al.*, Machine Learning Based Solar Photovoltaic Power Forecasting: A Review and Comparison. *IEEE Access*, 11, 40820–40845, 2023, doi: 10.1109/ACCESS.2023.3270041.
13. Ledmaoui, Y., *et al.*, Forecasting solar energy production: A comparative study of machine learning algorithms. *Energy Rep.*, 10, 1004–1012, 2023, doi: 10.1016/j.egyr.2023.07.042.
14. Kallio, S. and Siroux, M., Photovoltaic power prediction for solar microgrid optimal control. *Energy Rep.*, 9, 1, 594–601, 2023, doi: 10.1016/j.egyr.2022.11.081.
15. Zazoum, B., Solar photovoltaic power prediction using different machine learning methods. *Energy Rep.*, 8, 1, 19–25, 2022, doi: 10.1016/j.egyr.2021.11.183.

16. AlShafeey, M. and Csáki, C., Evaluating neural network and linear regression photovoltaic power forecasting models based on different input methods. *Energy Rep.*, 7, 7601–7614, 2021, doi: 10.1016/j.egyr.2021.10.125.
17. Narayanan, S., Kumar, R., Ramadass, S., Ramasamy, J., Hybrid Forecasting Model Integrating RNN-LSTM for Renewable Energy Production. *Electr. Power Compon. Syst.*, 52, 1–19, 2024, doi: 10.1080/15325008.2024.2316247.
18. du Plessis, A.A., Strauss, J.M., Rix, A.J., Short-term solar power forecasting: Investigating the ability of deep learning models to capture low-level utility-scale Photovoltaic system behaviour. *Appl. Energy*, 285, 116395, 2021, doi: 10.1016/j.apenergy.2020.116395.
19. Feng, C. and Zhang, J., SolarNet: A Deep Convolutional Neural Network for Solar Forecasting via Sky Images. *2020 IEEE Power & Energy Society Innovative Smart Grid Technologies Conference (ISGT)*, pp. 1–5, Washington, DC, USA, 2020, doi: 10.1109/ISGT45199.2020.9087703.
20. Fungtammasan, G. and Koprinska, I., Convolutional and LSTM Neural Networks for Solar Power Forecasting. *2023 International Joint Conference on Neural Networks (IJCNN)*, pp. 1–7, Gold Coast, Australia, 2023, doi: 10.1109/IJCNN54540.2023.10191813.
21. Hou, X., Ju, C., Wang, B., Prediction of solar irradiance using convolutional neural network and attention mechanism-based long short-term memory network based on similar day analysis and an attention mechanism. *Heliyon*, 9, 11, e21484, 2023, doi: 10.1016/j.heliyon.2023.e21484.
22. Ospina, J., Newaz, A., Faruque, M.O., Forecasting of PV plant output using hybrid wavelet-based LSTM-DNN structure model. *IET Renewable Power Gener.*, 13, 1087–1095, 2019, doi: 10.1049/iet-rpg.2018.5779.
23. Aghmadi, A. and Mohammed, O., Enhancing Solar Radiation Forecasting with Hybrid Ensemble Empirical Mode Decomposition and Machine Learning Techniques. *2023 IEEE Intl. Conf. on Environment and Electrical Engg. and 2023 IEEE Industrial and Commercial Power Systems Europe (EEEIC / I&CPS Europe)*, pp. 1–6, Madrid, Spain, 2023, doi: 10.1109/EEEIC/ICPSEurope57605.2023.10194879.
24. Trong, T.N., *et al.*, Short-term PV power forecast using hybrid deep learning model and Variational Mode Decomposition. *Energy Rep.*, 9, 10, 712–717, 2023, doi: 10.1016/j.egyr.2023.05.154.
25. Dragomiretskiy, K. and Zosso, D., Variational mode decomposition. *IEEE Trans. Signal Process.*, 62, 3, 531–544, Feb. 2014.
26. Hochreiter, S. and Schmidhuber, J., Long short-term memory. *Neural Comput.*, 9, 8, 1735–1780, 1997.

2

Location Analysis and Environmental Validation for Installation of Hybrid Solar-Wind Energy Generation System in Hilly Areas of Uttarakhand: Study Toward Forecasting

Paramjeet Singh Paliyal[1], Shyam Kumar Menon[1], Surajit Mondal[1]* and Vikas Thapa[2]

[1]*Electrical Cluster, School of Advanced Engineering, UPES, Dehradun, India*
[2]*School of Design, UPES, Dehradun, India*

Abstract

Although the sunshine and wind resources in Uttarakhand are abundant, they are not fully used. The power demand is currently rising quickly, and non-renewable energy sources cannot satisfy it on their own. Nature provides renewable energy sources like solar and wind power. Although it is a growing option to meet energy demand, renewable energy is unpredictable. Systems that combine two or more renewable energy sources, such as solar panels and wind turbine systems, are known as hybrid solar and wind renewable energy systems (HRESs). The battery output serves as the voltage for the load after the outputs of the solar and wind power conversion systems are combined, charging the same battery. This study's objective is to present the application of a combination arrangement and a thorough examination of numerous HRES-related issues. With less developed capacity, renewable energy in Uttarakhand has enormous potential. As a result, there is a significant difference between potential and installed power. Renewable energy from solar, wind, and hydropower installations is effective in Uttarakhand. The most rapidly expanding area of technology is in hybrid systems, which has increased energy-generating productivity. There are various locations in Uttarakhand where weather statistics can be researched to use hybrid energy systems, even though the resources to draw energy are not yet being employed

**Corresponding author*: surajitmondalee@gmail.com

Jai Govind Singh, Rupendra Kumar Pachauri and Sasidharan Sreedharan (eds.) Forecasting Methods for Renewable Power Generation, (21–48) © 2025 Scrivener Publishing LLC

there. The Uttarakhand weather should be taken into account while using this engineering in Uttarakhand's mountainous areas for small-scale loads like lighting systems in a particular village.

***Keywords*:** Hybrid system, PV system, wind mills, Uttarakhand, energy generation

2.1 Introduction

Both conventional and unconventional energy sources are employed to produce electricity. The main drawbacks are the connection to global warming and climate change, as they would vanish from the planet if a traditional source was still used. As a result, the hunt for many other kinds of energy ends with both unconventional types of energy. Due to their complementing nature, some renewable energy sources, like solar and wind energy, are frequently utilized to produce electrical power [3].

Two systems—solar and wind hybrid energy generation system—make up the energy systems that generate the electrical fuel. According to experts in renewable energy, compact hybrid systems that integrate wind and solar energy have some advantages for use in domestic applications. Energy will be the most crucial resource because, without electricity, daily existence will be impossible in the future. The concept behind a hybrid power generation is to obtain continuous electricity during the day and at night for both small-scale power usage and the battery system [4].

Solar wind hybrid energy system (SWHES) are used to generate the entire amount of power from the wind and solar dual generating units. Depending on the amount of required load, these systems enter service mode. The remaining time is spent charging this battery feeding system. We can operate a wind and solar power project simultaneously in this single unit to obtain continuous energy supply, increasing the overall system efficiency through this integrated mode of operation. With batteries serving as the storage option for household applications, the combined power generation would offer a constant supply of power.

The state of Uttarakhand has achieved the biggest advances over the past few years, yet the state is still developing in the country. Because of the state's steep terrain, the improvement graph is much clearer than in other states. Given that it requires the least amount of care, it appears that this project is being carried out to promote growth in the hilly region. There are thermal energy stations and hydroelectric power projects spread throughout the state of Uttarakhand that can handle the necessary demand [5].

2.1.1 Brief Introduction to the State of Uttarakhand

Following the dissolution of the state of Uttar Pradesh in 2000, Uttarakhand was created. Due to Uttar Pradesh's division, Uttarakhand is now included in the region's development plan. However, Uttarakhand was given the status of an autonomous region in response to the expanding political and economic activities, taking into account the territory's expansion and infrastructure requirements, as well as those of the capital Dehradun, the GIS map of Uttarakhand state is given in Figure 2.1. This has increased the state's energy needs as well but does not address the current power situation. According to the Central Electricity Authority, Uttarakhand was expected to use 8,363 MU of usable energy and 10,480 MU of electricity between 2019 and 2020. However, the state's overall demand during this period was 1,600 MW, while the available capacity was only 1,430 MW, leading to a significant energy shortfall. Although Uttarakhand has undoubtedly improved over the past few years, improvement has not been universal. Because the majority of the state's production units are located in its plain regions, Uttarakhand's growth cycle is limited to those areas, with the exception of its hilly areas. In some states, 40% of the people still struggle to make ends meet. Even if the unemployment rate in the state is too low and there are a lot of poor people, Uttarakhand is an odd instance of poverty, this refers to an economy's low rate of wages and profits, which results in a particular issue known as the primary problem of subpar working conditions. In the state of Uttarakhand, there is currently a need to provide the rural inhabitants with sustainable work. A rural community can do this by having efficient and secure access to power. Furthermore,

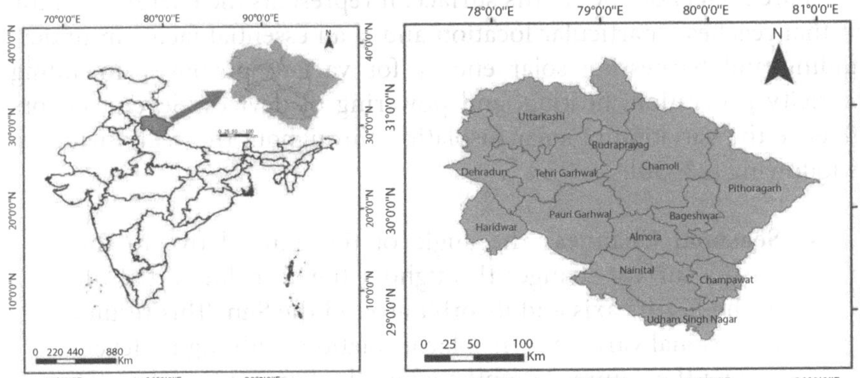

Figure 2.1 GIS Map of the Uttarakhand state.

construction methods are heavily dependent on the traditional energy sources that function in a plains setting, probably because of the challenging terrain that they require. These terrain types did not prove to be highly productive in the hills [6].

2.1.2 Solar and Wind Energy Availability in Uttarakhand State

A total integrated aggregate of 576.38 MW of renewable power (grid linked and decentralized power stations) is used in the state in addition to renewable sources. Small hydro and bioenergy have been employed to a maximum of 45.46% (262 MW), 34.31% (197.78 MW), and 20.23% (116.60 MW) total. The Tehri region of Uttarakhand has been prepared to be one of the first Himalayan locales, and the state government has also been given permission to construct a wind power facility on top of a hill in Bachelikhal. For the 2.4 MW of wind energy output, this impact has previously been announced in a circular. The Chennai Wind Energy Technology Centre laid the groundwork and conducted all other surveys on the effectiveness of wind energy in the Bachelikhal area (CWET). In order to understand the potential of wind power in the state's steep hills, Uttarakhand Renewable Energy Development Agency (UREDA) plans to install wind mapping equipment there. Through CWET, UREDA has begun to negotiate the mounting of particular equipment for such a purpose [7].

2.1.2.1 The Solar Statistics of the Uttarakhand State

Figures 2.2 and 2.3 show that solar insolation, also known as solar radiation, refers to the amount of solar energy received per unit area (typically per square meter) on the Earth's surface. It represents the energy from the Sun that reaches a particular location and is an essential factor in understanding and harnessing solar energy for various purposes, including electricity generation, heating, and powering of devices. Several factors influence the variation of solar insolation throughout the year, including the following [25–30]:

- **Seasonal Changes**: The angle of the Sun relative to the Earth's surface changes throughout the year due to the tilt of the Earth's axis and its orbit around the Sun. This results in seasonal variations in solar insolation, with higher levels during the summer months when the Sun is more directly overhead and lower levels during the winter months when the Sun is lower in the sky.

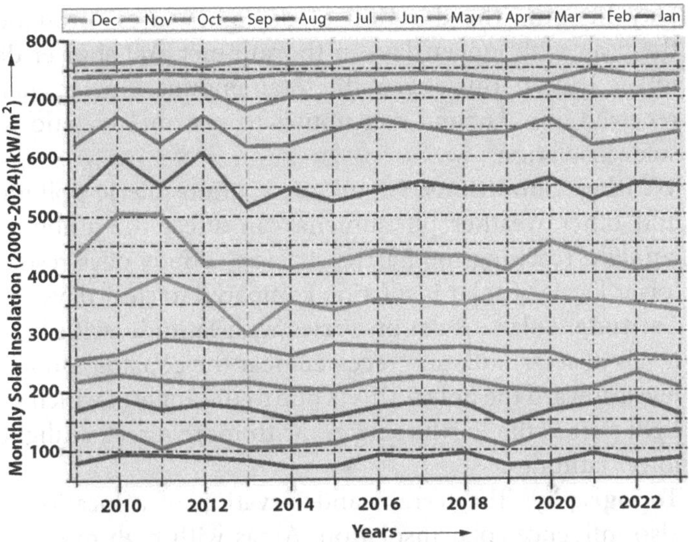

Figure 2.2 Monthly solar insolation from 2009 to 2023 in kilowatt hours per meter square.

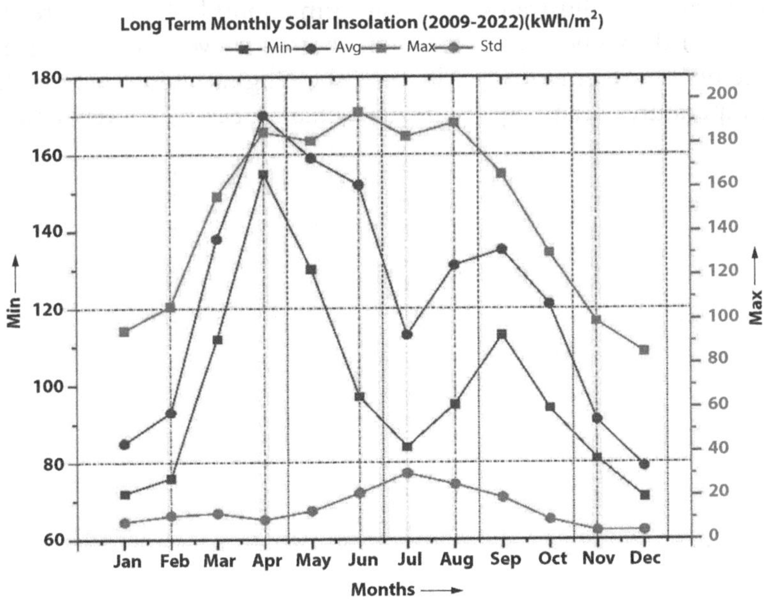

Figure 2.3 Long-term monthly solar insolation from year with minimum, maximum, average, and standard value.

- **Day Length**: The duration of daylight varies throughout the year, with longer days in the summer and shorter days in the winter. This affects the total amount of solar energy received in a day and contributes to seasonal variations in solar insolation.
- **Weather Conditions**: Cloud cover, atmospheric pollution, and other weather phenomena can affect the amount of sunlight reaching the Earth's surface. Cloudy days result in lower levels of solar insolation compared to clear days.
- **Latitude**: Solar insolation varies with latitude, with higher levels of solar radiation received near the equator and lower levels toward the poles. This is due to the angle at which sunlight strikes the Earth's surface, with more direct sunlight at lower latitudes.
- **Topography**: The terrain and elevation of a location can also influence solar insolation. Areas with high mountains or tall buildings may experience shading effects that reduce the amount of sunlight reaching the surface, whereas flat, unobstructed areas receive more direct sunlight.

The Figure 2.4 illustrates the long-term monthly temperature variation (in °C) throughout the year, with the red curve representing maximum temperatures and the black curve showing minimum temperatures. The temperatures gradually rise from January, peaking in May and June,

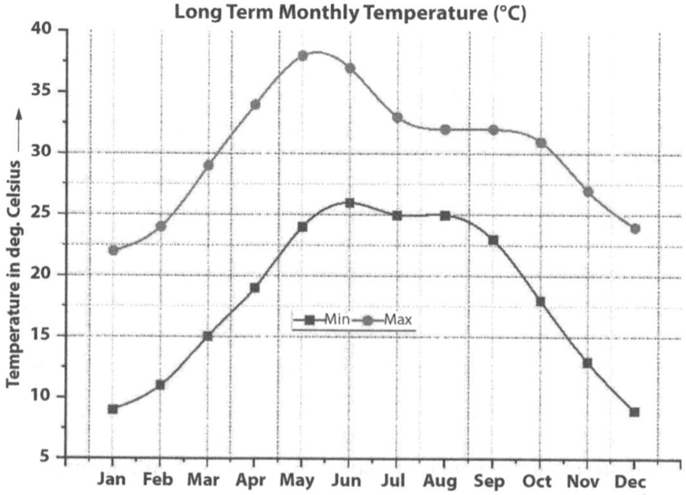

Figure 2.4 Long-term monthly temperature pattern in degrees celsius.

indicating the hottest period, and then decline steadily toward December, reflecting the coldest months. The x-axis displays the months from January to December, while the y-axis represents temperature in degrees Celsius. This graph highlights the seasonal temperature trends, with clear distinctions between summer and winter months.

Understanding the movement of the sun across the sky is fundamental to various fields, from renewable energy to agriculture and navigation (Figures 2.5 and 2.6). ISRO-VEDAS (Visualization of Earth Observation Data and Archival System) platform offers a powerful tool for visualizing and analyzing Earth observation data, including the sun's path. By leveraging VEDAS, students can explore the intricate dynamics of solar motion and its implications on Earth's systems. This study material aims to provide a comprehensive understanding of how to trace the sun's path using ISRO's VEDAS platform. Through theoretical explanations and practical demonstrations, students will gain insights into the factors influencing the sun's path, the significance of solar angles, and the real-world applications of this knowledge. Tracing the sun's path involves understanding the apparent motion of the sun as observed from the Earth's surface. This motion is a result of the Earth's rotation on its axis and its orbit around the sun.

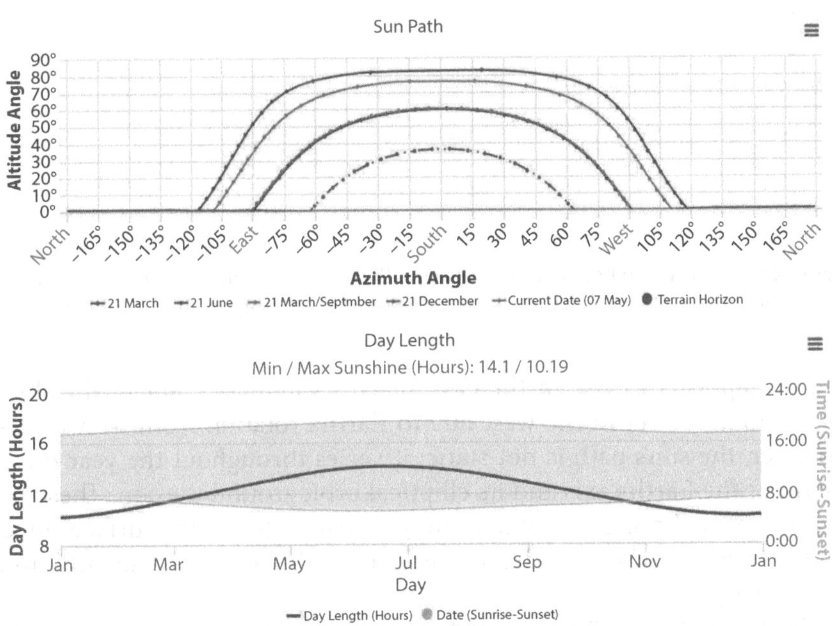

Figure 2.5 Tracing of the sun path to get the azimuth angle and day length by satellite applications [Data: VEDAS].

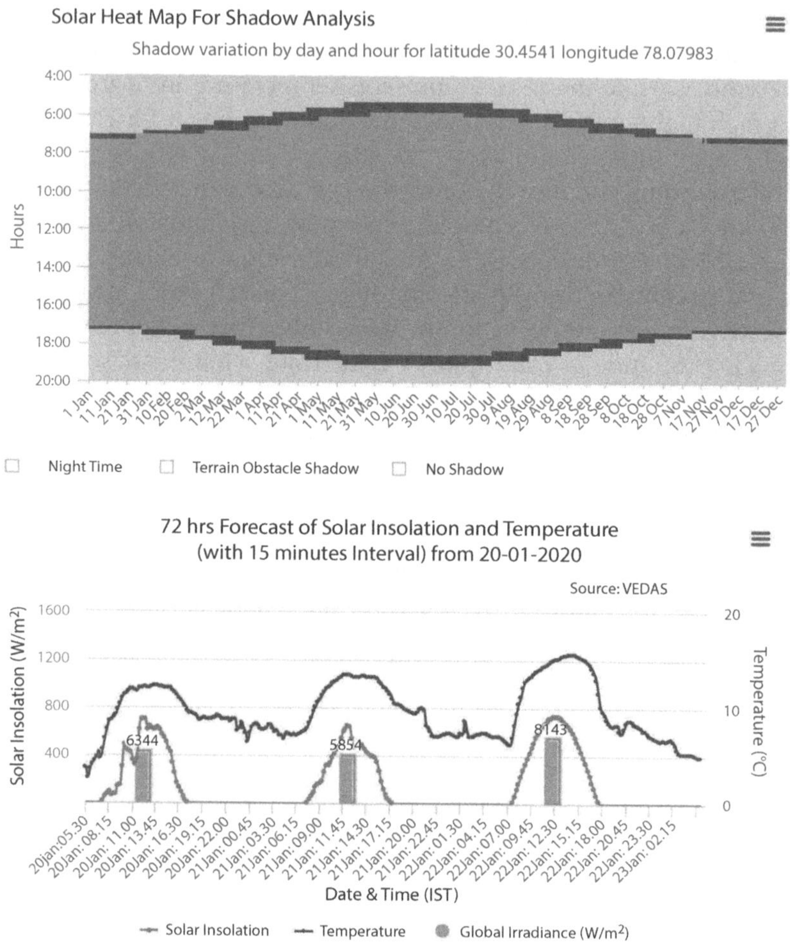

Figure 2.6 Seventy-two hours of forecasting of the solar isolation and temperature on the basis of pre-specified data parameters using the satellite applications [Data: VEDAS].

The sun appears to rise in the east, reach its highest point in the sky at solar noon, and set in the west due to Earth's rotation from west to east. However, the sun's path is not static; it varies throughout the year due to the tilt of the Earth's axis and its elliptical orbit around the sun. These factors influence the angle at which sunlight strikes the Earth's surface, affecting the intensity of solar radiation received at different latitudes and times of the year [9].

ISRO's VEDAS platform provides a user-friendly interface for visualizing the sun's path based on Earth observation data. Users can select a specific location (latitude and longitude) and date to analyze the sun's motion.

VEDAS offers tools to visualize solar angles such as solar azimuth (the angle between the sun's position and true north) and solar elevation (the angle between the sun and the horizon). By tracing the sun's path using VEDAS, students can observe how solar angles change throughout the day and over the course of the year, influencing factors such as daylight duration and solar energy availability [10].

To increase solar electricity output in the state of Uttarakhand, a renewable energy conservation organization in that state has decided to build more than 2,000 units that will produce 44 MW of electrical power using solar energy. Additionally, the state of Uttarakhand would have been required to purchase 8% of the electricity generated by solar power facilities under the new legislation. Previously, the state only bought 3% of the electricity generated by these kinds of plants. There are 23 solar power plants there, making it capable of supporting the 3% mandatory energy cost.

All around India, different photovoltaic energy production stations have been constructed, and, as part of this program, fresh specialists are interested in boosting the overall capacity to produce power. The state would have to work toward the development of diverse electricity infrastructures in order to comply with the requirements set forth by the Ministry of the Alliance. For this purpose, more than 2,000 units of various capacities will be installed in various locations. Programs run by the centralized government will sponsor every platform [11–16].

Everyone Needs Energy, Endayal Upadhyay The Gramin Jyti Yojana (DDUGGY) Integrated Framework for Energy Creation - IPDS, which is roughly Rs. 3332.37 crore, is set aside by the State Department of Energy to ensure last mile connectivity and power accessibility to rural communities in order to better serve rural customers. In fact, there is a means to follow R-APDRP with more funds. During the current 12th Five-Year Plan period (FYP 2012–17) and the upcoming 13th Five-Year Program, a goal of Rs 233.54 crore is set for the deployment of 24–7 efficient, safe, and stable power as well as all forms of connectivity to rural, semi–urban, and metropolitan centers (FYP 2017–22). Underneath the program, the household will reduce the corporation's AT&C losses by 14% by 31 March 2022 as well as eventually by 14% [2].

2.1.2.2 *The Wind Statistics of an Uttarakhand State*

For Uttarakhand state in India, wind statistics can vary based on the region's topography and elevation. Generally, Uttarakhand experiences diverse wind patterns influenced by its mountainous terrain and proximity to the Himalayas. In the lower regions and valleys, wind speeds may be relatively moderate and influenced by local factors such as valley breezes and terrain channeling. In contrast, higher-altitude areas and ridges might experience stronger and more variable winds, especially during certain seasons like winter and monsoon. Wind direction can also be influenced by local geography, with prevailing winds often originating from the northwest during winter due to the influence of western disturbances, whereas monsoon winds may arrive from the southwest. Detailed wind statistics for Uttarakhand would require location-specific measurements and analysis considering its diverse landscapes and microclimates [17].

Table 2.1 illustrates the wind speed and direction patterns across the year at a specific location situated at approximately 30.06675° latitude and 79.0193° longitude. Notably, the wind speeds tend to peak during the winter and early spring months, notably in February and March, reaching a maximum of 2.4 m/s, before gradually declining toward the summer,

Table 2.1 Wind speed and wind direction of latitude 30.06675 and longitude 79.0193 for 12 months.

Latitude	30.06675	
Longitude	**79.0193**	
Month	**Wind speed (m/s)**	**Wind direction (°)**
January	**2.1**	**246**
February	**2.4**	**250**
March	**2.4**	**239**
April	**2.3**	**227**
May	**2.2**	**223**
June	**1.8**	**223**
July	**1.2**	**246**
August	**1**	**261**
September	**1.1**	**245**
October	**1.6**	**223**
November	**2**	**233**
December	**1.9**	**236**

hitting a low of 1 m/s in August. Concurrently, the prevailing wind direction consistently originates from the southwest quadrant, ranging between 223° to 261° throughout the year, with minor fluctuations across the seasons. These insights offer valuable information for various sectors, including agriculture, energy, and infrastructure planning, facilitating informed decision-making processes and resource management strategies tailored to seasonal wind dynamics [18, 19].

2.2 Observations

This study using the ISRO VEDAS earth observation technique to examine the wind statistics of the hilly areas of Uttarakhand. The Pithoragarh region of Uttarakhand is considered in this study to be a hilly location where solar and wind statistics are collected. Using this information, a hybrid solar-wind energy model that may be used for street lights is created.

A brief description of a chosen hilly region in Uttarakhand is as follows:

- Pithoragarh, which is the municipality board for the district of Pithoragarh in the Indian state of Uttarakhand, is a town in the Western Himalayas [8].
- This is the largest organized cluster of semimurbanized to urbanized towns and villages in Uttarakhand's highlands.
- One of Uttarakhand's most populated districts, the city has all the amenities, including an airport.
- At 80.22°E, Pithoragarh is 29.58°N. It has an average altitude of about 1,514 meters (4,967 ft). It is situated in the Valley of the Tire. The valley occupies 50 square kilometers.
- In Pithoragarh, solar and wind resources are plentiful, so it is a favorable place for the installation of a hybrid solar wind power system [1].

2.2.1 Satellite Image of Pithoragarh District

Figures 2.7 and 2.8 show the satellite image of the district Pithoragarh, located in the northern state of Uttarakhand, India, which is known for its mountainous terrain as it lies in the Himalayan foothills. A topographic map of Pithoragarh would likely display the surrounding peaks, such as the Panchchuli Range and the Nanda Devi peak, rivers like the Kali River, as well as valleys and plateaus in the region. Elevation contours would show the variations in height, helping you understand the steepness of

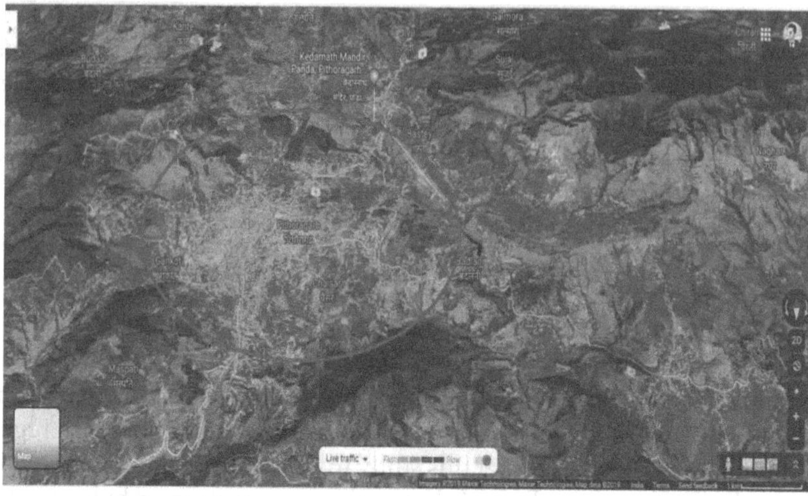

Figure 2.7 Satellite image of Pithoragarh district of Uttarakhand [21].

Figure 2.8 Bird-eye view of Pithoragarh district of Uttarakhand [21].

slopes and the ruggedness of the landscape. During certain times of the day, particularly in the afternoon and evening, you may experience valley winds or mountain winds. Valley winds occur when cooler air from higher elevations descends into the valleys due to the temperature difference between the valley and the surrounding mountains. Mountain winds, on the other hand, can occur when warm air rises along mountain slopes, creating localized wind patterns [20].

Pithoragarh's climate is influenced by regional weather systems, including monsoons, cyclonic circulations, and western disturbances. These weather systems can bring strong winds, especially during certain seasons or weather events. Pithoragarh district encompasses a wide range of elevations, from low-lying areas along river valleys to high mountain peaks. The altitude variation can lead to differences in air pressure and temperature, which can contribute to wind patterns and turbulence. Pithoragarh is located in a region prone to weather extremes, including thunderstorms, rain showers, and snowfall in higher elevations. These weather phenomena can be accompanied by strong winds, particularly during stormy conditions.

2.2.2 Solar Insolation of the District Pithoragarh Region in Uttarakhand

Figure 2.9 shows the solar insolation and temperature pattern, Pithoragarh is situated at a certain latitude on the Earth's surface, which affects the angle at which sunlight reaches the region.

Higher latitudes receive sunlight at a lower angle compared to equatorial regions, which can impact the intensity of solar radiation. Pithoragarh is situated at a certain latitude on the Earth's surface, which affects the angle at which sunlight reaches the region. Higher latitudes receive sunlight at a lower angle compared to equatorial regions, which can impact the intensity of solar radiation [22].

2.2.3 Wind Statistics of the District Pithoragarh Region in Uttarakhand

The wind pattern using the VEDAS is analyzed. Winds are primarily driven by the uneven heating of Earth's surface by the sun. The resulting pressure differences lead to the formation of large-scale wind patterns known as the global wind belts. These include the trade winds, westerlies, and polar easterlies, which flow in both hemispheres between the equator and the poles. In addition to global wind patterns, local geography, topography, and land-sea temperature differences also play significant roles in shaping wind patterns in specific regions [23]. For example, mountain ranges can deflect

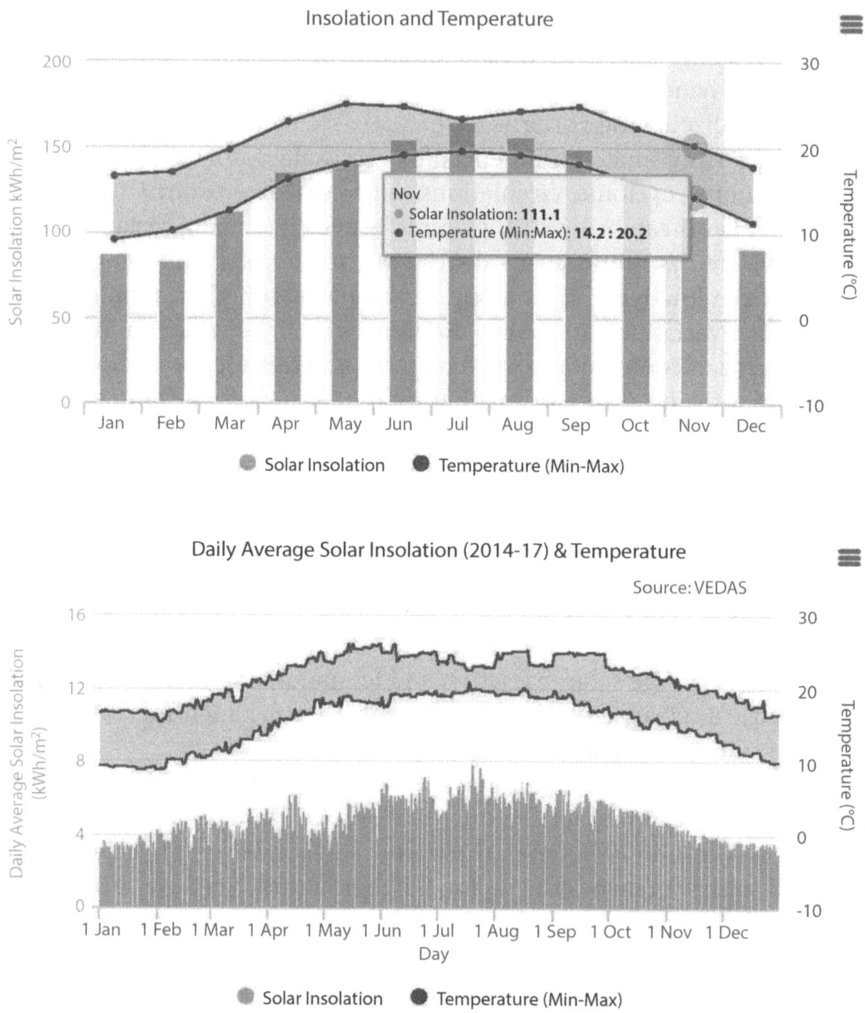

Figure 2.9 Solar insolation and temperature pattern of the Pithoragarh district.

and channel winds, creating localized wind systems such as valley breezes, mountain winds, and downslope winds. In regions like Uttarakhand, monsoon winds have a significant impact on weather patterns. The Indian summer monsoon brings moist air from the Indian Ocean to the Indian subcontinent, resulting in heavy rainfall during the monsoon season. The onset, duration, and intensity of the monsoon vary from year to year, affecting local wind patterns and weather conditions [24].

2.2.4 Monthly Speed Pattern of the Wind in the Study Area and Its Forecasting

Wind speed exhibits a seasonal pattern on average, the lower wind speeds (around 1 m/s) are observed during the warmer months (April to September). Conversely, the cooler months (October to March) experience higher wind speeds (around 2.5 m/s). The wind direction is predominantly from the southwest (around 230°) throughout the year. However, there is some variation observed monthly as shown in Table 2.2. The data shows a seasonal pattern in wind speed, with lower speeds (around 1 m/s) observed in the warmer months (April to September) and higher speeds (around 2.5 m/s) in the cooler months (October to March). Wind direction is predominantly from the southwest (around 230°) throughout the year, with some variation as shown in Figure 2.10.

Table 2.2 Wind speed and wind direction of latitude 30.14607 and longitude 80.34041 for 12 months.

Latitude	**30.14607**	
Longitude	**80.34041**	
Month	**Wind speed (m/s)**	**Wind direction (°)**
January	2.9	231
February	3	243
March	2.3	241
April	1.9	230
May	1.6	233
June	1.5	316
July	1	17
August	1	11
September	0.9	272
October	1	249
November	1.8	218
December	2.2	229

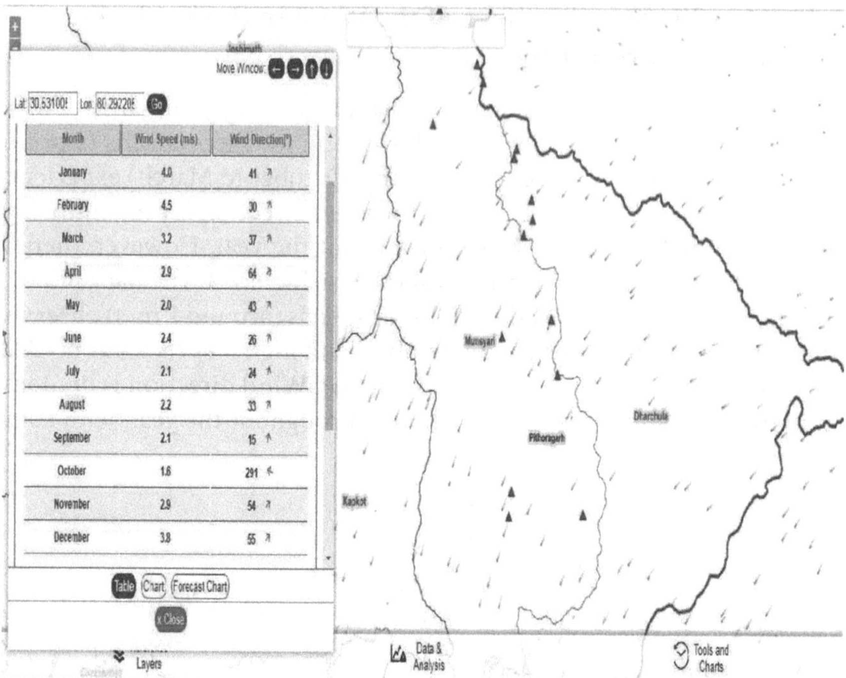

Figure 2.10 Wind pattern for the latitude 30.53 and longitude 80.29 showing the wind speed and wind direction.

Wind Forecasting: Figure 2.11 shows the 72 h of forecasting for the tracing of wind and its direction and intensity. Wind patterns in Pithoragarh, as in many regions, exhibit seasonal variations. During the summer months, for example, winds may be influenced by the Indian summer monsoon, whereas the winter months may experience different wind patterns due to changes in atmospheric pressure systems and temperature gradients. Local winds, such as valley breezes, mountain winds, and katabatic winds, are influenced by the specific geography and topography of the Pithoragarh region. For instance, during the day, warm air rises in the valleys, creating valley breezes that flow upslope. At night, the reverse occurs, with cool air descending from the mountains, resulting in downslope winds.

Wind Measurement: Wind speed and direction are typically measured using anemometers and wind vanes, respectively. Data from weather stations, remote sensing instruments, and numerical weather prediction models are used to analyze wind patterns and provide wind statistics for a given region.

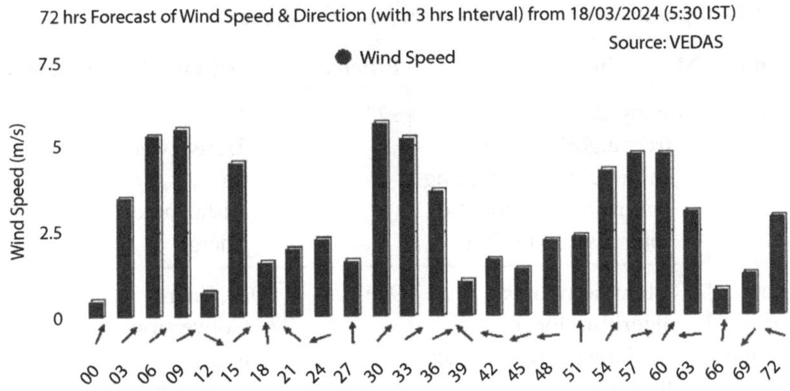

Figure 2.11 Seventy-two hours of forecasting of wind speed and wind direction.

2.3 Imperative of Machine Learning for Present Study

In Table 2.3 comparison of various machine learning (ML) methodologies used to optimize solar tracking systems. Researchers have explored a wide range of techniques, including artificial neural networks (ANNs), support vector machines (SVMs), decision trees, random forests, genetic algorithms, convolutional neural networks (CNNs), K-nearest neighbors

Table 2.3 Imperative of machine learning for the present study.

Authors	ML methodology	Efficiency	Year/journal	References
A. Smith *et al.*	Utilized artificial neural networks to predict the best value in the day length and the wind speed in a particular day.	98%	2018, Renewable Energy	[31]
B. Chen *et al.*	Applied support vector machines (SVM) for solar tracking optimization on the pre-specified data.	95%	2019, Solar Energy	[32]

(*Continued*)

Table 2.3 Imperative of machine learning for the present study. (*Continued*)

Authors	ML methodology	Efficiency	Year/journal	References
C. Kim *et al.*	Employed decision trees algorithm for automatic solar tracking optimization using the solar insolation data.	92%	2020, IEEE Transactions on Sustainable Energy	[33]
S. Gupta and R. Sharma	Used random forests algorithm for solar tracking control to trace the path of sun.	96%	2021, Energy Conversion and Management	[34]
M. Patel and S. Jain	Implemented genetic algorithms for optimization, by using techniques such as selection, crossover, and mutation to search for the optimal path for solar tracing.	94%	2019, Applied Energy	[35]
T. Nguyen and N. Tran	Deep Learning (CNN): Employed convolutional neural networks for optimization by using convolutional layers to extract input.	97%	2020, Solar Energy	[36]
H. Lee and J. Park	Used K-Nearest Neighbors (KNN), by finding the k nearest data points to a given query point in the feature space and using their labels or values to make predictions.	93%	2018, International Journal of Photoenergy	[37]

(*Continued*)

Table 2.3 Imperative of machine learning for the present study. (*Continued*)

Authors	ML methodology	Efficiency	Year/journal	References
Y. Wu *et al.*	Utilized Gaussian processes for tracking, they model the distribution over functions rather than individual data points and are characterized by a mean function and a covariance function.	95%	2022, Solar Energy	[38]
Q. Zhang *et al.*	Employed ensemble learning techniques, by training a group of diverse models and aggregating their predictions to make the final prediction.	94%	2019, Applied Energy	[39]
L. Tan and K. Lim	Used reinforcement learning for optimization by receiving feedback in the form of penalties based on its actions and uses this feedback to learn optimal behavior over time.	96%	2020, IEEE Access	[40]
X. Wang and Z. Li	Employed hidden Markov models for prediction, they consist of a set of hidden states, each of which emits observable symbols according to a probability distribution.	93%	2018, Solar Energy	[41]

(*Continued*)

Table 2.3 Imperative of machine learning for the present study. (*Continued*)

Authors	ML methodology	Efficiency	Year/journal	References
Y. Liu *et al.*	Applied naive Bayes for solar tracking, a simple probabilistic classifier based on Bayes' theorem with strong independence assumptions between the feature.	91%	2021, Renewable and Sustainable Energy Reviews	[42]
M. Yang and L. Zhang	Utilized fuzzy logic for solar tracking, for dealing with uncertainty and imprecision in data and knowledge representation.	94%	2022, Solar Energy	[43]
W. Hu and L. Chen	Used PCA (principal component analysis) for solar panel tracking optimization, by finding the principal components or directions of maximum variance in high-dimensional data and projecting the data onto a lower-dimensional subspace while preserving.	96%	2019, Energy Procedia	[44]
S. Park and Y. Kim	Employed extreme learning machines (ELMs) for solar tracking control, which utilizes a single-layer feedforward neural network trained using random weights and biases.	97%	2020, Solar Energy	[45]

(*Continued*)

Table 2.3 Imperative of machine learning for the present study. (*Continued*)

Authors	ML methodology	Efficiency	Year/journal	References
J. Li *et al.*	Utilized Bayesian networks for prediction, they model the conditional dependencies between variables using a directed acyclic graph, where nodes represent random variables and edges represent probabilistic dependencies.	94%	2018, IEEE Transactions on Industrial Electronics	[46]
Y. Wang and Z. Wu	Used CNN for solar panel tracking optimization, CNNs use convolutional layers to automatically learn hierarchical representations of features from raw input data.	98%	2021, Solar Energy	[47]
X. Chen and H. Zhang	Employed LSTM (long short-term memory) networks for solar tracking, to capture long-term dependencies and process sequential data.	95%	2019, Applied Energy	[48]
Q. Xu *et al.*	Used autoencoders for solar tracking control, they consist of an encoder network that maps input data to a latent representation and a decoder network that reconstructs the input data from the latent representation.	96%	2020, Renewable Energy	[49]

(*Continued*)

Table 2.3 Imperative of machine learning for the present study. (*Continued*)

Authors	ML methodology	Efficiency	Year/journal	References
H. Zhao and S. Liu	Recurrent neural networks (RNN) have connections that form directed cycles, allowing them to capture temporal dependencies and context information over time, which makes the solar tracking more efficient.	97%	2022, Solar Energy	[50]

(KNN), Gaussian processes, ensemble learning, reinforcement learning, hidden Markov models (HMMs), naive Bayes, fuzzy logic, principal component analysis (PCA), extreme learning machines (ELMs), Bayesian networks, long short-term memory (LSTM) networks, autoencoders, and recurrent neural networks (RNNs). The efficiency (percentage) and year/journal of publication are also provided for each study. Mainly, deep learning approaches like CNNs and RNNs achieved the highest efficiencies (around 97%–98%).

2.4 Conclusion

By getting the wind and solar statistics of the Pithoragarh district of Uttarakhand, the locations pointed in the observations can be set for the installment of solar wind hybrid model; this model can be installed for the street lighting applications; the particular village of that pointed locations in the map showed in observations can be selected for street lighting applications. The monthly solar insolation and wind flow speed graph obtained shown in the observations clearly illustrates, that the months in which the solar insolation is less, in that month the wind flow is more, this will provide 44 the continuous generation in day and night, in summer and winter the comparison of solar and wind resource obtained by the ISRO VEDAS. The planned hybrid PV wind system includes excellent implementations of wind turbines, and solar photovoltaic produces optimum energies from two sources. It creates a converter model from compacts, thus minimizing costs. The proposed system for wind-solar hybrid modules includes significant improvements in the performance of wind energy conversion systems (WECS) while addressing constraints for increasing generation capacity. The system integrates solar PV with an appropriate capacity enhancement and uses limited battery storage to optimize overall efficiency and reliability. The storage facility offers solutions to the problems of energy production during the conditions of low winds. Currently, three types of battery model are defined in particular: experiment analysis, and electrochemical and electrical circuit-based studies. The model is not well suited to reflect cell motility for the purpose of battery state of charging (SOC) predictions. Battery pack experimental and electrical analysis is conducted to reflect and evaluate the electrical characteristics of a battery, including parameters such as voltage, current, capacity, and efficiency. To reflect the electrical characteristics of a battery. However, an electrical circuitry-based design may be useful [51, 52]. The basic electrical models have an optimum source of voltage series with the internal resistances.

References

1. Finster, C., El heliostato de la Universidad santa maria. *Scientia*, 119, 5–20, 1962.
2. Sahu, S.K., Kopalakrishnaswami, A.S., Natarajan, S.K., Historical overview of power generation in solar parabolic dish collector system. *Environ. Sci. Pollut. Res.*, 29, 43, 64404–64446, 2022.
3. Muhammad, J.Y., Jimoh, M.T., Kyari, I.B., Gele, M.A., Musa, I., A review on solar tracking system: A technique of solar power output enhancement. *Eng. Sci.*, 4, 1, 1–11, 2019.
4. Fahad, H.M., Islam, A., Islam, M., Hasan, M.F., Brishty, W.F., Rahman, M.M., Comparative analysis of dual and single axis solar tracking system considering cloud cover, in: *2019 International Conference on Energy and Power Engineering (ICEPE)*, IEEE, pp. 1–5, 2019.
5. Wang, S., Shen, Y., Zhou, J., Li, C., Ma, L., Efficiency Enhancement of Tilted Bifacial Photovoltaic Modules with Horizontal Single-Axis Tracker—The Bifacial Companion Method. *Energies (Basel)*, 15, 4, 1262, 2022.
6. Paliyal, P. S., Mondal, S., Layek, S., Kuchhal, P., & Pandey, J. K., Automatic solar tracking system: a review pertaining to advancements and challenges in the current scenario. *Clean Energy*, 8, 6, 237–262, 2024.
7. Riedel-Lyngskær, N., Petit, M., Berrian, D., Poulsen, P.B., Libal, J., Jakobsen, M.L., A spatial irradiance map measured on the rear side of a utility-scale horizontal single axis tracker with validation using open source tools, in: *2020 47th IEEE Photovoltaic Specialists Conference (PVSC)*, IEEE, pp. 1026–1032, 2020.

8. Gutiérrez, S. and Rodrigo, P.M., Energetic analysis of simplified 2-position and 3-position North-South horizontal single-axis sun tracking concepts. *Sol. Energy*, 157, 244–250, 2017.
9. Rana, G., Thakur, N.S., Pandey, S., Performance evaluation of solar photovoltaic system using vertical single-axis solar tracker in Himachal Pradesh, India, in: *Application of Soft Computing Techniques in Mechanical Engineering*, pp. 23–41, CRC Press, Boca Raton, 2022.
10. Willockx, B. and Cappelle, J., A comparison study of the performance of vertical vs single axis tracking bifacial agrivoltaic systems in Belgium, in: *2022 IEEE 49th Photovoltaics Specialists Conference (PVSC)*, IEEE, p. 1162, 2022.
11. Mourad, A.-H.I., Shareef, H., Ameen, N., Alhammadi, A.H., Iratni, M., Alkaabi, A.S., A state-of-the-art review: Solar trackers, in: *2022 Advances in Science and Engineering Technology International Conferences (ASET)*, IEEE, pp. 1–5, 2022.
12. Karthik, M., Vishnu, R., Vigneshwar, M., Logaeshwar, M., Arduino based Dual Axis Smart Solar Tracking System, in: *2023 Third International Conference on Artificial Intelligence and Smart Energy (ICAIS)*, IEEE, pp. 169–174, 2023.
13. Pawar, V.R. and Gitanjali Vasant, L., Solar-Wind Hybrid Energy System Using MPPT. *IEEE International Conference on Intelligent Computing and Control Systems ICICCS 2017.*
14. Prashanth, B.N., Pramod, R., Veeresh Kumar, G.B., Design and Development of Hybrid Wind and Solar Energy System for Power Generation. *International Conference on Materials Manufacturing and Modelling (ICMMM - 2017)*, pp. 2214–7853, © 2017 Elsevier Ltd.
15. Kanathipan, K. and Lam, J., A New Modified Maximum Power Extraction Technique for Wind and Hybrid Renewable Energy Systems. *2017 IEEE Electrical Power and Energy Conference (EPEC).*
16. Georges, S. and Slaoui, F.H., Case Study of Hybrid Wind-Solar Power Systems for Street Lighting. *IEEE 2011 21st International Conference on Systems Engineering.*
17. Anuradha, Sinha, S.K., Yadav, A., Sharma, A., Tiwari, A., scope of solar and wind energy in uttarakhand. *TEJAS Eng. J. Appl. Scopes*, 1, 2, July 2016.
18. Pandey, A., Sharma, M., Kumar, M., renewable energy scenario, policies and development barriers in the 27th state of India: Uttarakhand. *Int. Res. J. Eng. Technol. (IRJET)*, 5, 07 July 2018.
19. Chang, J. and Jia, S.-Y., Modeling and application of wind-solar energy hybrid power generation system based on multi-agent technology. *Proceedings of the Eighth International Conference on Machine Learning and Cybernetics*, Baoding, 12-15 July 2009.
20. Elsayed, I., Nassar, I., Mostafa, F., Optimization and economic evaluation of small scale hybrid solar/wind power for remote areas. *2017 Nineteenth International Middle East Power Systems Conference (MEPCON)*, Menoufia University, Egypt, 19-21 December 2017.

21. Salmi, T., Bouzguenda, M., Gastli, A., Masmoudi, A., MATLAB/Simulink Based Modelling of Solar Photovoltaic Cell. *Int. J. Renew. Energy Res.*, 2, 2, 213–218, 2012.
22. Chen, J. and Jiang, D., Study on Modeling and Simulation of Non-gridconnected Wind Turbine, 978-1-4244-4702-2/09/$25.00, ©2009 IEEE.
23. Kuo, C.-W., Du, S.-H., Juang, F.-S., Ho, Y.-Y., Solar Cell Module Installation Optimization and Evaluation of Power Conversion Simulation. *PEDS2009.*
24. Bhattacharjee, T., Ghosh, P., Mondal, S., Hazards and Environmental Issues in Biodiesel Industry. Green Technologies for Sustainable Production, 1, 383, 2023.
25. Mondal, S., Dalvi, A., Mondal, A.K., Thapa, V., A novel approach for aero-generators-triple state absorption-based cooling system. *Mater. Today Proc.*, 99, 44–48, 2023.
26. Sunilkumar, A., Singh, S.K., Mondal, A.K., Ghosh, P., Mondal, S., Aggravation of CoVID-19 infections due to air pollutant concentrations in Indian cities. *Spatial Inf. Res.*, 31(5), 525–536, 1–12, 2023.
27. Kuchhal, P., Kumar, A., Upadhyay, P., Mondal, S., Opportunities and challenges of solar thermal technologies in the Indian context. *Heat Transfer*, 51, 8, 7139–7173, 2022.
28. Mukherjee, M., Singhal, S., Saraswat, J., Prashanth Kumar, S., Mondal, S., Sharma, A.K., Enabling Resource Efficiency Through Reduce, Reuse, and Recycling—A Perspective on "Industrial Networking", in: *International Conference on Advances and Innovations in Recycling Engineering*, 2021, October, Springer Nature Singapore, Singapore, pp. 35–50.
29. Manna, S., Bobde, P., Roy, D., Sharma, A.K., Mondal, S., Separation of iodine using neem oil-cashew nut shell liquid based-phenol formaldehyde resin modified lignocellulosic biomatrices: Batch and column study. *J. Taiwan Inst. Chem. Eng.*, 122, 98–105, 2021.
30. Johnson, J., Mondal, S., Mondal, A.K., Rana, S., Pandey, J.K., Feasibility study of a 200 kW solar wind hybrid system. *Appl. Sol. Energy*, 54, 376–383, 2018.
31. Mondal, S., Mondal, A.K., Sharma, A., Devalla, V., Rana, S., Kumar, S., Pandey, J.K., An overview of cleaning and prevention processes for enhancing efficiency of solar photovoltaic panels. *Curr. Sci.*, 115, 6, 1065–1077, 2018.
32. Mukherjee, M., Adhikary, I., Mondal, S., Mondal, A.K., Pundir, M., Chowdary, V., A vision of IoT: applications, challenges, and opportunities with Dehradun perspective, in: *Proceeding of International Conference on Intelligent Communication, Control and Devices: ICICCD 2016*, Springer Singapore, pp. 553–559, 2017.
33. Gheibi, A., Mohammadi, S.M.A., Maghfoori Farsangi, M., 2016. A proposed maximum power point tracking by using adaptive fuzzy logic controller for photovoltaic systems. *Scientia Iranica*, 23, 3, 1272–1281.
34. Shamshirband, S., Kasra M., Hossein K., Lip Yee, M.L., Petković, D., Zalnezhad, E., Estimating the diffuse solar radiation using a coupled support

vector machine–wavelet transform model. *Renew. Sustain. Energy Rev.*, 56, 428–435, 2016.

35. Rambhowan, Y. and Oree, V., Improving the dual-axis solar tracking system efficiency via drive power consumption optimization. *Appl. Solar Energy*, 50, 74–80m 2014.
36. Paliyal, P.S., Mondal, S., Pandey, J.K., An Analysis of Springwater Parameter Status for Different Locations in the Rural Hilly Garhwal Region of Uttarakhand. *Environ. Claims J.*, 36, 3, 243–277, 2024. https://doi.org/10.1080/10406026.2024.2423964.
37. Gairola, M., Paliyal, P.S., Kashyap, M., Bahuguna, N.K., Modeling of Solar Emergency System in Power Station to Mitigate Blackout, in: *Proceedings of Integrated Intelligence Enable Networks and Computing*, S. Mer, K.K., Semwal, V.B., Bijalwan, V., Crespo, R.G. (eds) Algorithms for Intelligent Systems. Springer, Singapore, 2021. https://doi.org/10.1007/978-981-33-6307-6_3.
38. Menon, S.K., Mondalm S., Singh Paliyal, P., The Role of Hydrogen as a Sustainable Energy Source and its Integration with the Grid, in: *Advancements in Communication and Systems*, Ashish Kumar Tripathi and Vivek Shrivastava (eds), SCRS, India, pp. 397–406, 2024. https://doi.org/10.56155/978-81-955020-7-3-35.
39. Welepe, H.J.N., Günerhan, H., Bilir, L., Theoretical performance assessment of a parabolic trough humidifying solar collector-based solar still, *Isı Bilimi ve Tekniği Dergisi*, 44, 1, 163–189, 2024.
40. Majeed, M.H., Alwan, N.T., Salih, S.A., Aljashaami, B.A., Shcheklein, S.E., Velkin, V.I., An experimental study of the effect of solar tracking technology on the performance of a single phase open thermosyphon: a case study. *Energy Sources, A: Recovery, Utilization, and Environmental Effects*, 46, 1, 6049–6063, 2024.
41. Alkhalaf, Q., Chandel, S.S., Suri, A.R.S., Thermal performance analysis of a novel Conical-Cylindrical-Conical cavity receiver for a Scheffler concentrated solar system. *Appl. Therm. Eng.*, 245, 122798, 2024.
42. Guillet, G., Gaspar, J., Barbosa, S., Fasquelle, T., Kadoch, B., Experimental evaluation of the concentrated solar heat flux distribution provided by an 8 m2 Scheffler reflector. *Renew. Energy*, 223, 119958, 2024.
43. Abed, A.A., El-Marghany, M.R., El-Awady, W.M., Hamed, A.M., Recent advances in parabolic dish solar concentrators: Receiver design, heat loss reduction, and nanofluid optimization for enhanced efficiency and applications. *Solar Energy Mater. Solar Cells*, 273, 112930, 2024.
44. Wadate, P. and Dharmadhikari, H., Thermal performance evaluation of solar paraboloidal dish concentrator, in: *AIP Conference Proceedings*, AIP Publishing, 2023.
45. Kailaswar, O., Lakhe, S., Mahant, S., Joshi, S., Patil, S., Performance investigation of concentrated solar thermal system, in: *AIP Conference Proceedings*, AIP Publishing, 2024.

46. Gupta, M.K. *et al.*, Energy, environmental and economic evaluation of decentralized solar distillation and wood based distillation for extraction of organoleptically superior essential oils: A comparative analysis. *Sustain. Energy Technol. Assess.*, 69, 103895, 2024.
47. Sagade, A.A., Mawire, A., Palma-Behnke, R., Sagade, N.A., Low-cost solar concentrating collector-receiver system as an effective enabler for clean cooking and heating in urban areas. *Solar Energy*, 263, 111813, 2023.
48. Phate, M., Toney, S., Phatem V., Modelling and simulation of portable solar Scheffler reflector water heater using soft computing techniques. *Int. J. Comput. Appl. Technol.*, 73, 2, 91–103, 2023.
49. Herrería-Alonso, S., Suárez-González, A., Rodríguez-Pérez, M., Rodríguez-Rubio, R.F., López-García, C., A solar altitude angle model for efficient solar energy predictions. *Sensors*, 20, 5, 1391, 2020.
50. Wei, D. *et al.*, Optimum tilt and azimuth angles of heat pipe solar collector, an experimental approach. *Case Stud. Thermal Eng.*, 55, 104083, 2024.
51. Umar, S., Qureshi, M.S., Nawaz, M.U., Thermal Imaging and AI in Solar Panel Defect Identification. *Int. J. Adv. Eng., Technol., Innov.*, 1.3, 73–95, 2024.
52. Sheikholeslami, M. and Khalili, Z., Enhancing photovoltaic solar panel performance with integration of PCM-based spectral filter and self-cleaning coating. *J. Build. Eng.*, 94, 110019, 2024.

3

Harnessing Wind Energy: Ontological Frameworks for Optimizing Wind Turbine Lifecycle Management and Performance

Gaurav Jaglan[1], Aman Jolly[2], Vikas Pandey[3]*, Shashikant[3] and Priyanka Sharma[4]

[1]*Semantics & AI Departmet, Knowledge Learning Infotech (Opc) Pvt. Ltd., Delhi, India*
[2]*Computer Science Department, KIET Group of Institutions, Uttar Pradesh, India*
[3]*Electrical Engineering Departmet, Babu Banarasi Das University, Lucknow, UP, India*
[4]*Department of Mathematics, Shivalik College of Engineering, Dehradun, Uttarakhand, India*

Abstract
This study examines the use of ontological frameworks in wind energy, specifically their impact on improving the management and performance of wind turbines throughout their lifespan. This work offers a succinct examination of the underlying principles of ontology, their practical application in design and development, and real-life examples that showcase their influence in standardizing understanding and enhancing decision-making in wind energy sector. The text classifies and evaluates several wind energy ontologies, providing valuable perspectives for scholars and practitioners. Furthermore, it tackles obstacles and emphasizes possibilities for utilizing ontologies to stimulate advancement in wind turbine technology. This survey aims to be a guide for those interested in the intersection of ontologies and wind energy.

Keywords: Ontology, knowledge graphs, wind energy, sustainability, renewable energy, data management, interoperability

**Corresponding author*: 04.pandey@gmail.com

Jai Govind Singh, Rupendra Kumar Pachauri and Sasidharan Sreedharan (eds.) Forecasting Methods for Renewable Power Generation, (49–66) © 2025 Scrivener Publishing LLC

3.1 Introduction

Wind energy has been a crucial factor in the worldwide shift toward clean and renewable energy sources. The importance of wind energy resides in its capacity to utilize the force of the wind and transform it into electrical energy, providing a sustainable and plentiful substitute for conventional fossil fuels. The deployment of wind turbines is a significant factor in the success of wind energy as they are instrumental in turning wind power into useful electricity [1].

Wind turbines consist of expensive blades affixed to a tower, which harness the kinetic energy of the wind and transform it into rotational energy. Subsequently, this rotational energy is converted into electrical energy by use of a generator. The exponential expansion of wind energy installations globally highlights the growing acknowledgment of wind power as a dependable and environmentally acceptable energy resource [2]. The crucial significance of wind turbines in the generation of renewable energy is apparent in their ability to produce substantial quantities of electricity without releasing greenhouse gases or depending on limited resources. Wind turbines have become a fundamental component of sustainable energy strategy as governments work toward achieving ambitious renewable energy goals and decreasing their carbon emissions. Nevertheless, the process of managing the lifespan of wind turbines presents substantial difficulties. The challenges are discussed below.

Operational efficiency is of utmost importance in order to maximize the energy output of wind turbines. Factors such as equipment malfunctions, maintenance requirements, and subpar performance can have a detrimental effect on the overall effectiveness of a wind farm [3]. Wind turbines endure severe environmental conditions, resulting in deterioration and damage. Efficient maintenance planning is crucial for swiftly addressing faults and preventing expensive periods of inactivity. End-of-life considerations: As wind turbines mature, they reach the conclusion of their functional lifespan. The process of decommissioning and disposing of turbine components has environmental and logistical concerns.

Data Management: As wind turbine systems get more intricate, the issue of effectively handling the extensive data produced by sensors and monitoring equipment becomes increasingly important. Effective data management is essential for predictive maintenance and optimizing performance.

In order to tackle these difficulties, ontologies present a viable resolution. Ontologies are hierarchical systems that establish the connections and attributes of entities within a certain field. Ontologies can be utilized in wind turbine lifecycle management to the following:

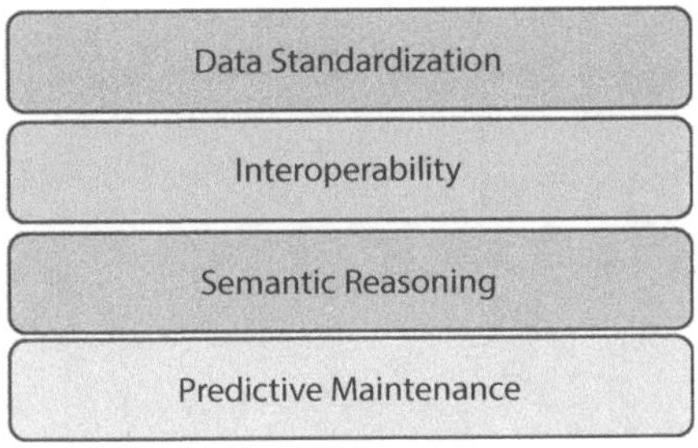

Figure 3.1 Use cases for ontology-based wind turbine lifecycle management.

- Standardizing Data: Ontologies facilitate the standardization of data formats and definitions, enabling smooth communication and integration of information among various systems and stakeholders [4].
- Enhance Interoperability: Ontologies improve interoperability among various software programs and systems used in wind turbine management by establishing a shared understanding of the domain.
- Activate Semantic Reasoning: Ontologies facilitate semantic reasoning, enabling higher order analytics, decision assistance, and the extraction of significant insights from intricate data sets [5].

To improve predictive maintenance, ontologies can be employed to represent the connections among different aspects that impact wind turbine performance, facilitating the creation of methods for anticipating maintenance needs [6]. Figure 3.1 shows the use cases for ontology-based wind turbine lifecycle management as follows.

3.2 Fundamentals of Ontologies

Ontologies function as formal representations of knowledge within a particular domain, offering a systematic framework for organizing information and establishing connections between entities [7]. They have a vital function in promoting a collective comprehension of a specific field among various systems, applications, and users. Ontologies serve as a means to represent and describe concepts, their attributes, and the connections between them in a manner that is comprehensible and interpretable by

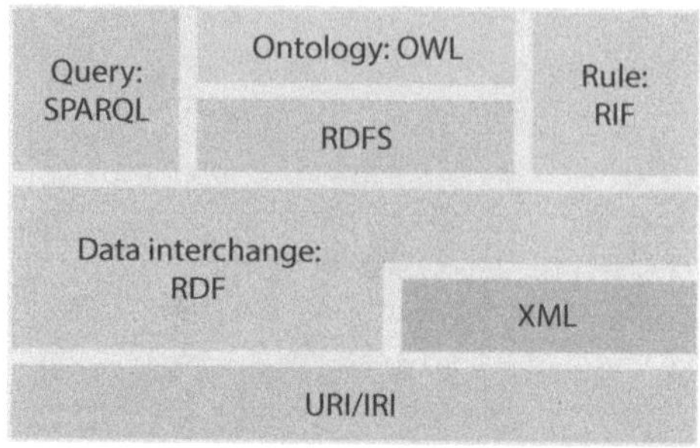

Figure 3.2 W3C standard technologies encompassing semantic web stack.

both humans and machines. Ontologies are part of semantic web [8] stack build on top of W3C standards as depicted below in Figure 3.2.

The following are the fundamental elements of ontologies:

Ontologies employ a formal language to precisely articulate concepts, attributes, and relationships. By formalizing information, we can achieve a clear and consistent definition of the knowledge framework in a certain field. Ontologies often consist of classes or ideas that represent entities within a specific domain, properties that specify the attributes of these things, and relationships that establish links between entities. In a medical oncology, concepts could encompass terms such as "disease," characteristics could refer to "symptoms," and relationships could involve "causes" or "treats."

Hierarchical Structure: Ontologies commonly have a hierarchical arrangement, where classes are organized in a tree-like manner. This hierarchical structure facilitates the organization of concepts according to their larger or narrower categories, resulting in a taxonomy that accurately represents the intrinsic relationships within the domain. Interoperability is enhanced through the use of ontologies, which establish a shared vocabulary and structure that enables various systems to effectively communicate and comprehend information. This is especially crucial in situations where various programs or databases must communicate efficiently. Ontologies facilitate reasoning by enabling systems to deduce new information through the use of rules defined in the ontology. It is imperative to make informed decisions, respond to inquiries, and derive conclusions based on the existing facts.

The Web Ontology Language (OWL) [9] is a major ontology language. OWL is a web standard established by the World Wide Web Consortium (W3C) with the purpose of representing ontologies on the internet [10]. The system offers a comprehensive range of elements for expressing classes, characteristics, and connections in a formal and easily interpreted format. OWL has three variations with escalating levels of expressiveness—OWL

Lite, OWL DL, and OWL Full—enabling users to select a complexity level that suits their individual needs.

The functions and purposes of OWL-related ontology languages are as follows:

- **Expressivity:** OWL enables the depiction of intricate relationships and limitations within a specific field. The system offers several elements, including classes, individuals, characteristics, and limits, to accurately represent the intricacies of diverse domains.
- **Inference:** OWL [10] facilitates automated inference by utilizing the axioms and rules specified in the ontology. This inference is crucial for deriving implicit information, verifying coherence, and drawing logical conclusions.
- **Standardization:** The utilization of ontology languages such as OWL facilitates the establishment of uniformity in the representation and dissemination of knowledge. Standardization improves the ability for different systems to work together and makes it easier to combine information from many sources.

The Semantic Web is an expanded version of the World Wide Web that assigns clear meaning to data. OWL plays a crucial role in this framework. OWL ontologies enhance the semantic richness of the web, facilitating robots to comprehend and manipulate information in a more intelligent fashion.

3.3 Wind Turbine Life Cycle Overview

The lifespan of a wind turbine can be categorized into distinct stages, each accompanied by its own unique difficulties. Below is a summary of the difficulties connected with each stage [11].

- **Design Phase:** The design phase encompasses the development of blueprints for the wind turbine, encompassing its structural elements, aerodynamic properties, and electrical systems. Difficulties may develop while attempting to optimize the design to enhance efficiency, dependability, and cost-effectiveness. Achieving equilibrium between these parameters while taking into account the precise environmental conditions at the installation location is a multifaceted undertaking.

- **Manufacturing Phase:** The manufacturing phase involves the fabrication of turbine components on a large scale while ensuring exceptional quality. Ensuring consistency in production processes, obtaining resources in a sustainable manner, and adhering to industry standards are of utmost importance. Moreover, the shipping logistics involved in handling big turbine components might pose significant difficulties.
- **Installation Phase [12]:** The installation phase encompasses the transportation and assembly of the turbine components at the designated location. The challenges encompass the coordination of transporting sizable and weighty components, guaranteeing construction safety, and managing diverse environmental circumstances. The installation process can become more challenging when it involves accessing remote or offshore places.
- **Operation Phase:** During operations, turbines encounter issues pertaining to the optimization of performance and the assurance of dependability. Potential issues that may occur include device malfunctions, variations in wind conditions, and difficulties in integrating with the power system. Monitoring and overseeing the turbine's performance to optimize energy production while minimizing periods of inactivity are crucial elements of the operational phase [13].
- **Maintenance Phase:** Maintenance is crucial for maintaining the continuous operation of wind turbines. Challenges encompass the formulation of efficient predictive maintenance techniques to proactively tackle prospective problems before they result in operational interruptions. Managing and preserving components, particularly for offshore turbines, can pose significant logistical and financial difficulties [14].
- **Decommissioning Phase:** Wind turbines have a limited duration of operation, and decommissioning entails the secure disassembly and proper disposal of their components. Challenges encompass the ecological consequences of decommissioning, appropriate disposal of wastes, and resolving any conflicts with local residents. Formulating economical decommissioning solutions and complying with regulatory mandates are additional notable obstacles.

Furthermore, there are overarching issues that extend throughout numerous stages of the lifespan in addition to the challenges specific to each phase.

3.3.1 Technological Progress

The swift advancement of wind turbine technology presents difficulties in maintaining current infrastructure. Integrating new technology into turbines while maintaining compatibility with older components can provide difficulties. Environmental and social factors must be taken into account while considering wind farms, particularly those located offshore, as they may encounter difficulties associated with their ecological consequences and relationships with nearby communities. Mitigating issues regarding noise, visual impact, and animal problems necessitates meticulous planning and active involvement.

The wind energy operations are affected by dynamic regulatory and policy changes. Government incentives, environmental laws, and grid interconnection requirements have the potential to significantly affect the whole lifecycle of a project. These factors can influence design decisions, project feasibility, and the decommissioning process.

To effectively tackle these difficulties, it is crucial to foster collaboration among stakeholders, conduct continuous research and development, adhere to established best practices, and demonstrate a steadfast commitment to sustainability throughout the entire lifespan of wind turbines. Addressing these problems is crucial for assuring the enduring success and sustainability of wind power as the wind energy sector expands [14].

3.4 Ontologies in Wind Turbine Design and Development

Ontologies are essential in the wind turbine lifespan as they offer a systematic and standardized representation of knowledge. Ontologies play a significant role in several stages of the wind turbine lifetime.

- **Creation and Implementation:** The ontologies can specify the concepts and relationships associated with different components of a wind turbine, including blades, towers, and control systems. This aids in precisely defining the needs and attributes of each component throughout the design stage.
- **Standards Compliance:** Utilizing ontologies that align with industry standards, such as the IEC 61400 series, can help guarantee that the design and development processes conform to established rules and regulations.

- **Production:** Logistics and operations management: Ontologies are utilized to represent the connections and interdependencies within the supply chain, guaranteeing that the materials and components employed in manufacturing adhere to the necessary criteria. Ontologies play a role in the establishment of thorough quality assurance systems in manufacturing by defining quality standards and criteria [15].
- **Installation and Commissioning:** Site-Specific Conditions: Ontologies can incorporate data regarding the geographic and environmental factors at the installation site, facilitating the enhancement of the turbine's efficiency according to its precise location.
- **Integration With Power Grids:** Ontologies can streamline the incorporation of wind turbines into the broader power grid by describing electrical attributes and communication protocols.
- **Execution and Upkeep:** Ontologies can be utilized to represent the connections among various components and their possible failure modes, facilitating problem identification and the formulation of proactive maintenance policies.
- **Performance Monitoring:** Ontologies enable the representation of crucial performance indicators, facilitating the efficient monitoring of the turbine's operational efficiency and energy production [6].
- **Data Integration:** Data Harmonization: Ontologies facilitate the integration of data from multiple sources by giving a consistent vocabulary andstructure. This is particularly critical in the wind turbine lifecycle, as data originates from multiple sensors, monitoring systems, and maintenance records.
- **Interoperability:** By adhering to common ontologies, different software systems and platforms engaged in the wind turbine lifespan can connect and exchange data more seamlessly.

End-of-Life considerations ontologies can model the environmental impact of decommissioning and recycling operations, helping to ensure that the disposal of components is carried out in an environmentally appropriate manner. In summary, ontologies contribute to the wind turbine lifecycle by facilitating communication, standardizing information, improving decision-making processes, and supporting the integration of multiple data sources across the many stages of a turbine's life.

3.5 Different Ontologies Used for Wind Energy and Wind Turbine

Here are some examples of ontologies used in the field of wind energy:

- **Wind Ontology (Platoon Project):** Platoon project is an EU initiative for bringing renewable sources of energy together under common standards. Wind turbine ontology under this initiative is developed on OWL standards and inherits/extends entities defined in OntoWind, SAREF, SAREF4BLDG, as well as SEAS ontology extending structural as well as performance metadata [16].
- **Ontology for Wind Energy:** OntoWind represents and applies semantic information by providing a model that defines the structural components for wind, i.e., blade, generator, sensors plant, and turbine information. Prediction and forecast models for weather as well as wind power have also been covered in scope of ontology. Lastly, various organization types involving public and private organizations have been modeled [4].
- **IEC 61400-25 Ontology:** The International Electro technical Commission (IEC) has produced standards for wind turbines, including the IEC 61400 series. An ontology based on these standards can capture information relating to the design, operation, and maintenance of wind turbines [17].
- **Weather Ontologies:** Weather conditions have a vital impact on wind energy production. Ontologies linked to meteorological data, atmospheric conditions, and climate can be merged with wind energy ontologies to provide a full perspective of the environmental elements affecting wind turbines [18].

It is worth mentioning that the creation and implementation of ontologies in the field of wind energy are ongoing, and new ontologies may arise over time. Additionally, interoperability standards such as the Open Platform for Energy Data (OPEn) are being created to simplify data interchange and integration in the energy industry, including wind generation. Table 3.1 is showing the comparison of ontologies supporting wind turbine life cycle management as follows.

Table 3.1 Comparison of ontologies supporting wind turbine life cycle management.

Ontology	Definition and purpose	Scope	Integration with WindIO ontology	Applications	Use in wind turbine design	Collaboration and benchmarking	Information models defined	References and documentation
Wind Task 37 Blade Ontology [19]	Part of IEA Wind Task 37, standardizes information for MDAO workflows in wind turbine and plant design, operation, and control.	Wind turbine and plant design, operation, and control	Yes, part of WindIO ontology for sharing design information across MDAO practitioners	Standardizing information flow in MDAO workflows, supporting benchmarking, comparison, and streamlined workflows	Used to standardize representation of information for MDAO workflows in wind turbine design	Facilitates transparent benchmarking, comparison, and integration of models for improved efficiency and performance	Blade ontology tailored for wind turbine blades, considering factors like aerostructural design, materials, and airfoils	Various sources, including official IEA Wind Task 37 and documentation links.
IEC 61400-25 Ontology [17]	Part of IEC 61400 series, provides uniform information exchange for monitoring and control of wind power plants	Communication between wind power plant components, focusing on monitoring and control	Yes, used in conjunction with WindIO ontology.	Uniform information exchange for monitoring and control of wind power plants	Used as part of IEC 61400 series for communication between wind power plant components	Addresses real wind power business needs for communication, standardizing crucial part of information exchange in wind power plants	Defines compatible logical node names and data names for communication between wind power plant components	Various sources, including Scientific. Net, iec.ch, ResearchGate, and documentation links

(Continued)

Table 3.1 Comparison of ontologies supporting wind turbine life cycle management. (*Continued*)

Ontology	Definition and purpose	Scope	Integration with WindIO ontology	Applications	Use in wind turbine design	Collaboration and benchmarking	Information models defined	References and documentation
WindIO [20]	Comprehensive ontology for sharing design information across MDAO practitioners.	Comprehensive ontology for design information sharing	N/A	Sharing design information across MDAO practitioners	Used for sharing design information in MDAO workflows	Supports collaboration among MDAO practitioners for design information sharing	Comprehensive ontology for design information sharing	Documentation available on readthedocs.io
Systems Modeling Framework and Ontology for Wind Turbines and Plants [21]	Focuses on developing a modeling framework and ontology for wind turbines and plants	Modeling framework and ontology for wind turbines and plants	N/A	Modeling and ontology for wind turbines and plants	Part of the modeling framework and ontology for wind turbines and plants	Supports modeling and benchmarking in the context of wind turbines and plants	Modeling framework and ontology for wind turbines and plants	Documentation available for the systems modeling framework and ontology for wind turbines and plants
Data Structuring for the Ontological Modeling of Wind Energy Systems [5]	Focuses on data structuring for ontological modeling of wind energy systems	Data structuring for ontological modeling of wind energy systems	N/A	Data structuring for ontological modeling of wind energy systems	Used for data structuring in ontological modeling of wind energy systems	Supports ontological modeling for data structuring in the context of wind energy systems	Data structuring for ontological modeling of wind energy systems	Documentation available for data structuring for the ontological modeling of wind energy systems

(*Continued*)

Table 3.1 Comparison of ontologies supporting wind turbine life cycle management. *(Continued)*

Ontology	Definition and purpose	Scope	Integration with WindIO ontology	Applications	Use in wind turbine design	Collaboration and benchmarking	Information models defined	References and documentation
Weather Ontologies [18, 22]	Broad category encompassing various ontologies related to weather in the context of wind energy	Encompasses various ontologies related to weather in the context of wind energy	N/A	Various ontologies related to weather and wind energy	Various ontologies related to weather, specific to the context of wind energy	Supports various aspects of ontologies related to weather in the context of wind energy	Various ontologies related to weather in the context of wind energy	Documentation for various weather ontologies may vary based on specific ontologies

3.6 Challenges and Opportunities

The challenges and opportunities in incorporating semantic data and potential research work for utilizing ontologies for wind energy are discussed below.

3.6.1 Dynamic and Evolving Environments

- **Challenge:** Wind turbine systems operate in dynamic environments with changing weather patterns, grid conditions, and maintenance requirements. Traditional ontologies may struggle to adapt to these dynamic settings.
- **Opportunity for Improvement:** Explore adaptive ontologies that can dynamically evolve based on real-time data and changing conditions. Incorporate machine learning techniques for continual learning and updating of ontological information.

3.6.2 Semantic Interoperability

- **Challenge:** Achieving semantic interoperability between diverse wind turbine systems, databases, and stakeholders can be challenging, preventing efficient communication and collaboration.
- **Opportunity for Improvement:** Develop ontological standards and best practices for the wind energy domain to increase semantic interoperability. Explore the usage of ontologies to bridge the gap between diverse industry standards and protocols.

3.6.3 Scale and Complexity

- **Challenge:** Large-scale wind farms with several turbines create large amounts of data. Managing the scale and complexity of ontologies for such systems can be resource-intensive.
- **Opportunity for Improvement:** Investigate scalable ontology design concepts and optimization strategies. Explore distributed ontology designs and technologies to handle large-scale wind energy systems efficiently.

3.6.4 Human-Computer Interaction

- **Challenge:** The usability of ontologies for non-experts in the wind energy sector may be constrained. Interacting with complex ontologies could provide issues for operators and maintenance people.
- **Opportunity for Improvement:** Integrate natural language processing and visualization tools with ontological systems to increase human-computer interaction. Develop user-friendly interfaces and technologies that allow domain experts to engage with ontologies without requiring deep ontology knowledge.

3.6.5 Real-Time Decision Support

- **Challenge:** Traditional ontologies may not be suited for real-time decision support, which is critical for controlling the dynamic nature of wind turbine operations.
- **Opportunity for Improvement:** Investigate AI-driven ontologies that can perform real-time analysis and decision-making. Explore the integration of streaming data and complicated event processing with ontologies for immediate response to changing conditions.

3.6.6 Security and Privacy Problems

- **Challenge:** As ontologies handle sensitive data relating to wind turbine operations, security and privacy problems become crucial. Ensuring the secure processing of data within ontologies is critical.
- **Opportunity for Improvement:** Research strategies to strengthen the security of ontological systems, including encryption, access controls, and secure communication protocols. Address privacy concerns with anonymization and data protection techniques.

3.6.7 Future Research Opportunities

a) **Explainable AI in Ontologies:** Explore the integration of explainable AI approaches into ontologies to promote transparency and interpretability, notably in decision-making processes linked to wind turbine operations and maintenance [23].

b) **Dynamic Ontologies for Adaptive Systems:** Investigate the construction of dynamic ontologies that can adapt and evolve in real-time, incorporating feedback from AI algorithms, IoT sensors, and operational experiences [24].

c) **Ontologies for Hybrid Energy Systems:** Extend ontologies to incorporate hybrid energy systems, considering the integration of wind energy with other renewable sources and storage technologies. Explore how ontologies might help optimum energy management in hybrid systems [25].

d) **Interdisciplinary Collaboration:** Promote multidisciplinary collaboration between ontology experts, wind energy engineers, data scientists, and AI researchers to ensure a holistic and domain-specific approach to ontology creation and implementation.

e) **Ontology-Based Standards:** Advocate for the establishment of common ontologies within the wind energy industry to enhance consistency, interoperability, and information exchange across different stakeholders and systems.

f) **Ethical Considerations in Ontologies:** Investigate the ethical implications of employing ontologies in wind turbine systems, addressing topics such as data ownership, transparency, and the appropriate use of AI in decision-making processes [26].

g) **Benchmarking and Evaluation:** Develop established benchmarks and evaluation criteria for ontologies in the context of wind energy. This includes examining the performance, scalability, and efficiency of ontologies in enhancing many elements of wind turbine lifecycle management.

By solving these problems and pursuing these future research prospects, the application of ontologies to wind turbines can be further refined, contributing to the creation of more intelligent, adaptive, and sustainable wind energy systems.

3.7 Conclusion and Future Work

To conclude, the increasing importance of wind energy highlights the necessity for efficient management of wind turbine lifecycles. Ontologies, utilizing languages such as OWL, have a vital function in constructing organized models that formalize information within a certain domain. These improve communication, compatibility, and logical abilities, making them valuable tools in several sectors such as artificial intelligence, information integration, and knowledge representation. Thus, ontologies provide a systematic and standardized method to tackle the difficulties related to the operation, maintenance, and decommissioning of wind turbines. This helps ensure the ongoing success and sustainability of wind energy as a significant contributor to the global energy sector.

References

1. Advantages and Challenges of Wind Energy, accessed on Nov-23, 2023 https://www.energy.gov/eere/wind/advantages-and-challenges-wind-energy.
2. https://www.irena.org/Energy-Transition/Technology/Wind-energy
3. Renewable Energy Fact Sheet: Wind Turbines, United States Environmental Protection Agency, August 2013.
4. Küçük, D. and Küçük, D., OntoWind: An improved and extended wind energy ontology, arXiv preprint arXiv:1803.02808, 2018.
5. Groza, A., Data structuring for the ontological modelling of wind energy systems, arXiv preprint arXiv:1510.03179, 2015.
6. Papadopoulos, P. and Cipcigan, L., Wind turbines' condition monitoring: an ontology model, in: *2009 International Conference on Sustainable Power Generation and Supply*, IEEE, pp. 1–4, 2009, April.
7. Gómez-Pérez, A. and Corcho, O., Ontology languages for the semantic web. *IEEE Intell. Syst.*, 17, 1, 54–60, 2002.
8. Berners-Lee, T., Hendler, J., Lassila, O., The semantic web. *Sci. Am.*, 284, 5, 34–43, 2001. Grimm, S., Abecker, A., Völker, J., Studer, R., Ontologies and the semantic web, in: *Handbook of semantic web technologies*, 2011.
9. McGuinness, D.L., Owl web ontology language overview w3c recommendation. 10 february 2004. http://www.w3.org/TR/2004/REC-owl-features-20040210/. 2004.
10. Antoniou, G. and Harmelen, F.V., Web ontology language: Owl, in: *Handbook on ontologies*, pp. 91–110, 2009.
11. Beauson, J., Laurent, A., Rudolph, D.P., Jensen, J.P., The complex end-of-life of wind turbine blades: A review of the European context. *Renew. Sustain. Energy Rev.*, 155, 111847, 2022.
12. Mendecka, B. and Lombardi, L., Life cycle environmental impacts of wind energy technologies: A review of simplified models and harmonization of the results. *Renew. Sustain. Energy Rev.*, 111, 462–480, 2019.
13. Jiang, Z., Installation of offshore wind turbines: A technical review. *Renew. Sustain. Energy Rev.*, 139, 110576, 2021.

14. Yeter, B. and Garbatov, Y., Structural integrity assessment of fixed support structures for offshore wind turbines: A review. *Ocean Eng.*, 244, 110271, 2022.
15. Mohamed, E., Seresht, N.G., AbouRizk, S., Context-driven ontology-based risk identification for onshore wind farm projects: A domain-specific approach. *Adv. Eng. Inf.*, 56, 101962, 2023.
16. Vidal, M.E., Sakor, A., Iglesias, E., Bechara, M., Deliverable D5. 3 Harmonization and Knowledge Extraction Service, 2022.
17. Nguyen, T.H., Prinz, A., Friisø, T., Nossum, R., Smart grid for offshore wind farms: Towards an information model based on the IEC 61400-25 standard, in: *2012 IEEE PES Innovative Smart Grid Technologies (ISGT)*, IEEE, pp. 1–6, 2012, January.
18. Kaladevi, R., Geetha, G., Narayanasamy, P., Ontological based interoperability and integration framework for heterogeneous weather systems. *Rev. Tec. Ing. Univ. Zulia*, 39, 1, 185–192, 2016.
19. Dykes, K. and Barter, G., IEA Wind TCP Task 37.
20. Scott, S., Greaves, P., Macquart, T., Pirrera, A., Comparison of blade optimisation strategies for the IEA 15MW reference turbine. *J. Phys. Conf. Ser.*, 2265, 3, 032029, IOP Publishing, 2022, May.
21. Bortolotti, P., Bay, C., Barter, G., Gaertner, E., Dykes, K., McWilliam, M., Friis-Moller, M., Molgaard Pedersen, M., Zahle, F., System Modeling Frameworks for Wind Turbines and Plants: Review and Requirements Specifications, 2022.
22. Staroch, P., A weather ontology for predictive control in smart homes (Doctoral dissertation), 2013.
23. de Sousa Ribeiro, M. and Leite, J., Aligning artificial neural networks and ontologies towards explainable AI. *Proc. AAAI Conf. Artif. Intell.*, 35, 6, 4932–4940, 2021, May.
24. Wang, Y., Herrling, D., Stroganov, P., Rausch, A., Ontology-based Automatic Adaptation of Component Interfaces in Dynamic Adaptive Systems, in: *Proc. 8th Int. Conf. on Adaptive and Self-Adaptive Systems and Applications (ADAPTIVE2016)*, 2016.
25. Saba, D., Laallam, F.Z., Belmili, H., Abanda, F.H., Bouraiou, A., Development of an ontology-based generic optimisation tool for the design of hybrid energy systems. *Int. J. Comput. Appl. Technol.*, 55, 3, 232–243, 2017.
26. Rovetto, R.J., The ethics of conceptual, ontological, semantic and knowledge modeling. *AI Soc.*, 39, 1547–1568, 2024. https://doi.org/10.1007/s00146-022-01563-3.

4

Statistical Forecasting Model for Solar Power Generation Under Different Environmental Conditions

Varun Pratap Singh[1,2]* and Bharti Sharma[3]

[1]Solar Thermal Energy Research Group, Department of Mechanical and Mechatronics Engineering, Stellenbosch University, Stellenbosch, South Africa
[2]Department of Mechanical Engineering, School of Advanced Engineering, UPES, Dehradun, Uttarakhand, India
[3]School of Computing, DIT University, Dehradun, Uttarakhand, India

Abstract
Accurate solar power forecasting is essential for the reliable operation of energy grids, energy markets, and renewable energy systems. This chapter explores statistical forecasting models for solar power generation under varying environmental conditions. It begins with an introduction to solar power forecasting, discussing its significance and the challenges posed by intermittency and weather variability. The fundamentals of solar energy are reviewed, focusing on the key environmental and geographical factors that influence power output. A detailed analysis of statistical forecasting techniques is provided, covering time-series methods and machine learning approaches, with a specific focus on neural networks and artificial neural network models for solar forecasting. The chapter also examines the impacts of weather variability and geographical location on solar power generation. Future directions emphasize the potential of emerging technologies and methodologies, such as artificial intelligence and big data integration, to improve forecasting accuracy. This chapter offers insights into the evolving landscape of solar forecasting, supporting energy operators and researchers in developing more precise and adaptive models for sustainable energy management.

Keywords: Solar power forecasting, statistical forecasting models, time-series methods, machine learning in solar forecasting, artificial neural networks (ANN), environmental impact, weather variability

**Corresponding author*: ray_varun@yahoo.com

Jai Govind Singh, Rupendra Kumar Pachauri and Sasidharan Sreedharan (eds.) Forecasting Methods for Renewable Power Generation, (67–94) © 2025 Scrivener Publishing LLC

4.1 Introduction

4.1.1 Overview of Solar Power Forecasting

Solar power forecasting plays a crucial role in integrating renewable energy sources (RESs), particularly solar power, into the energy grid. This process is essential for managing grid efficiency, enhancing the reliability of power supplies, and maximizing the use of solar energy. Effective forecasting helps predict energy production levels, aiding utility companies in managing load balancing and reducing dependence on costly and environmentally harmful Peaker plants. Accurate predictions of solar output are vital for minimizing the costs associated with energy generation and distribution, ensuring that solar power fluctuations do not destabilize the grid [1].

RESs are inherently dynamic and greatly influenced by geographic and meteorological conditions. The integration of large-scale RES, such as solar photovoltaic (PV) systems, into existing energy grids presents significant challenges. These systems are crucial in smart grids for their ability to generate electricity. Issues such as system stability, reliability, electric power balance, reactive power compensation, and frequency response are prominent when integrating large-scale grid-connected solar PV plants [2].

The modeling stage, which includes sizing, designing, and simulating solar PV plants, is vital. Researchers focus on optimizing solar PV power generation by considering the number of PV modules, storage capacity, and inverter and controller types. Effective energy storage scheduling improves the operation of grids powered by renewable energy. Sizing is a critical aspect of system design, emphasizing the need for careful planning of PV plants under varying meteorological conditions. The intermittent nature of solar resources can create significant challenges in balancing power generation and demand. Scheduling solar PV plants to align closely with anticipated load profiles is essential. Reliable forecasting information helps grid operators manage potential shortages or surpluses of solar power, enhancing system stability and reducing the costs of maintaining ancillary devices [3].

Forecasting models depend heavily on spatial and temporal resolutions, meteorological variability, selection of input parameters, and training algorithms. The effectiveness of these models is influenced by the chosen training methodologies, such as online and offline modes, which are applied to various data-driven models to assess their adaptability and performance. Solar PV power forecasting is a powerful tool that aids system operators and designers in efficiently managing and modeling solar PV plants [4].

This chapter aims to provide researchers and engineers involved in the modeling and planning of solar PV plants with valuable information on forecasting techniques. It reviews, classifies, and compares different solar PV power generation forecasting techniques, including statistical-time series methods, physical methods, and ensemble methods. Statistical approaches, which are subdivided into artificial neural networks (ANNs), support vector machines (SVMs), Markov chains, autoregressive models, and regression models, rely heavily on historical data for time-series forecasting. Physical methods, including numerical weather prediction, sky imagery, and satellite-imaging models, depend on the dynamics of solar radiation interactions within the atmosphere. Ensemble methods combine various models to leverage their unique features, addressing the limitations of individual models and enhancing overall forecasting performance [5].

Solar power forecasting has evolved from basic statistical methods to complex machine learning techniques that analyze large datasets for more accurate power output predictions. Initially reliant on simple time-series analyses, the field now incorporates sophisticated models like ANNs, SVMs, and ensemble methods. These advanced techniques use inputs from weather forecasts, satellite imagery, and real-time sensor data to predict solar irradiance and power generation more effectively. The trend is moving toward hybrid models that merge physical and statistical approaches, capturing the strengths of both to improve predictive performance across different geographical areas and temporal scales [6].

4.1.2 Importance and Challenges of Accurate Forecasting

One of the main challenges in solar forecasting is the variability and unpredictability of weather conditions, which directly influence solar irradiance. Factors such as cloud movements, atmospheric conditions, and other local environmental influences can significantly alter the amount of sunlight reaching solar panels. Moreover, the varying angles of sunlight throughout the year and the diverse geographic features of installation sites further complicate accurate model creation. These challenges are compounded by the need for real-time data integration and the computational demands of complex models, which require substantial processing power and robust data-handling capabilities to be effective [7].

4.2 Fundamentals of Solar Power

PV technology converts sunlight directly into electricity using semiconductor materials that exhibit the PV effect. When photons from sunlight strike a PV cell, they may be absorbed by the semiconductor material, typically silicon. The energy from the absorbed light excites electrons, allowing them to flow freely. PV cells have electric fields that force these free electrons into a direct current. This current can be drawn off for external use by placing metal contacts on the top and bottom of the PV cell [8, 9].

4.2.1 Key Factors Influencing Solar Power Output

A typical solar panel, or module, consists of multiple PV cells connected in series and parallel to increase voltage and current output respectively. These panels are encased in protective materials and are equipped with anti-reflective coatings to maximize light absorption and minimize losses. The entire assembly includes a glass front, a protective back sheet, and a frame that provides structural stability. Solar panels are often combined into larger units called arrays, which can be customized to meet specific power needs by altering the number and configuration of the panels [7, 10].

Physical and Environmental Parameters Affecting Solar Energy: Several factors influence the efficiency and output of solar power systems:

- **Sunlight Intensity:** Solar power output is directly proportional to the intensity of sunlight, which varies according to the time of day, season, and weather conditions. Cloud cover, fog, and pollution can significantly reduce the amount of sunlight reaching the panels.
- **Temperature:** Contrary to popular belief, solar panels are less efficient at higher temperatures. Increased temperatures can reduce the voltage and overall efficiency of a solar panel. Thus, cooler conditions are generally more favorable for solar power generation [11].
- **Angle of Incidence:** The angle at which sunlight strikes the solar panels affects their ability to absorb light. Panels oriented directly toward the sun at a perpendicular angle absorb the maximum amount of energy. The optimal angle of installation depends on the geographic location and the specific environmental conditions of the site [12].
- **Shading:** Even partial shading of a solar panel can significantly reduce its output. Shading from trees, nearby buildings, or other obstructions can cause a drop in power production, as most panels are connected in series within a

module, and the weakest-performing cell can affect the output of the entire system.

- **Panel Degradation:** Over time, solar panels suffer from degradation due to exposure to environmental conditions such as UV light and weather cycles. This degradation reduces the output efficiency of the panels gradually.

Understanding these fundamental aspects of solar power is crucial for optimizing the design and location of PV installations. By considering the various physical and environmental parameters, engineers and designers can maximize the efficiency and performance of solar power systems, adapting their strategies to meet the energy demands effectively and sustainably [13].

4.3 Statistical Forecasting Techniques

Statistical models in solar forecasting are mathematical frameworks used to predict future solar energy output based on historical data. These models analyze patterns and relationships between variables, such as solar irradiance, temperature, and time, to estimate future values. The utility of statistical models lies in their ability to provide quantitative assessments that can help in planning and decision-making processes for energy management. They are crucial for optimizing the operation of solar power systems, managing energy supply in grids, and facilitating effective integration of solar energy into existing power systems [14, 15].

4.3.1 Time Series Forecasting Methods

Time series forecasting involves using historical data to predict future values. This method is extensively used in various fields such as finance, meteorology, and energy production, including the prediction of solar PV generation as mentioned in your document. Let us explore the different categories of time series statistical methods, their basic concepts, formulas, applications, limitations, and potential modifications [16, 17].

4.3.1.1 Autoregressive Models (AR)

Autoregressive models predict future **behavior** based on past **behavior**. They are used to forecast future solar power output by expressing the next value in the series as a linear combination of previous values. This approach is particularly useful for handling time-series data where past solar output significantly influences future outputs [18].

AR models predict future values based on past values. The model assumes that past data points are useful predictors of the future.

$$Xt = \alpha + \phi_1 X_{t-1} + \phi_2 X_{t-2} + \ldots + \phi_p X_{t-p} + \epsilon_t \quad (4.1)$$

where:

α is a constant.

ϕ are the parameters of the model.

p is the number of lag observations included.

ϵ_t is white noise.

Ar models used in economics for predicting stock prices, in meteorology for weather forecasting, and in engineering for signal processing. Assumes a linear relationship between past and future values. Sensitive to non-stationary data which can lead to misleading results. Adding seasonality and integrating differencing steps can help stabilize the mean of the time series, leading to the development of ARIMA models [19, 20].

4.3.1.2 *Moving Average Models (MA)*

Moving average models use the average of past data points to smooth the series and forecast future points, helping to identify trends. In solar power forecasting, MA models can help smooth out short-term fluctuations due to cloud cover or other temporary environmental changes, making the underlying trends in solar power generation more apparent [21].

MA models use past forecast errors in a regression-like model. Unlike AR models, MA models do not directly depend on past values of the series.

$$X_t = \mu + \epsilon_t + \theta_1 \epsilon_{t-1} + \ldots + \theta_q \epsilon_{t-q} \quad (4.2)$$

where:

μ is the mean of the series.

θ are the parameters of the model.

q is the number of lagged forecast errors in the prediction equation.

ϵt is white noise.

MA forecasting stock returns, economic indicators, and demand forecasting in retail. It is effective only when the series is stationary and difficult to interpret when combined with AR models. MA is often combined with AR models to form ARMA or ARIMA models for better adaptability to real-world data [22, 23].

4.3.1.3 *Autoregressive Integrated Moving Average (ARIMA) Models*

ARIMA models combine autoregressive models and moving average models and integrate differencing to make the time series stationary. This method is highly effective for non-seasonal patterns with trends and is widely used for forecasting solar power output where both seasonality and non-linear trends may be present [24].

ARIMA models are used to describe and forecast stationary time series data through differencing non-stationary data.

$$(1-\sum_{i=1}^{p} \phi_i L^i)(1-L)^d X_t = (1+\sum_{i=1}^{q} \theta_i L_i)\epsilon_t \tag{4.3}$$

where:

L is the lag operator.

d is the degree of differencing.

p and q are the orders of the AR and MA parts of the model.

ARIMA is widely used in financial time series prediction, economic forecasting, and environmental modeling. The limitation of ARIMA assumes the linearity and stationarity of data after differencing and involves intensive computational work to identify the right parameters. Seasonal ARIMA (SARIMA) models incorporate seasonal effects, which are particularly useful in fields like retail and climatology [25, 26].

4.3.1.4 *Exponential Smoothing Models*

Exponential smoothing models are a popular and effective set of forecasting techniques used especially in time series analysis where data exhibit a trend or seasonal patterns. These models are highly flexible, making them well suited to making short-term forecasts, which is especially important in fields such as inventory management, demand forecasting, and weather prediction [27]. These models forecast by estimating the level, trend, and seasonality in the data. Exponential smoothing models apply weighted averages of past observations to make forecasts. The weights decrease exponentially as the observations get older, meaning that more recent observations have more influence on the forecast. This method is particularly useful in updating forecasts to reflect more recent changes [28–30].

Simple Exponential Smoothing:

$$\hat{x}_{t+1} = \alpha x_t + (1-\alpha)\hat{x}_t \tag{4.4}$$

Holt's Linear Trend Model:

$$\hat{x}_{t+1} = l_t + b_t \tag{4.5}$$

$$l_t = \alpha x_t + (\mathbf{1} - \alpha)(l_{t-1} + b_{t-1}) \tag{4.6}$$

$$b_t = \beta^*(l_t - l_{t-1}) + (1-\beta^*)b_{t-1} \tag{4.7}$$

where:

lt is the estimated level.

bt is the estimated trend.

Exponential Smoothing Models are generally used in forecasting airline passengers, electricity demand, and inventory studies. The limitations of Exponential Smoothing Models include simplistic assumptions that can lead to inaccuracies in the presence of complex patterns and sensitivity to the smoothing parameters chosen. Advanced models like Holt-Winters add seasonal adjustments to capture repeated cycles of behavior over time [31].

Time series statistical methods are powerful tools for forecasting, offering insights into future trends based on historical data. Each method has its own set of strengths and limitations, which makes it essential to choose the appropriate model based on the specific characteristics of the data. Modifications and integrations of different methods often provide better performance by addressing the limitations of single models. Understanding these methods deeply enables effective and efficient planning and decision-making across various applications [32, 33].

4.3.2 Machine Learning Approaches in Solar Forecasting

4.3.2.1 Neural Networks

Neural networks are powerful machine learning models that can model complex nonlinear relationships. In solar forecasting, neural networks analyze historical data on solar irradiance, weather patterns, and other relevant factors to learn and predict solar energy output. Their ability to learn from large datasets and capture non-linear dependencies makes them highly effective for this application [34].

4.3.2.1.1 Artificial Neural Network (ANN) Models for Solar Power Forecasting

Artificial neural networks (ANNs) are computational models inspired by the human brain's structure, capable of handling complex and nonlinear relationships between inputs and outputs [35]. This capability makes ANNs particularly suitable for solar power forecasting, where the variability in weather conditions and solar irradiance presents challenging data patterns.

Structure and Working of ANNs:

ANNs consist of multiple layers of interconnected nodes or neurons, which include the following:

- **Input Layer:** Receives raw data inputs, such as solar irradiance levels, temperature, and time of day.
- **Hidden Layers:** One or more layers that transform the input into something the output layer can use. Neurons in these layers apply weights to the inputs and pass them through an activation function to introduce non-linearities into the output.
- **Output Layer:** Produces the final prediction output, such as the expected power output from a solar PV system.

The connections between neurons have associated weights that adjust as the network learns from data during the training phase. The most common type of ANN used in solar forecasting is the multilayer perceptron (MLP), which includes at least three layers (input, hidden, and output) and utilizes backpropagation for learning as shown in the below mentioned Figure 4.1.

Modeling Process

The modeling of ANNs involves several key stages as discussed below and shown in figure:

1. **Design Stage:**
    - **Network Architecture:** Selection of the type of ANN, number of neurons in each layer, and the number of layers.
    - **Parameter Setting:** Establishing the input and output parameters for the network.

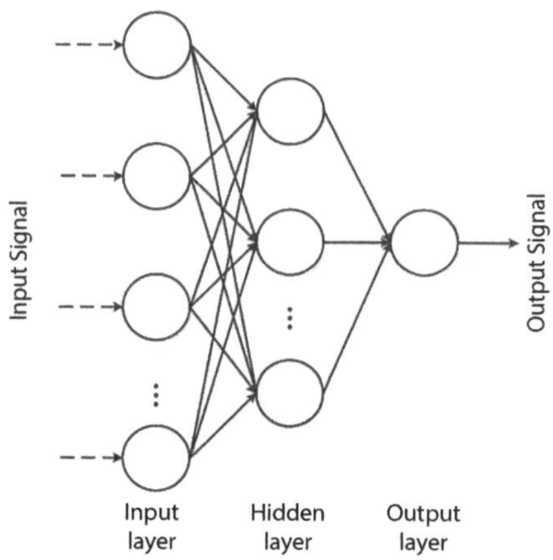

Figure 4.1 Basic ANN structures.

- **Validation:** Using a separate dataset to validate the model ensures that it generalizes well to new data.

2. **Training Phase:**
   - Data is fed into the network, and the model iteratively adjusts the weights of the connections to minimize the difference between the predicted output and the actual data.
   - This phase continues until a predetermined performance metric is met or no significant improvement is noted.

3. **Validation Phase:**
   - The trained model is tested against a new set of data that was not used during the training phase.
   - This step is crucial to confirm that the model is robust and performs well on unseen data.

Artificial neural networks (ANNs) are widely utilized in solar power forecasting to estimate the power output from PV systems, leveraging inputs such as solar radiation, temperature, and cloud coverage. The ability to precisely forecast solar power is essential for efficient grid management and optimizing the utilization of RESs [36]. ANNs are particularly adept at short-term and intraday forecasting, where they provide valuable predictions a few minutes to several hours ahead, aiding operational decisions and load balancing. Additionally, they play a crucial role in day-ahead forecasting, which is instrumental in planning and managing energy supply more

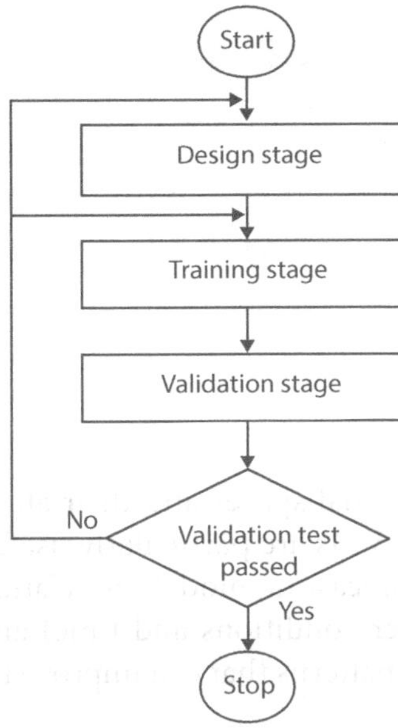

Figure 4.2 The modeling process flowchart for ANN.

effectively, thus reducing reliance on non-RESs [37–39]. Figure 4.2 shows the modeling process flowchart for ANN used in solar energy forcasting.

Despite their significant advantages in forecasting accuracy and adaptability, ANNs come with their set of challenges. The performance of an ANN heavily depends on the quantity, quality, and variety of the training data, making it sensitive to the data it is trained on. There is also the risk of overfitting, where without proper regularization and validation, ANNs may become overly complex and start to capture noise as if they were actual patterns, which can degrade their performance when applied to new data [40]. Additionally, training ANNs, especially those with many deep layers, can be computationally demanding. Nevertheless, the flexibility and computational power of ANNs make them an asset in enhancing the reliability and efficiency of solar power systems. By continuously refining these models and incorporating real-time data, the accuracy of solar power forecasts can be significantly improved, supporting the broader adoption and efficient integration of solar energy into the power grid [41–43].

4.3.2.2 Support Vector Machines (SVMs)

Support vector machines are supervised learning models that analyze data used for classification and regression analysis. Due to their effectiveness

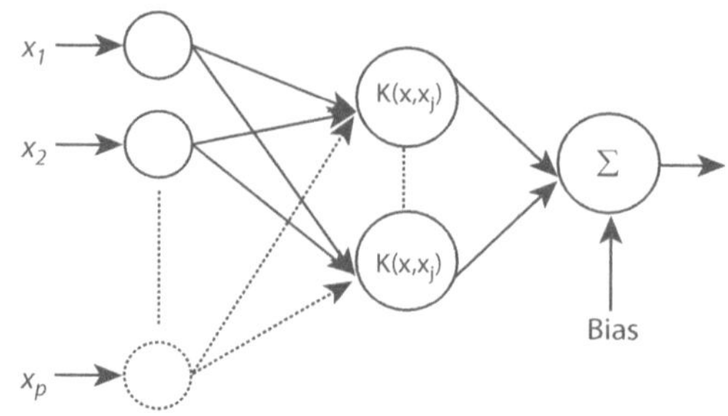

Figure 4.3 SVM architecture.

in handling high-dimensional spaces and their ability to model complex nonlinear relationships, SVMs are particularly useful in solar power forecasting. SVMs in solar forecasting model the relationships between input variables (such as weather conditions and time) and solar output, effectively capturing complex patterns that can improve the accuracy of predictions [44].

SVM operates based on the principle of finding a hyperplane that best divides a dataset into classes with maximum margin [45, 46]. The support vectors are the data points nearest to the hyperplane, and they directly influence the position and orientation of the hyperplane. This methodology helps in achieving a generalized model with good performance on unseen data. A basic SVM architecture is shown in the mentioned Figure 4.3.

Mathematical Formulation:

The SVM model seeks to solve the following optimization problem:

$$\min_{w,b,\xi\cdot} \frac{1}{2} \| w \|^2 + C \sum_{i=1}^{n} \xi_i \tag{4.8}$$

Subject to the constraints for each data point :

$$y_i(w \cdot \phi(x_i) + b) \geq 1 - \xi_i, \quad \xi_i \geq 0 \tag{4.9}$$

here:

w represents the weights of the hyperplane.

b is the bias term.

$\phi(xi)$ is the function mapping *xixi* into a higher-dimensional space.

C is the regularization parameter that balances the margin maximization and loss minimization.

ξi are the slack variables allowing for the violation of the margin for non-separable cases.

yi are the labels associated with each data point.

Kernel Trick:

The kernel trick is employed to solve non-linear problems by using a kernel function $K(x_i, x_j)$ to compute the dot product in high-dimensional space, which allows SVM to create non-linear boundaries in the original space.

$$K(x_i, x_j) = \phi(x_i) \cdot \phi(x_j) \tag{4.10}$$

Common kernels include the following:

- Linear: $K(x_i, x_j) = x_i^T x_j$
- Polynomial: $K(x_i, x_j) = (\gamma x_i^T x_j + r)^d$
- Radial Basis Function (RBF): $K(x_i, x_j) = e^{-\gamma \|x_i - x_j\|^2}$

Support vector machines (SVMs) are extensively used in solar power forecasting to predict solar radiation and power output. They excel in handling a variety of inputs such as historical solar radiation data, temperature, and cloud cover, which are crucial for making accurate predictions about solar power. SVMs are utilized not only for predictive modeling—forecasting hourly or daily solar power output based on historical weather conditions and solar irradiance data—but also for classification tasks, such as identifying weather patterns that significantly impact solar power generation, like distinguishing between cloudy and sunny days. While SVMs are known for their high generalization performance, which makes them less prone to overfitting, and offer the flexibility to model a wide range of non-linear relationships due to the choice of the kernel function, they do present challenges [47–49]. Scalability can become an issue as SVMs can become computationally intensive with increasing training samples, particularly with the large datasets typical in solar energy applications. Additionally, the performance of SVM heavily depends on the meticulous tuning of the kernel and regularization parameters, which can be a demanding process. Support vector machines provide a powerful toolkit for forecasting solar power output. Their ability to handle complex, non-linear relationships in the data makes them ideal for predicting variables influenced by multiple environmental factors, such as solar power generation. Properly tuned SVM models can significantly enhance the accuracy of solar energy forecasts, contributing to more efficient energy management and grid integration of solar power resources [50].

4.3.3 Ensemble Methods

Ensemble methods are advanced machine learning techniques that combine predictions from multiple models to achieve higher accuracy than any individual model could on its own. This approach is particularly useful in complex predictive tasks like solar power forecasting, where the objective is to minimize the prediction errors that come from various sources of uncertainty such as weather variability and sensor inaccuracies [51].

Principle of Ensemble Methods
The fundamental idea behind ensemble methods is to build a predictive model by integrating the outputs of multiple models. This integration helps to reduce both bias and variance in predictions:

- **Bias** is reduced because ensemble methods can combine simpler models to form a complex model that captures a broad range of data complexities.
- **Variance** is reduced because averaging several models' predictions tends to cancel out some of the noise.

Ensemble methods are commonly divided into two main types:

1. **Bagging (Bootstrap Aggregating):** This method involves training multiple models on different subsets of the original dataset, selected with replacement (bootstrap samples). The final prediction is typically the average (for regression) or the majority vote (for classification) of all model predictions. An example of bagging is the random forest algorithm, which combines multiple decision trees.

Equation for Bagging:

$$\hat{y} = \frac{1}{N}\sum_{i=1}^{N} f_i(x) \tag{4.11}$$

where N is the number of models, $f_i(x)$ is the prediction of the ith model, and x is the input feature set.

2. **Boosting:** This method involves sequentially training models, each focusing on the errors of the previous ones. The final prediction is a weighted sum of all the model predictions, where more accurate predictors are given more weight.

Equation for Boosting:

$$\hat{y}(x) = \sum_{i=1}^{N} \alpha_i f_i(x) \tag{4.12}$$

where α_i represents the weight of the ith model, $f_i(x)$ is the prediction of the ith model, and N is the total number of models.

Ensemble methods in solar power forecasting leverage the strengths of various predictive models to enhance accuracy and robustness. These methods integrate models like ANNs, which excel in capturing nonlinear relationships in data, SVMs, which are effective in high-dimensional spaces, and decision trees, which can handle both linear and nonlinear relationships without needing transformations [52, 53]. By combining these diverse models, ensemble methods can mitigate the individual weaknesses of each model [54]. For instance, while ANNs may be highly accurate, they are prone to overfitting; decision trees provide excellent interpretability but might be too simplistic to capture complex data patterns; SVMs are robust but may struggle with very large datasets. Thus, an ensemble of these models provides a composite model that is both robust and capable of making highly accurate forecasts [55].

The practical application of ensemble methods in solar power forecasting includes both cooperative and competitive ensemble techniques. Cooperative ensemble forecasting involves dividing the forecasting task into sub-tasks, each addressed by a separate model or a subset of the data. This approach can involve techniques like pre-processing, where the dataset is split into smaller chunks that are each modeled separately, and post-processing, which might involve applying different models to different features of the data, such as using ARIMA models for linear time series data. Competitive ensemble forecasting, on the other hand, uses multiple predictive models, each with different parameters to develop individual forecasts. These forecasts are then aggregated, typically by averaging, to produce a final ensemble forecast. This method leverages diversity in kernel, parameter, and data to enhance the forecasting accuracy, ensuring that the ensemble model outperforms any single predictive model in terms of accuracy and robustness, particularly useful in the variable and complex field of solar power forecasting [56, 57].

In the realm of solar power forecasting, ensemble methods utilize a combination of diverse predictive models to enhance forecast accuracy and robustness. This approach often integrates various types of models such as neural networks (ANNs), which are proficient in handling nonlinear data relationships; SVMs, which excel in high-dimensional spaces; and decision trees, which adeptly capture both linear and non-linear relationships without the need for data transformation [58]. By harnessing the strengths of these models while compensating for their individual weaknesses, ensemble methods can create a more robust forecasting tool. For instance, while ANNs are highly accurate, they are also prone to overfitting; decision trees provide excellent interpretability but might oversimplify complex patterns; and SVMs, although robust, can struggle with large datasets. The combination of these models through ensemble techniques can yield a forecasting tool that is both accurate and efficient. Different model classification based on temporal and spatial resolutions is shown in Figure 4.4 [59].

The use of ensemble methods in solar power forecasting comes with notable advantages and challenges. One major advantage is the improved accuracy obtained from integrating multiple models, which typically outperforms any single model approach. This integration also helps reduce the risk of overfitting, as the ensemble tends to average out the noise to which

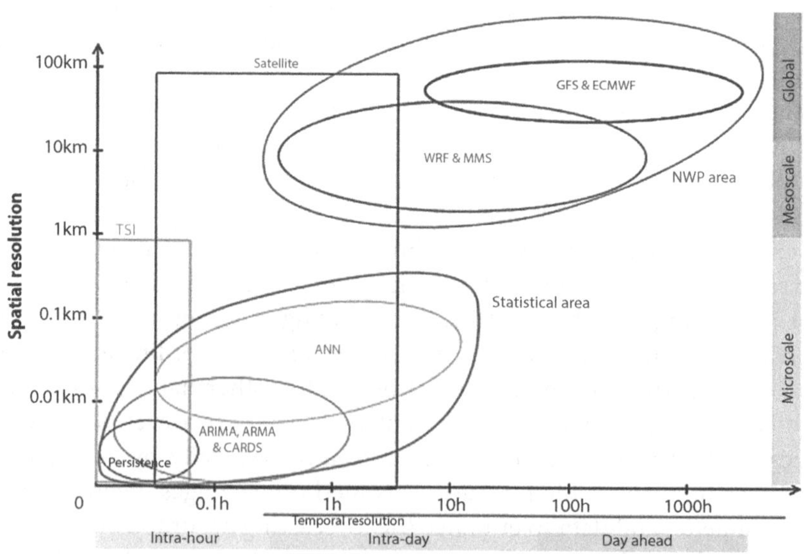

Figure 4.4 Model classification based on temporal and spatial resolutions [61].

individual models might overreact. Furthermore, the flexibility of ensemble methods allows them to be adapted to various data characteristics, making them versatile tools in solar power forecasting. However, these benefits are accompanied by challenges such as increased complexity in managing and combining multiple models, which can be computationally intensive. Additionally, the selection and tuning of models within the ensemble—such as deciding which models to include and how to optimally combine them (e.g., setting weights in boosting)—require meticulous planning and careful tuning. Despite these hurdles, ensemble methods stand as a potent set of techniques in solar power forecasting, providing a practical solution to navigate the complexities of predicting solar irradiance and power output [60].

These statistical and machine learning techniques provide a comprehensive toolkit for accurately forecasting solar power generation, each offering unique strengths that can be tailored to specific characteristics of the data and forecasting needs. By leveraging these models, stakeholders can enhance the predictability and stability of solar power supply, contributing to more efficient and sustainable energy systems.

4.4 Environmental Impacts on Solar Power Generation

4.4.1 Influence of Weather Variabilities

Weather conditions have a profound impact on the efficiency and output of solar power systems. Various weather elements affect solar generation as follows [62, 63]:

- **Sunlight Availability:** The most direct factor is the amount of sunlight reaching the solar panels. Clear, sunny days provide optimal conditions for solar power generation, while cloudy or rainy days significantly reduce the output. Even a small amount of cloud cover can lead to sharp drops in power generation.
- **Temperature:** Solar panels operate more efficiently at cooler temperatures. High temperatures can reduce the efficiency of PV cells, thereby decreasing the electricity output. This effect is particularly pronounced during hot summer days when the sun is brightest.

- **Humidity:** High humidity can affect solar panel efficiency. Moisture in the air can scatter and absorb sunlight, reducing the direct irradiance reaching the panels. Additionally, high humidity can lead to corrosion and degradation of panel materials over time if not properly sealed.
- **Wind:** Wind can have both positive and negative impacts on solar power. While the cooling effects of wind may enhance panel efficiency by lowering their temperature, strong winds can carry dust and debris that accumulate on the panels, obstructing sunlight and reducing their operational efficiency.
- **Snow and Ice:** In colder climates, snow and ice can cover panels and completely halt electricity production. However, the albedo effect of snow can increase irradiance due to reflected sunlight under clear conditions post-snowfall.

4.4.2 Geographical Impact on Solar Radiation

Latitude and Its Effects on Solar Incidence: Latitude plays a crucial role in the amount of solar energy received at a particular location. Areas closer to the equator receive more direct sunlight year-round, making them ideal for solar power installations. The sun's rays hit these regions more perpendicularly, minimizing energy loss that occurs when sunlight strikes at an angle. In contrast, regions at higher latitudes experience significant seasonal variations in solar incidence due to the tilt of the Earth's axis. During winter months, these areas receive less sunlight due to the lower angle of the sun above the horizon, reducing solar panel efficiency [64].

Local Weather Patterns and Their Influence on Solar Power: Local weather patterns, driven by geographical features such as mountains, bodies of water, and urban structures, significantly influence solar power generation:

- **Mountainous Regions:** These areas can experience rapid weather changes that can affect solar output. Mountains can enhance cloud formation and lead to increased precipitation, which may reduce sunlight availability.
- **Proximity to Bodies of Water:** Locations near large bodies of water can experience more cloud cover and fog, which reduces solar radiation. However, these areas might also benefit from more stable temperatures, which can help maintain solar panel efficiency.

- **Urban Heat Islands:** Urban areas can experience higher temperatures due to the heat absorbed by roads and buildings. This urban heat island effect can decrease the efficiency of solar panels installed in densely populated areas [65].

Understanding the environmental impacts on solar power generation is crucial for predicting energy output and planning the placement and management of solar power installations. It ensures that systems are designed and operated to maximize efficiency and adapt to changing environmental conditions [66].

4.5 Future Directions and Innovations

4.5.1 New Technologies and Methodologies Improving Forecasting Accuracy

Advancements in solar forecasting technologies are key to optimizing the integration of solar energy into the power grid [67]. Some emerging technologies and methodologies include the following:

- **Sky Imaging Systems:** Ground-based sky imagers can now predict solar irradiance for the short term (a few minutes to a few hours ahead) by analyzing the movement and type of clouds. These systems allow for real-time adjustments in grid management, helping to accommodate sudden changes in solar power generation.
- **Satellite Data Utilization:** Enhanced satellite imaging technologies provide comprehensive data on cloud cover and atmospheric conditions over large geographical areas. By analyzing this data, forecasters can improve the accuracy of solar irradiance predictions over longer periods and larger scales.
- **Quantum Dot Solar Cells:** While not a direct forecasting technology, the development of quantum dot solar cells may significantly impact forecasting needs by potentially offering higher efficiency and broader spectral absorption, which could mitigate some variability issues by making solar power generation more consistent across different weather conditions.
- **Radiative Cooling Panels:** Innovations like radiative cooling solar panels that can emit heat into space and reduce panel temperature can help maintain the efficiency of solar

cells even under strong sunlight, reducing the impact of temperature on forecasting models.

4.5.2 Integrating AI and Big Data into Solar Energy Systems

The integration of artificial intelligence (AI) and big data analytics into solar energy systems represents a transformative shift in how solar energy is forecasted, managed, and integrated into power grids [68].

- **Machine Learning Models:** AI and machine learning algorithms can analyze vast amounts of data from weather stations, satellites, and existing solar farms to learn patterns and predict solar output more accurately. These models can adapt to changing patterns in weather and environmental conditions, improving their accuracy over time.
- **Big Data Analytics:** Big data technologies allow for the processing and analysis of huge datasets that include historical weather data, real-time sensor data from solar panels, and other relevant information. By analyzing these data, predictive models can be refined to improve accuracy and dynamically adapt to changing environmental conditions.
- **Internet of Things (IoT):** IoT devices connected across solar power systems can provide real-time data collection and analysis, enhancing the responsiveness of forecasting models. This network of connected sensors and controllers can monitor and adjust to local changes in sunlight and weather conditions instantaneously, optimizing power output and reducing inefficiencies.
- **Advanced Simulation Models:** Using high-performance computing, simulation models can now run multiple scenarios based on different weather conditions and system configurations. These simulations help in understanding potential impacts on solar power generation and in developing strategies to mitigate risks associated with solar power variability [69].

4.6 Conclusion

The exploration of statistical forecasting models for solar power generation under varying environmental conditions underscores the complexity and necessity of accurate solar forecasting. This chapter provided an in-depth analysis of the fundamental mechanisms of solar power through PV systems, discussing how components such as solar panels convert sunlight into electricity and the factors that influence their efficiency. The role of environmental variables like temperature, sunlight intensity, and geographic factors was highlighted, elucidating how these elements impact the performance of solar energy systems.

In delving into statistical forecasting techniques, the discussion spanned several models, from traditional time series models like ARIMA to advanced machine learning approaches including neural networks and SVMs. These methods are integral to developing predictive models that are not only accurate but also capable of handling the inherent variability associated with solar power. The introduction of ensemble methods further exemplifies the industry's move toward more robust and reliable forecasting solutions, integrating multiple predictive models to mitigate individual weaknesses.

The influence of environmental conditions on solar power generation was analyzed, emphasizing the significance of weather variabilities and geographic impacts. It was discussed how local weather patterns and geographic peculiarities like latitude and proximity to water bodies can drastically alter solar irradiance levels, thereby affecting power output.

Looking forward, the prospects for research and technological advancements in solar power forecasting are promising and diverse. One area ripe for further exploration is the enhancement of short-term forecasting accuracy through the integration of real-time data from IoT devices and advanced sky imaging systems. These technologies offer the potential to dynamically adjust forecasts based on immediate environmental changes, such as sudden cloud cover or atmospheric conditions, which are currently challenging to predict with high accuracy.

Another promising direction involves the application of hybrid models that combine physical and statistical forecasting methods. These models could leverage the strengths of each approach, potentially offering superior prediction accuracy over models that rely solely on one type or the other. Research could focus on developing adaptive algorithms that can seamlessly integrate data from various sources and update their predictions in real time, a critical capability for managing the integration of solar power into smart grids.

The development of quantum dot solar cells and radiative cooling technology also presents an intriguing avenue for reducing the dependency on precise weather forecasts by making solar panels more efficient and less susceptible to environmental variables. Further research into these materials and technologies could revolutionize the way solar panels are manufactured and operated, potentially reducing the impact of temperature and irradiance variability on solar output.

The role of big data analytics and machine learning in solar forecasting is expected to expand, with future research likely to explore deeper integrations of these technologies. This could involve the use of more sophisticated machine learning models that can process and analyze larger datasets with greater complexity, including unstructured data types that are increasingly available but currently underutilized. The demand for renewable energy continues to grow, so too does the need for advanced forecasting techniques that can ensure the efficient and effective integration of solar power into the global energy market. The continued advancement of forecasting technologies, alongside innovations in solar panel efficiency and energy management systems, will be crucial for maximizing the potential of solar energy as a major component of sustainable energy strategies worldwide. The journey toward highly reliable solar forecasting is just beginning, and the horizon is bright with potential.

References

1. Ahmed, R., Sreeram, V., Mishra, Y., Arif, M.D., A review and evaluation of the state-of-the-art in PV solar power forecasting: Techniques and optimization. *Renew. Sustain. Energy Rev.*, 124, 109792, 2020, doi: https://doi.org/10.1016/j.rser.2020.109792.
2. Yang, D., Li, W., Yagli, G.M., Srinivasan, D., Operational solar forecasting for grid integration: Standards, challenges, and outlook. *Sol. Energy*, 224, 930–937, 2021, doi: https://doi.org/10.1016/j.solener.2021.04.002.
3. Sharma, A.K., Nayal, M., Jain, S., Singh, V.P., Chapter 07 Optimization Techniques of Solar Thermal and Hybrid Energy Systems, in: *Highly Efficient Thermal Renewable Energy Systems Design, Optimization and Applications*, 1st ed., R.W., V. Verma, S. Thangavel, N. Dutt, A. Kumar (Eds.), pp. 1–14, CRC Press, Boca Raton, FL, USA, 2024, [Online]. Available: https://www.routledge.com/Highly-Efficient-Thermal-Renewable-Energy-Systems-Design-Optimization/Verma-Thangavel-Dutt-Kumar-Weerasinghe/p/book/9781032595641.
4. Nayal, M., Sharma, A.K., Jain, S., Singh, V.P., Chapter 06 Design and Modelling of Solar, Geothermal and Hybrid Energy Systems.pdf, in: *Highly Efficient Thermal Renewable Energy Systems Design, Optimization and Applications*, 1st ed., R.W., V. Verma, S. Thangavel, N. Dutt, A. Kumar (Eds.), pp. 1–15, CRC Press, Boca Raton, FL, USA, 2024, [Online]. Available: https://www.routledge.com/Highly-Efficient-Thermal-Renewable-Energy-Systems-Design-Optimization/Verma-Thangavel-Dutt-Kumar-Weerasinghe/p/book/9781032595641.

5. Lovegrove, K. and Pye, J., 2- Fundamental principles of concentrating solar power (CSP) systems, in: *Woodhead Publishing Series in Energy*, K. Lovegrove and W. B. T.-C. S. P. T. Stein (Eds.), pp. 16–67, Woodhead Publishing, Cambridgeshire, United Kingdom, 2012, doi: https://doi.org/10.1533/9780857096173.1.16.
6. Singh, V.P. and Jain, S., Economic analysis of a large scale solar updraft tower power plant. *Sustain. Energy Technol. Assess.*, 58, 103325, 2023, doi: https://doi.org/10.1016/j.seta.2023.103325.
7. Khosrojerdi, F., Gagnon, S., Valverde, R., Identifying Influential Factors Affecting the Shading of a Solar Panel, in: *2021 IEEE Electrical Power and Energy Conference (EPEC)*, pp. 255–260, 2021, doi: 10.1109/EPEC52095.2021.9621688.
8. Kumar, A., Gori, Y., Meena, C.S., Singh, V.P., Sharma, V.K., Design and analysis of heavy vehicle medium duty transmission gearbox system, in: *Additive Manufacturing in Industry 4.0: Methods, Techniques, Modeling, and Nano Aspects*, pp. 67–86, 2022, doi: 10.1201/9781003360001-4.
9. Bawankar, S., *et al.*, Environmental impact assessment of lithium ion battery employing cradle to grave. *Sustain. Energy Technol. Assess.*, 60, 1–10, May, 2023, doi: 10.1016/j.seta.2023.103530.
10. Li, W., Ren, H., Chen, P., Wang, Y., Qi, H., Key Operational Issues on the Integration of Large-Scale Solar Power Generation—A Literature Review. *Energies (Basel)*, 13, 22, 1–25, 2020, doi: 10.3390/en13225951.
11. Kumar, A., Singh, V.P., Meena, C.S., Dutt, N., Preface of Thermal Energy Systems, in: *Thermal Energy Systems: Design, Computational Techniques, and Applications*, pp. x–xiii, CRC Press, Technical Education Department, Kanpur, Uttar Pradesh, India, 2023, [Online]. Available: https://www.scopus.com/inward/record.uri?eid=2-s2.0-85161660089&partnerID=40&md5=056f377c943550d48151824d45dcb030.
12. Rathore, G.S., S.R., A.N., Karn, A., Singh, V.P., Recent Progress in Solar Dryers using Phase Changing Material: A Review. *Int. J. Energy Resour. Appl. (IJERA)*, 2, 1, 57–77, 2023, doi: 10.56896/IJERA.2023.2.1.005.
13. Kumar, A., Singh, V.P., Meena, C.S., Dutt, N., Aim and Scope of Thermal Energy Systems: Design, Computational Techniques, and Applications, in: *Thermal Energy Systems: Design, Computational Techniques, and Applications*, A. Kumar, V.P. Singh, C.S. Meena, N. Dutt (Eds.), pp. 9–10, CRC Press, Boca Raton, FL, USA, 2022.
14. Dash, S., *et al.*, A Novel Algorithmic Forex Trade and Trend Analysis Framework Based on Deep Predictive Coding Network Optimized with Reptile Search Algorithm. *Axioms*, 11, 8, 1–29, Aug. 2022, doi: 10.3390/axioms11080396.
15. Wu, Y.-K., Huang, C.-L., Phan, Q.-T., Li, Y.-Y., Completed Review of Various Solar Power Forecasting Techniques Considering Different Viewpoints. *Energies*, 15, 9, 1–22, 2022, doi: 10.3390/en15093320.

16. Kim, E., Akhtar, M.S., Yang, O.-B., Designing solar power generation output forecasting methods using time series algorithms. *Electr. Power Syst. Res.*, 216, 109073, 2023, doi: https://doi.org/10.1016/j.epsr.2022.109073.
17. Sharadga, H., Hajimirza, S., Balog, R.S., Time series forecasting of solar power generation for large-scale photovoltaic plants. *Renew. Energy*, 150, 797–807, 2020, doi: https://doi.org/10.1016/j.renene.2019.12.131.
18. Sharma, B., Kumar, S., Tiwari, P., Yadav, P., Nezhurina, M.I., ANN based short-term traffic flow forecasting in undivided two lane highway. *J. Big Data*, 5, 1, 1–16, Dec. 2018, doi: 10.1186/s40537-018-0157-0.
19. Pant, G., Meena, C.S., Saxena, A., Kumar, A., Singh, V.P., Dutt, N., Study the Temperature Variation in Alternate Coils of Insulated Condenser Cum Storage Tank: Experimental Study, in: *Lecture Notes in Mechanical Engineering*, B.S. Sikarwar, S.K. Sharma, A. Jain, K.M. Singh (Eds.), pp. 627–638, Springer Nature Singapore, Singapore, 2023, doi: https://doi.org/10.1007/978-981-99-2382-3_52.
20. Dutt, N., Jageshwar, A., Ashwani, H., Mukesh, K., Awasthi, K., Pratap, V., Thermo - hydraulic performance of solar air heater having discrete D - shaped ribs as artificial roughness. *Environ. Sci. Pollut. Res.*, 30, 28, 1–22, 2023, doi: 10.1007/s11356-023-28247-9.
21. Zheng, J., *et al.*, Time series prediction for output of multi-region solar power plants. *Appl. Energy*, 257, 114001, 2020, doi: https://doi.org/10.1016/j.apenergy.2019.114001.
22. Singh, V.P. and Dwivedi, G., Technical Analysis of a Large-Scale Solar Updraft Tower Power Plant. *Energies (Basel)*, 16, 1, 103325, 2023, doi: 10.3390/en16010494.
23. Saxena, A., Prajapati, A.N., Pant, G., Meena, C.S., Kumar, A., Singh, V.P., Water Consumption Optimization of Hybrid Heat Pump Water Heating System, in: *Lecture Notes in Mechanical Engineering*, A.K. Shukla, B.P. Sharma, A. Arabkoohsar, P. Kumar (Eds.), pp. 721–732, Springer Nature Singapore, Singapore, 2023, doi: https://doi.org/10.1007/978-981-99-1894-2_61.
24. Prema, V. and Rao, K.U., Development of statistical time series models for solar power prediction. *Renew. Energy*, 83, 100–109, 2015, doi: https://doi.org/10.1016/j.renene.2015.03.038.
25. Kundu, A., Kumar, A., Singh, V.P., Meena, C.S., Dutt, N., Chapter 1: Introduction to Thermal Energy Resources and Their Smart Applications, in: *Thermal Energy Systems: Design, Computational Techniques, and Applications*, 1st Edition, V. P. S. C. S. M. N. D. A. Kumar (Ed.), pp. 1–15, CRC Press, Boca Raton, 2023, doi: 10.1201/9781003395768-1.
26. Kumar, A., Singh, V.P., Meena, C.S., Dutt, N., *Thermal Energy Systems: Design, Computational Techniques, and Applications*, 1st ed., CRC Press, Boca Raton, FL, USA, 2023, doi: https://doi.org/10.1201/9781003395768.
27. Kumar, S., Sharma, B., Nezhurina, M.I., Responsible Factors of Powered Two Wheeler Accidents: A Review, in: *International Conference on Research in*

Intelligent and Computing in Engineering (RICE), pp. 116–122, 22-24 Aug. 2018.
28. Sivasakthivel, T., Verma, V., Tarodiya, R., Meena, C.S., Singh, V.P., Kumar, R., Chapter 11: Analysis of Optimum Operating Parameters for Ground Source Heat Pump System for Different Cases of Building Heating and Cooling Mode Operations, in: *Thermal Energy Systems: Design, Computational Techniques, and Applications*, 1st ed., A. Kumar, N. Dutt, V. Singh, C. Meena (Eds.), pp. 183–207, CRC Press, Boca Raton, FL, USA, 2023, doi: 10.1201/9781003395768-11.
29. Kundu, A., Kumar, A., Dutt, N., Singh, V.P., Meena, C.S., Chapter 7: Modelling and Simulation of Thermal Energy System for Design Optimization, in: *Thermal Energy Systems: Design, Computational Techniques, and Applications*, A. Kumar, N. Dutt, V. Singh, C. Meena (Eds.), pp. 103–137, CRC Press, Boca Raton, FL, USA, 2023, doi: 10.1201/9781003395768-07.
30. Singh, A.R., Singh, S.K., Singh, V.P., Process parameters optimization of carbon nano tube based catalytic transesterification of algal oil. *Mater. Today Proc.*, 72, 5, 684–691, 2023, doi: 10.1016/j.matpr.2023.01.418.
31. Sharma, B., Katiyar, V.K., Kumar, K., Traffic Accident Prediction Model Using Support Vector Machines with Gaussian Kernel, in: *Proceedings of Fifth International Conference on Soft Computing for Problem Solving*, M. Pant, K. Deep, J.C. Bansal, A. Nagar, K.N. Das (Eds.), pp. 1–10, Springer Singapore, Singapore, 2016.
32. Singh, V.P., Kumar, A., Meena, C.S., Dwivedi, G., *Energy Efficient Vehicles*, 1st ed., CRC Press, Boca Raton, 2024, doi: 10.1201/9781003464556.
33. Singh, V.P. and Kumar, A., Techno-Economic and Future Aspects of the HEV-EV-FCV: Decarbonization, Digitalization, and Sustainability, in: *Energy Efficient Vehicles*, 1st ed., vol. 1, pp. 212–231, CRC Press, Boca Raton, FL, USA, 2024, doi: https://doi.org/10.1201/9781003464556-11.
34. Munawar, U. and Wang, Z., A Framework of Using Machine Learning Approaches for Short-Term Solar Power Forecasting. *J. Electr. Eng. Technol.*, 15, 2, 561–569, 2020, doi: 10.1007/s42835-020-00346-4.
35. Kumar, V., Jain, V., Sharma, B., Chatterjee, J.M., Shrestha, R., *Smart City Infrastructure: The Blockchain Perspective*, Wiley, United States, 2022, ISBN: 9781119785545. [Online]. Available: https://books.google.co.in/books?id=iM1gEAAAQBAJ.
36. Agarwal, V. and Sharma, S., Q Learning Algorithm for Network Resource Management in Vehicular Communication Network, in: *Autonomous Vehicles Volume 2*, pp. 259–274, 2022, doi: https://doi.org/10.1002/9781394152636.ch13.
37. Kumar, R., *et al.*, Experimental and RSM-Based Process-Parameters Optimisation for Turning Operation of EN36B Steel. *Materials*, 16, 1, 1–18, 2023, doi: 10.3390/ma16010339.
38. Singh, V.P., Karn, A., Dwivedi, G., Alam, T., Kumar, A., Experimental Assessment of Variation in Open Area Ratio on Thermohydraulic

Performance of Parallel Flow Solar Air Heater. *Arabian J. Sci. Eng.*, 41, 12, 1–17, 2022, doi: 10.1007/s13369-022-07525-7.

39. Singh, V.P., *et al.*, Heat transfer and friction factor correlations development for double pass solar air heater artificially roughened with perforated multi-V ribs. *Case Stud. Therm. Eng.*, 39, September, 102461, 2022, doi: https://doi.org/10.1016/j.csite.2022.102461.
40. Noori, M.A.R., Sharma, B., Mehra, R., Feature selection from disaster tweets using Spark-based parallel meta-heuristic optimizers. *Soc. Netw. Anal. Min.*, 12, 1, 93, 2022, doi: 10.1007/s13278-022-00930-8.
41. Singh, V.P., *et al.*, Nanomanufacturing and Design of High-Performance Piezoelectric Nanogenerator for Energy Harvesting, in: *Nanomanufacturing and Nanomaterials Design: Principles and Applications*, 1st ed., S. K. B. A. K. K. V. S. Singh (Ed.), pp. 241–272, CRC Press, Boca Raton, 2022, doi: 10.1201/9781003220602-15.
42. Singh, V.P., Jain, S., Gupta, J.M.L., Analysis of the effect of perforation in multi-v rib artificial roughened single pass solar air heater: - Part A. *Exp. Heat Transfer*, 36, 2, 1–20, Oct. 2021, doi: 10.1080/08916152.2021.1988761.
43. Ashwani Kumar, Y.G., Gangwar, A.K.S., Kumar, A., Meena, C.S., Singh, V.P., Dutt, N., Prasad, A., Biomedical study of femur bone fracture and healing, in: *Advanced Materials for Biomedical Applications*, 1st edition, Y. G. A. K. C. S. M. N. D. A. Kumar (Ed.), pp. 283–298, CRC Press, Boca Raton, 2022, doi: 10.1201/9781003344810-14.
44. Zeng, J. and Qiao, W., Short-term solar power prediction using a support vector machine. *Renew. Energy*, 52, 118–127, 2013, doi: https://doi.org/10.1016/j.renene.2012.10.009.
45. Shubham Srivastava, R.K., Verma, D., Thusoo, S., Kumar, A., Singh, V.P., Nanomanufacturing for Energy Conversion and Storage Devices, in: *Nanomanufacturing and Nanomaterials Design: Principles and Applications*, 1st ed., S. K. B. A. K. K. V. S. Singh (Ed.), pp. 165–174, CRC Press, Boca Raton, 2022, doi: 10.1201/9781003220602-10.
46. Datta, A., Kumar, A., Kumar, A., Kumar, A., Singh, V.P., Advanced materials in biological implants and surgical tools, in: *Advanced Materials for Biomedical Applications*, 1st ed., Ashwani Kumar, N.D., Gori, Y., Kumar, A., Meena, C.S. (Eds.), pp. 283–298, 2022, doi: 10.1201/9781003344810-2.
47. Dutt, N., Binjola, A., Hedau, A.J., Kumar, A., Singh, V.P., Meena, C.S., Comparison of CFD Results of Smooth Air Duct with Experimental and Available Equations in Literature. *Int. J. Energy Resour. Appl. (IJERA)*, 1, 1, 40–47, 2022, doi: https://doi.org/10.56896/IJERA.2022.1.1.006.
48. Yadav, A.K., Singh, V.P., Kumar, A., Gori, Y., Kumar, A., Study the Properties of High-Density Polyethylene with Nano Particles Silica Composite. *Int. J. Mater. Manuf. Sustain. Technol. (IJMMST)*, 1, 1, 29–36, 2022, doi: https://doi.org/10.56896/IJMMST.2022.1.1.005.

49. Singh, V.P., Jain, S., Karn, A., Kumar, A., Dwivedi, G., Mathematical Modeling of Efficiency Evaluation of Double Pass Parallel Flow Solar Air Heater. *Sustainability*, 14, 17, 1–22, 2022, doi: https://doi.org/10.3390/su141710535.
50. Jang, H.S., Bae, K.Y., Park, H.-S., Sung, D.K., Solar Power Prediction Based on Satellite Images and Support Vector Machine. *IEEE Trans. Sustain. Energy*, 7, 3, 1255–1263, 2016, doi: 10.1109/TSTE.2016.2535466.
51. Ren, Y., Suganthan, P.N., Srikanth, N., Ensemble methods for wind and solar power forecasting—A state-of-the-art review. *Renew. Sustain. Energy Rev.*, 50, 82–91, 2015, doi: https://doi.org/10.1016/j.rser.2015.04.081.
52. Sharma, B., Katiyar, V.K., Gupta, A.K., Singh, A., The Automated Vehicle Detection of Highway Traffic Images by Differential Morphological Profile. *J. Transp. Technol.*, 04, 02, 150–156, 2014, doi: 10.4236/jtts.2014.42015.
53. Sharma, B. and Kumar, S., Delay Optimization Using Genetic Algorithm at the Road Intersection. *Int. J. Inf. Retr. Res.*, 9, 2, 1–10, Apr. 2019, doi: 10.4018/IJIRR.2019040101.
54. Singh, V.P., Jain, S., Kumar, A., Establishment of correlations for the Thermo-Hydraulic parameters due to perforation in a multi-V rib roughened single pass solar air heater. *Exp. Heat Transfer*, 35, 5, 1–20, 2022, doi: 10.1080/08916152.2022.2064940.
55. Zhang, X., Li, Y., Lu, S., Hamann, H.F., Hodge, B.-M., Lehman, B., A Solar Time Based Analog Ensemble Method for Regional Solar Power Forecasting. *IEEE Trans. Sustain. Energy*, 10, 1, 268–279, 2019, doi: 10.1109/TSTE.2018.2832634.
56. Verma, M. and Singh, V.P., Boundary layer flow in aerospace applications, in: *Computational Fluid Flow and Heat Transfer*, pp. 66–94, CRC Press, Boca Raton, FL, USA, 2024, doi: https://doi.org/10.1201/9781003465171-5.
57. Verma, M. and Singh, V.P., Design optimization and sensitivity analysis in aerospace engineering, in: *Computational Fluid Flow and Heat Transfer*, pp. 112–132, CRC Press, Boca Raton, FL, USA, 2024, doi: https://doi.org/10.1201/9781003465171-7.
58. Upadhyay, R.K., Singh, V.P., Kumar, A., Sustainable Transportation: Policy, Planning, and Implementation, in: *Energy Efficient Vehicles*, 1st ed., pp. 74–95, CRC Press, Boca Raton, FL, USA, 2024, doi: https://doi.org/10.1201/9781003464556-5.
59. Diagne, H.M., Lauret, P., David, M., Solar irradiation forecasting: state-of-the-art and proposition for future developments for small-scale insular grids, in: *WREF 2012-World Renewable Energy Forum*, 2012.
60. Singh, V.P., Kumar, A., Meena, C.S., Dwivedi, G., Introduction to Energy-Efficient Vehicles for Sustainable Transportion: Transitions and Challenges, in: *Energy Efficient Vehicles*, pp. 1–18, CRC Press, Boca Raton, FL, USA, 2024, doi: https://doi.org/10.1201/9781003464556-1.
61. Diagne, M., David, M., Lauret, P., Boland, J., Schmutz, N., Review of solar irradiance forecasting methods and a proposition for small-scale insular grids.

Renew. Sustain. Energy Rev., 27, 65–76, 2013, doi: https://doi.org/10.1016/j.rser.2013.06.042.
62. Hosenuzzaman, M., Rahim, N.A., Selvaraj, J., Hasanuzzaman, M., Malek, A.B.M.A., Nahar, A., Global prospects, progress, policies, and environmental impact of solar photovoltaic power generation. *Renew. Sustain. Energy Rev.*, 41, 284–297, 2015, doi: https://doi.org/10.1016/j.rser.2014.08.046.
63. Chauhan, K. and Singh, V.P., Prospect of biomass to bioenergy in India : An overview. *Mater. Today Proc.*, 72, 5, 941–953, 2023, doi: 10.1016/j.matpr.2023.01.419.
64. Rabaia, M.K.H., *et al.*, Environmental impacts of solar energy systems: A review. *Sci. Total Environ.*, 754, 141989, 2021, doi: https://doi.org/10.1016/j.scitotenv.2020.141989.
65. Yang, L., Gao, X., Lv, F., Hui, X., Ma, L., Hou, X., Study on the local climatic effects of large photovoltaic solar farms in desert areas. *Sol. Energy*, 144, 244–253, 2017, doi: https://doi.org/10.1016/j.solener.2017.01.015.
66. Prăvălie, R., Patriche, C., Bandoc, G., Spatial assessment of solar energy potential at global scale. A geographical approach. *J. Clean. Prod.*, 209, 692–721, 2019, doi: https://doi.org/10.1016/j.jclepro.2018.10.239.
67. Choudhary, P. and Srivastava, R.K., Sustainability perspectives- a review for solar photovoltaic trends and growth opportunities. *J. Clean. Prod.*, 227, 589–612, 2019, doi: https://doi.org/10.1016/j.jclepro.2019.04.107.
68. Li, J., Herdem, M.S., Nathwani, J., Wen, J.Z., Methods and applications for Artificial Intelligence, Big Data, Internet of Things, and Blockchain in smart energy management. *Energy AI*, 11, 100208, 2023, doi: https://doi.org/10.1016/j.egyai.2022.100208.
69. Said, Y. and Alanazi, A., AI-based solar energy forecasting for smart grid integration. *Neural Comput. Appl.*, 35, 11, 8625–8634, 2023, doi: 10.1007/s00521-022-08160-x.

5

Understanding Forecasting Models for Renewable Energy Generation and Market Operation

Varun Pratap Singh[1,2]*, Ashwani Kumar[3], Chandan Swaroop Meena[4,5] and Nitesh Dutt[6]

[1]*Solar Thermal Energy Research Group, Department of Mechanical and Mechatronics Engineering, Stellenbosch University, Stellenbosch, South Africa*
[2]*Department of Mechanical Engineering, School of Advanced Engineering, UPES, Dehradun, India*
[3]*Technical Education Department, Uttar Pradesh, Kanpur, India*
[4]*CSIR-Central Building Research Institute, Roorkee, India*
[5]*Academy of Scientific and Innovative Research (AcSIR), Ghaziabad, India*
[6]*Department of Mechanical Engineering, COER University, Roorkee, India*

Abstract
Forecasting renewable energy is essential for optimizing energy generation and market functionality by anticipating the fluctuations of sources like wind and solar power. This chapter offers a thorough examination of forecasting models employed in renewable energy systems. It begins with an introduction to the importance of accurate forecasting and explores different model categories, including physical, statistical, hybrid, and specialized approaches. In-depth analyses focus on forecasting methodologies for wind and solar energy production, emphasizing their distinct difficulties and strategies. The chapter also explores the utilization of forecasting in market activities, including energy trading, grid stabilization, and demand regulation. Complex subjects, including machine learning–based predictions and probabilistic techniques, are addressed to showcase the most recent advancements in the discipline. The chapter ultimately discusses contemporary problems, such as data constraints and model ambiguities, and delineates prospective research avenues to improve forecasting precision and integration with advancing energy markets. This synthesis provides essential insights for researchers, aiming to enhance the utilization of renewable energy sustainably and economically.

**Corresponding author*: ray_varun@yahoo.com

Jai Govind Singh, Rupendra Kumar Pachauri and Sasidharan Sreedharan (eds.) Forecasting Methods for Renewable Power Generation, (95–130) © 2025 Scrivener Publishing LLC

Keywords: Energy forecasting, wind energy prediction, solar energy forecasting, physical models, statistical models, hybrid models, market operations, energy trading optimization

5.1 Introduction to Renewable Energy Forecasting

Renewable energy forecasting has emerged as a pivotal field over the past few decades, driven by the global shift toward renewable energy sources in response to climate change and the depletion of fossil fuels. The history of renewable energy forecasting can be traced back to the early attempts at understanding weather patterns and their impacts on wind and solar power generation. With the advent of more sophisticated numerical weather prediction (NWP) models in the late 20th century, the foundation was laid for today's advanced forecasting techniques. These developments were paralleled by the growth of statistical methods, leveraging historical data to predict future energy output. The integration of renewable energy sources into power grids necessitated the evolution of forecasting models to manage their inherent variability and intermittency, marking a significant milestone in the operational planning of energy systems [1].

Renewable energy forecasting offers numerous advantages, primarily enabling more efficient grid management and integration of variable energy sources such as wind and solar power. It enhances grid reliability, optimizes the dispatch of energy resources, reduces operational costs, and supports the economic viability of renewable projects by minimizing the need for excess reserve capacity. The application of forecasting models spans various aspects of energy system operations, including load management, maintenance scheduling, and investment planning. A brief review of the literature reveals a focus on improving the accuracy and reliability of forecasts through the development of hybrid models that combine physical and statistical approaches. Recent studies have explored the use of machine learning algorithms and big data analytics to refine predictions, highlighting the ongoing innovation in the field. The continual advancement in renewable energy forecasting models not only supports the increasing integration of renewables into the energy mix but also contributes to the global pursuit of sustainable and resilient energy systems [2].

5.1.1 Importance of Renewable Energy Forecasting

The importance of renewable energy forecasting is deeply rooted in the global shift toward more sustainable energy sources. As the adoption of renewable energy—such as wind, solar, and hydropower—grows, the need for accurate forecasting becomes increasingly critical. The primary goal of renewable energy forecasting is to predict the amount of energy that will be generated by these resources at any given time. This forecasting is crucial for several reasons [3].

Grid Stability and Reliability: Renewable energy sources like wind and solar power are inherently variable; their output depends on weather conditions. Accurate forecasting helps grid operators anticipate changes in power production and take necessary actions to maintain a balance between supply and demand, ensuring the stability and reliability of the power grid [4].

Optimization of Resources: By predicting how much power will be generated, operators can optimize the mix of energy sources used at any given time. This includes efficiently integrating renewable energy with more traditional forms of power generation to meet demand without over-reliance on fossil fuels [5].

Economic Benefits: Forecasting allows energy producers and grid operators to reduce operational costs by minimizing the need for expensive, fast-acting backup power sources. Accurate predictions can lead to more efficient use of resources, lowering costs for both energy providers and consumers [6–8].

Enhanced Integration of Renewables: As the world aims to reduce carbon emissions, forecasting is indispensable in integrating higher percentages of renewable energy into the grid without compromising its performance. This supports global efforts toward a more sustainable and environmentally friendly energy future [9].

Improved Energy Storage Management: Accurate renewable energy forecasting is vital for managing energy storage systems effectively. With precise predictions, energy storage can be optimized to store excess energy during periods of high production and low demand and release it when the demand is high but renewable generation is low. This not only ensures a constant supply of energy but also maximizes the economic value of stored energy [10].

Facilitation of Energy Trading and Market Operations: Renewable energy forecasting plays a crucial role in energy trading and market operations. By providing accurate predictions of renewable energy production, it enables energy producers to make informed decisions about when to sell their energy and at what price, optimizing their profits. Similarly, it helps buyers in the energy market to plan their purchases and secure energy at the most favorable rates, promoting a more competitive and efficient market [11].

Enhanced Policy and Investment Decisions: Forecasting is essential for policymakers and investors as it provides insights into the future of energy trends and the viability of renewable energy projects. Accurate forecasting can guide investment decisions, ensuring that funds are directed toward the most promising and efficient renewable energy technologies. It also helps in shaping policies that support the growth of renewable energy, aligning with national and international goals for sustainability and carbon reduction [12, 13].
Support for Research and Development: Renewable energy forecasting drives innovation in the energy sector by highlighting areas where improvements are needed. It identifies the challenges of integrating renewable energy into the grid, spurring research and development efforts aimed at overcoming these challenges. This includes the development of advanced forecasting models, more efficient energy storage solutions, and technologies that enhance the flexibility and resilience of the power grid [14].
Consumer Empowerment and Participation: With the advent of smart grids and real-time data analytics, renewable energy forecasting empowers consumers to become active participants in the energy ecosystem. By having access to forecasting information, consumers can adjust their energy usage patterns to take advantage of times when renewable energy production is high, and costs are lower. This not only leads to cost savings for the consumer but also encourages a more sustainable consumption pattern [15].
Climate Change Mitigation: Finally, accurate renewable energy forecasting is instrumental in the fight against climate change. By enabling a higher penetration of renewable energy sources into the energy mix, forecasting supports the reduction of greenhouse gas emissions from fossil fuel-based power generation. This contributes to global efforts to mitigate the effects of climate change, aiming for a healthier planet for future generations [16].

5.1.2 Overview of Forecasting Models

Renewable energy forecasting models serve as critical tools for predicting the output from various renewable sources like wind, solar, and hydropower. These models, categorized into physical, statistical, and hybrid types, each possess unique capabilities to tackle the inherent unpredictability of renewable energy generation. Physical models, leveraging NWP systems, are adept at simulating atmospheric conditions to forecast critical factors like wind speed and solar irradiance [17, 18]. Within this category, mesoscale models are invaluable for regional weather predictions, offering a bridge between global weather patterns and localized conditions. Microscale models go further, offering predictions at a highly localized level, considering the specific geographical features of a site.

Satellite-based Models, on the other hand, harness earth observation satellite data to forecast solar energy generation, focusing on solar irradiance and cloud cover [19].

Statistical models complement this approach by utilizing historical data to discern patterns and relationships between weather conditions and energy generation, employing sophisticated machine learning techniques to predict future output. Techniques range from time-series forecasting, which utilizes models like ARIMA and exponential smoothing for short-term forecasts, to regression models that establish relationships between weather variables and energy output. The advent of machine learning algorithms, including neural networks, decision trees, random forests, and support vector machines, has significantly enhanced the accuracy of these statistical models [20]. Hybrid models aim to synergize the strengths of both physical and statistical models, employing techniques like ensemble methods, integrated physical-statistical models, and multi-model fusion to refine forecasting accuracy. Additionally, specialized models like persistence models, probabilistic models, and seasonal and cyclical models address specific forecasting challenges, offering solutions tailored to short-term conditions and long-term accuracy enhancement. This rich tapestry of forecasting models underscores the complexity and dynamism of renewable energy forecasting, highlighting ongoing advancements in the field aimed at integrating renewable energy more effectively into our energy systems. A detailed description of these models is discussed in the next heading [21].

5.1.3 Challenges in Renewable Energy Forecasting

Expanding upon the intricacies of renewable energy forecasting, several additional challenges are noteworthy. These further complicate the process but also offer avenues for innovation and improvement in the field [22].

Inherent Variability and Uncertainty: The primary challenge in renewable energy forecasting is the natural variability and unpredictability of weather conditions. Sudden weather changes can significantly affect the accuracy of energy generation forecasts, particularly over short time frames [23].

Data Quality and Accessibility: High-quality, granular meteorological and power generation data are crucial for developing accurate forecasting models. However, collecting and accessing comprehensive and reliable data sets can be challenging, particularly in regions with limited monitoring infrastructure [23].

Model Complexity and Integration: Developing models that accurately capture the complex interactions between weather conditions and energy generation requires sophisticated algorithms and significant computational resources. Additionally, integrating these forecasts into energy market operations and grid management systems poses technical and regulatory challenges [24].

Scalability and Regional Specificity: Renewable energy resources and their potential vary significantly by region. Forecasting models must be adaptable and scalable to different geographic and climatic conditions, requiring localized data and customized modeling approaches.

Scaling-Up Forecasting for Wide Area Networks: As renewable energy generation becomes more widespread, forecasting models must scale up to accommodate wide area networks. This involves accurately predicting energy production across vast and geographically diverse areas where weather patterns can vary significantly. The challenge lies in aggregating these forecasts to inform grid-level decisions, necessitating advanced modeling techniques that can handle large datasets and provide insights at both local and regional scales [25, 26].

Real-Time Data Processing: Renewable energy forecasting increasingly relies on real-time data to improve accuracy, particularly for short-term predictions. This necessitates robust systems capable of processing large volumes of data from diverse sources, including satellites, weather stations, and IoT devices, in real time. Developing such systems poses technical challenges, including data transmission, storage, and analysis, requiring significant computational resources and advanced data processing algorithms [27, 28].

Forecasting for Emerging Renewable Technologies: As new renewable energy technologies emerge, such as floating solar panels or high-altitude wind power, forecasting models need to adapt to these novel sources. Each technology has unique characteristics and dependencies on environmental conditions, demanding bespoke forecasting approaches. The challenge lies in developing and validating new models that can accurately predict the energy output of these emerging technologies [23, 29].

Integration with Energy Markets: Accurate forecasts are critical for renewable energy's participation in energy markets, where they influence bidding strategies and operational planning. The challenge is to integrate forecasting models seamlessly with market operations, ensuring that forecasts are timely, reliable, and actionable for energy traders and plant operators. This requires not only technical integration but also regulatory frameworks that recognize the value of forecasts in market mechanisms [30].

Addressing Climate Change Impacts: Long-term forecasting models must consider the impacts of climate change on renewable energy resources. For instance, changing precipitation patterns could affect hydropower generation, whereas rising temperatures might influence solar panel efficiency. The challenge is to incorporate climate projections into renewable energy forecasting models, ensuring they remain accurate over the long term. This requires interdisciplinary collaboration between climate scientists, energy modelers, and policymakers to develop models that can adapt to a changing climate.

Addressing these challenges is essential for the continued integration of renewable energy sources into the global energy mix. It requires a multidisciplinary approach, combining advancements in meteorology, data science, energy technology, and policy to create forecasting models that are robust, scalable, and adaptable to the dynamic nature of renewable energy generation [31].

5.2 Types of Forecasting Models for Renewable Energy

Renewable energy forecasting models are essential tools for predicting the output of renewable energy sources. Beyond the basic classification into physical, statistical, and hybrid models, a deeper exploration reveals several subcategories and specialized models designed to address specific forecasting challenges. This enriched classification provides a more comprehensive understanding of the methodologies employed in renewable energy forecasting [32].

5.2.1 Physical Models

Physical models, underpinned by NWP systems, are indispensable for simulating atmospheric conditions to forecast renewable energy generation. These models excel in predicting wind speed and solar irradiance, crucial parameters for wind and solar power generation. By leveraging complex atmospheric dynamics, physical models provide insights into how weather patterns influence renewable energy outputs [33].

5.2.1.1 Mesoscale Models

Mesoscale models serve a vital function in forecasting by focusing on regional weather predictions. They bridge the gap between global weather patterns and local conditions, making them especially valuable for wind and solar forecasting across extensive areas. These models are adept at capturing weather phenomena over a range of 2 to 2,000 kilometers, offering a detailed understanding of how regional weather impacts renewable energy production [34].

5.2.1.2 Microscale Models

Microscale models are designed to predict conditions at a highly localized level, such as around a wind farm or solar plant. They take into consideration local geographical features, including terrain and vegetation, to provide accurate predictions of energy generation potential in specific environments [35]. This precision is critical for optimizing the placement and operation of renewable energy installations.

5.2.1.3 Satellite-Based Models

Satellite-based models utilize data from earth observation satellites to predict solar irradiance and cloud cover, fundamental for solar energy forecasting. By analyzing satellite imagery and radiometric data, these models offer comprehensive insights into solar potential across different geographies and times, aiding in the planning and operation of solar energy projects [36].

5.2.2 Statistical Models

Statistical models use historical data to identify patterns and relationships between weather conditions and energy generation. These models often employ machine learning techniques, such as neural networks or regression analysis, to predict future energy production based on past trends.

5.2.2.1 Time-Series Forecasting

Time-series forecasting, utilizing models like ARIMA and exponential smoothing, is predominantly employed for short-term predictions. By analyzing historical data sequences, these models identify temporal patterns to forecast future energy outputs, crucial for day-ahead or hour-ahead planning.

5.2.2.2 Regression Models

Regression models, both linear and nonlinear, establish relationships between weather variables (such as wind speed and temperature) and energy outputs. They are fundamental in translating meteorological data into actionable energy production forecasts, facilitating more reliable energy supply planning.

5.2.2.3 *Machine Learning Algorithms*

Machine Learning Algorithms have revolutionized the field of renewable energy forecasting by offering advanced methods to analyze and predict energy outputs from variable sources such as wind and solar power. These algorithms can process vast amounts of historical weather and energy production data to uncover complex patterns and relationships that are not immediately apparent. Neural networks, particularly recurrent neural networks (RNNs) and long short-term memory (LSTM) networks excel in capturing the temporal dynamics of energy production data, making them highly effective for both short-term and long-term forecasting [37, 38]. Decision trees and random forests provide a structured approach to prediction, enabling the identification of the most influential factors affecting energy generation. Support vector machines (SVMs), on the other hand, are adept at finding the optimal separation between different data classes, making them useful for both classification and regression tasks in scenarios with high-dimensional data. By leveraging these machine learning techniques, energy forecasters can significantly improve the accuracy of their predictions, optimize the integration of renewable energy sources into the power grid, and make informed decisions about energy management and policy development [39, 40].

5.2.2.3.1 Neural Networks

Neural networks, particularly RNNs and LSTM networks stand at the forefront of machine learning technologies applied to renewable energy forecasting. These networks are distinguished by their ability to process sequences of data, making them ideally suited for analyzing time-series data such as energy production records and weather patterns [41, 42].

Recurrent neural networks (RNNs) are designed to recognize patterns in sequences of data by maintaining a “memory” of previous inputs in their internal state. This capability allows RNNs to exhibit temporal dynamic behavior, making them exceptionally effective for tasks where context from previous data points must be considered to accurately predict future events. In the context of renewable energy forecasting, RNNs can analyze sequences of weather conditions and historical energy production data to predict future energy outputs [43].

Long short-term memory (LSTM) networks, a special kind of RNN, are specifically designed to overcome the limitations of traditional RNNs related to long-term dependencies. LSTMs are capable of learning long-term relationships in data by employing a sophisticated gating mechanism that regulates the flow of information. This allows LSTMs to retain important historical information over long sequences while discarding irrelevant data. This feature is particularly useful in renewable energy forecasting, where understanding both recent and more distant past conditions is crucial for making accurate predictions [44].

5.2.2.3.2 Decision Trees and Random Forests

Decision trees are a type of supervised learning algorithm that is used for both classification and regression tasks. The model segments the dataset into smaller subsets while simultaneously developing an associated decision tree. The result is a tree with decision nodes and leaf nodes, where each leaf node corresponds to a decision outcome or predicted output. Decision trees are intuitive and easy to interpret, making them an excellent tool for identifying the key variables that impact energy production [45].
Random forests are an ensemble learning method that operates by constructing multiple decision trees during the training phase. The output for regression tasks is the average prediction of the individual trees. Random Forests improve upon the predictive accuracy and control over-fitting by averaging or "bagging" multiple decision trees, thereby reducing the variance without significantly increasing bias. In renewable energy forecasting, random forests can handle complex interactions and non-linear relationships between a wide array of input variables, from weather conditions to plant operational status, to predict energy output more accurately [46].

5.2.2.3.3 Support Vector Machines (SVM)

Support vector machines (SVMs) are a set of supervised learning methods used for classification, regression (SVR), and outliers' detection. SVMs are particularly renowned for their ability to handle high-dimensional data efficiently. In the context of renewable energy forecasting, SVMs can be used to delineate the complex relationships between various meteorological factors and energy outputs [47]. By constructing a hyperplane or set of hyperplanes in a high-dimensional space, SVMs work to separate different classes with a clear gap that is as wide as possible. For regression tasks, SVMs aim to fit the error within a certain threshold, making them suitable for scenarios where precision is crucial, such as predicting the exact energy output from a solar PV system under varying weather conditions [48].

Together, these machine learning techniques—neural networks (especially RNNs and LSTMs), decision trees and random forests, and SVMs—provide powerful tools for improving renewable energy forecasting. By leveraging the unique strengths of each method, forecasters can enhance the accuracy of their predictions, contributing to more efficient grid operation, better integration of renewable sources, and ultimately, the advancement of clean and sustainable energy systems [49].

5.2.3 Hybrid Models

Hybrid models in renewable energy forecasting represent an innovative approach that synergizes the predictive strengths of both physical and statistical models. The core idea behind these models is to harness the detailed, physics-based understanding of weather and atmospheric conditions provided by NWP systems, alongside the pattern recognition and learning capabilities of statistical methods, particularly machine learning algorithms [50, 51]. This combination aims to mitigate the limitations inherent in each approach when used in isolation, offering a more robust and adaptable forecasting tool.

The adoption of hybrid models has been driven by the need for more accurate and reliable forecasts in the face of the complex, dynamic nature of renewable energy sources. By integrating diverse modeling approaches, hybrid models can account for a wider range of variables and conditions, enhancing their ability to predict energy outputs under various weather scenarios and operational contexts. This adaptability makes them especially valuable for managing the integration of renewable energy into power grids, facilitating more efficient grid operation, energy storage, and load balancing [52].

5.2.3.1 Ensemble Methods

Ensemble methods are a cornerstone of hybrid modeling, designed to improve prediction accuracy by combining the forecasts from multiple models. These methods operate on the principle that a collective decision made by integrating multiple predictions is generally more accurate than any single model's output [53, 54]. Techniques such as bagging, boosting, and stacking represent different strategies for aggregating model predictions:

- **Bagging (bootstrap aggregating)** reduces variance and helps avoid overfitting by generating multiple subsets of data

from the training set with replacement, training a model on each subset, and then averaging the outputs.
- **Boosting** sequentially trains models, where each new model focuses on the errors of the previous ones, adjusting the weights of incorrectly predicted instances to improve overall accuracy.
- **Stacking** involves training a meta-model to optimally combine the predictions of several base models, thereby learning the best way to blend different forecasts based on their performance and characteristics.

5.2.3.2 Integrated Physical-Statistical Models

Integrated physical-statistical models take a complementary approach by directly using the outputs from physical models as inputs into statistical models. This method leverages the precision of NWP systems in capturing atmospheric and weather dynamics while also utilizing the flexibility of statistical models to adapt to historical data trends and patterns. The integration enables these models to effectively capture both the causative factors of energy production variability and the empirical relationships observed in historical data, offering nuanced forecasts that are grounded in both theory and observed reality [55].

5.2.3.3 Multi-Model Fusion

Multi-model fusion takes hybrid modeling a step further by integrating various types of models, such as mesoscale NWP models with machine learning algorithms. This approach not only combines physical and statistical methods but also blends different scales of modeling, from regional weather patterns to localized conditions affecting energy production. By doing so, multi-model fusion captures both the macro- and micro-level dynamics influencing renewable energy generation, providing a comprehensive and finely detailed forecasting tool [56].

Hybrid models, through ensemble methods, integrated physical-statistical models, and multi-model fusion, represent the forefront of innovation in renewable energy forecasting. These models' ability to draw upon the strengths of diverse forecasting approaches offers significant advantages in accuracy, reliability, and adaptability, making them indispensable tools in the ongoing effort to optimize the use of renewable energy sources [57].

5.2.4 Specialized Models

Specialized models in renewable energy forecasting cater to specific aspects of prediction challenges, offering tailored approaches that range from simplistic to complex methodologies. These models include persistence models, probabilistic models, and seasonal and cyclical models, each addressing unique facets of forecasting needs [58].

5.2.4.1 Persistence Models

Persistence models operate on a straightforward principle: they assume that future short-term conditions will mirror the current state. This model type is particularly useful for very short-term forecasts, often spanning from a few minutes to a few hours ahead. Despite their simplicity, persistence models can be remarkably effective in situations where weather conditions change slowly, making them a reliable tool for imminent forecasting needs [59, 60]. Their utility is most pronounced in applications requiring immediate decision-making, such as real-time grid management and operational adjustments in renewable energy systems. The persistence approach provides a baseline against which the performance of more sophisticated forecasting models can be compared [61].

5.2.4.2 Probabilistic Models

Probabilistic models mark a departure from deterministic forecasting by providing forecasts in terms of probability distributions rather than single-point estimates. This approach is instrumental in quantifying the uncertainty inherent in renewable energy forecasts. By presenting a range of possible outcomes with associated probabilities, probabilistic models enable decision-makers to assess risks and make informed choices under uncertainty. This method is particularly valuable in energy trading, grid operation, and capacity planning, where understanding the likelihood of various generation scenarios helps in devising strategies that hedge against potential shortfalls or excesses in energy production. Moreover, probabilistic forecasts are crucial for integrating renewable energy sources into power systems, facilitating more robust and resilience-oriented planning [57, 61].

5.2.4.3 Seasonal and Cyclical Models

Seasonal and cyclical models are designed to capture and predict the impact of seasonal variations and cyclical patterns on weather conditions and, consequently, on renewable energy output. These models recognize the inherent periodicities in climate and environmental factors—such as changes in solar irradiance levels across seasons or predictable wind patterns—that significantly affect renewable energy generation [62]. By incorporating seasonal indices or cyclical algorithms, these models enhance the accuracy of long-term forecasts, making them invaluable for strategic planning, maintenance scheduling, and investment analysis in the renewable energy sector. Seasonal and cyclical models ensure that the temporal dynamics of energy production are accurately reflected in forecasts, aiding in optimizing operations and financial planning over longer time horizons. A broader classification of the renewable energy forecasting models is also shown in below mentioned Figure 5.1.

The selection of a forecasting model or a combination of models is guided by factors such as the type of renewable energy source, the forecasting horizon (short-term, medium-term, and long-term), the geographical area of interest, and the specific requirements of the energy system or market operation. As the field of renewable energy forecasting evolves, the development of more sophisticated and specialized models is expected to continue, driven by advances in computational capabilities, data availability, and understanding of atmospheric and environmental dynamics [61].

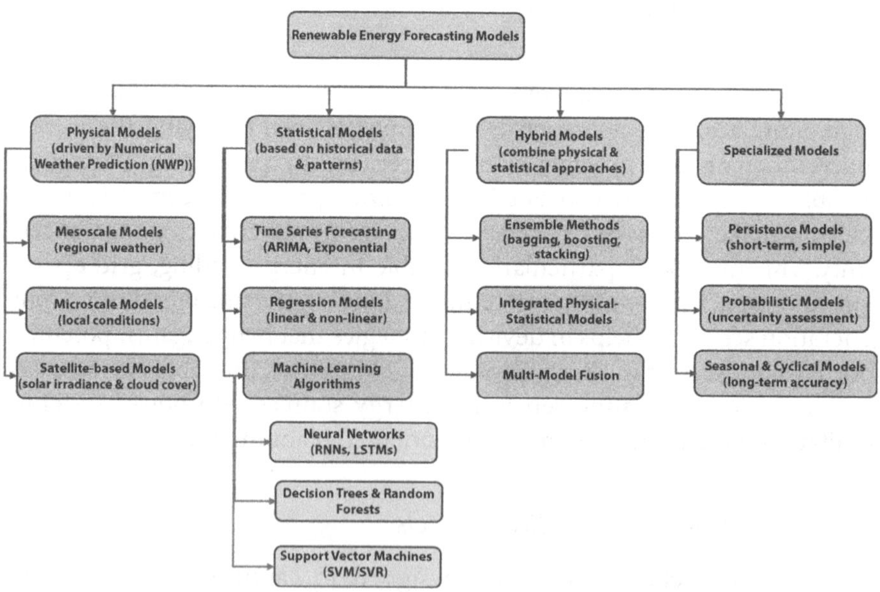

Figure 5.1 Classification of the renewable energy forecasting model.

5.3 Forecasting Wind and Solar Energy Generation

Forecasting wind and solar energy generation is a critical aspect of integrating renewable resources into the energy grid. Given the variable and intermittent nature of wind and solar power, accurate forecasting is essential for maintaining grid stability, optimizing energy dispatch, and ensuring the efficient use of renewable resources. This task involves predicting how much electricity wind turbines and solar panels will produce, which is inherently challenging due to the dependence on fluctuating weather conditions. Here is a detailed overview of forecasting wind and solar energy generation [63].

5.3.1 Wind Energy Forecasting Techniques

Wind energy forecasting is a critical component of integrating wind power into the electrical grid. As wind power generation depends heavily on the variability of wind speed and direction, accurate forecasting techniques are essential to predict how much power will be generated at wind farms. This forecasting enables grid operators to balance energy supply with demand, reduce reliance on reserve power sources, and increase the overall efficiency and reliability of the power system. Here is a detailed exploration of wind energy forecasting techniques, focusing on wind speed and direction forecasting and turbine output prediction [63].

5.3.1.1 Wind Speed and Direction Forecasting

Forecasting wind speed and direction is the foundational step in predicting wind energy production. Wind forecasts are generally derived from NWP models, which simulate the physical processes of the atmosphere to predict future weather conditions. These models can vary in complexity and spatial resolution, from global models covering large areas with coarse resolution to mesoscale models focusing on smaller regions with higher resolution [64].

Key Aspects:

- **Data Inputs:** Wind speed and direction forecasting relies on various meteorological data, including atmospheric pressure, temperature, humidity, and historical wind data.
- **Modeling Techniques:** Techniques range from simple statistical models that use historical data to predict future conditions, to complex physical models that simulate atmospheric

dynamics. Machine learning algorithms are also being increasingly applied to improve forecasting accuracy.
- **Temporal and Spatial Resolution:** The usefulness of wind forecasts depends on their temporal (how far into the future they predict) and spatial (the size of the geographic area they cover) resolutions. Higher-resolution models can provide more detailed and accurate forecasts but require more computational resources.

5.3.1.2 *Turbine Output Prediction*

Once wind speed and direction are forecasted, the next step is to predict the actual power output of wind turbines [65]. This process involves translating wind forecasts into expected electricity generation using power curves and considering the specific characteristics of the turbines and the wind farm.

Key Aspects:

- **Power Curves:** A power curve is a graphical representation of a wind turbine's power output at different wind speeds. It is a crucial tool for translating wind speed forecasts into power generation predictions.
- **Turbine Characteristics:** The specific design and operational characteristics of a turbine, such as rotor diameter, hub height, and cut-in/cut-out wind speeds, significantly influence its power output.
- **Environmental Factors:** Factors like air density, which can vary with temperature and elevation, also affect turbine performance, and are considered in output predictions.

Factors Involved in Wind Energy Forecasting

The Table 5.1 below outlines the key factors involved in wind energy forecasting, highlighting their impact on both wind speed and direction forecasting and turbine output prediction.

Wind energy forecasting techniques, encompassing wind speed and direction forecasting and turbine output prediction, are fundamental for optimizing wind power integration into the energy grid. Continuous advancements in meteorological modeling, data analysis, and machine learning are enhancing the accuracy and reliability of these forecasts, contributing to the sustainable expansion of wind energy as a major renewable power source [66].

Table 5.1 Factors involved in wind energy forecasting.

Factor	Description	Impact on wind forecasting	Impact on turbine output prediction
Atmospheric Pressure	High and low-pressure systems influence wind patterns	Critical for predicting wind speed and direction	Indirect impact through wind speed changes
Temperature	Affects air density and pressure gradients	Influences wind speed and direction	Affects air density, which influences turbine power output
Humidity	Impacts air density and temperature	Can influence local wind patterns	Indirectly affects turbine output through air density changes
Historical Wind Data	Past wind speed and direction records	Used in statistical models for future wind condition predictions	Provides data for calibrating power curves
Turbine Power Curves	Graph showing power output at various wind speeds	-	Essential for translating wind forecasts into power generation estimates
Turbine Characteristics	Specific design features of turbines	-	Directly influences the conversion of wind energy into electrical power
Environmental Factors	Conditions like air density and terrain	Influences local wind patterns	Affects the efficiency of turbine power generation

5.3.2 Solar Energy Forecasting Techniques

Solar energy forecasting is pivotal for the effective integration of solar power into the energy grid. With solar power's contribution to the energy mix growing rapidly, accurate forecasting methods are essential to manage its inherent variability and intermittency. These techniques aim to predict how much power solar panels will generate, considering the fluctuating availability of sunlight due to weather conditions and other influencing factors.

5.3.2.1 Solar Irradiance Models

Solar irradiance models are used to predict the amount of solar radiation that will reach the Earth's surface, a key determinant of solar power generation potential. These models range from simple empirical models to complex NWP models that account for various atmospheric conditions [63].

Key Aspects:

- **Data Inputs:** Inputs include satellite imagery, cloud cover data, atmospheric conditions, and historical irradiance records.
- **Modeling Approaches:** Approaches vary from statistical methods, which use historical data to predict future conditions, to physical models that simulate atmospheric processes affecting solar irradiance.
- **Temporal and Spatial Resolution:** The accuracy of solar irradiance forecasts can depend on the model's resolution. High-resolution models provide more detailed and localized predictions but require more computational resources.

5.3.2.2 Photovoltaic Output Prediction

Once solar irradiance is forecasted, the next step involves predicting the actual electricity output from photovoltaic (PV) panels. This process translates solar irradiance forecasts into expected power generation, considering the efficiency of solar panels and their operational characteristics [67, 68].

Key Aspects:

- **PV System Efficiency:** The conversion efficiency of PV systems, which can vary with temperature, irradiance level, and the angle of sunlight incidence.

- **Panel Orientation and Tilt:** The orientation (direction) and tilt (angle) of solar panels relative to the sun significantly influence their productivity.
- **Shading and Environmental Factors:** Local environmental factors, such as shading from nearby buildings or trees and soiling of the panels, can impact the amount of power generated.

Factors Involved in Solar Energy Forecasting

The Table 5.2 below outlines the critical factors involved in solar energy forecasting, emphasizing their impact on both solar irradiance models and PV output prediction.

Table 5.2 Factors involved in solar energy forecasting.

Factor	Description	Impact on solar irradiance forecasting	Impact on photovoltaic output prediction
Cloud Cover	The amount and type of clouds in the sky	Directly influences the amount of solar irradiance reaching the surface	Indirect impact through changes in received irradiance
Atmospheric Conditions	Includes aerosols, water vapor, and air pollutants	Affects solar irradiance by scattering and absorbing sunlight	Indirectly affects output by modifying the irradiance level
Temperature	Ambient temperature around PV panels	Can influence cloud formation and atmospheric clarity	Affects PV efficiency; higher temperatures generally decrease efficiency
Sun Position	The angle of the sun relative to the panels	-	Directly impacts the irradiance level received by panels, affecting output

(Continued)

Table 5.2 Factors involved in solar energy forecasting. (*Continued*)

Factor	Description	Impact on solar irradiance forecasting	Impact on photovoltaic output prediction
Panel Orientation and Tilt	Direction and angle of panels relative to the ground	-	Crucial for maximizing irradiance capture and hence power output
Shading	Obstructions that block sunlight from reaching panels	-	Reduces the effective irradiance, decreasing potential power output
Soiling	Accumulation of dirt and debris on panel surfaces	-	Diminishes irradiance reception, lowering output efficiency

Solar energy forecasting techniques, encompassing both solar irradiance models and PV output prediction, are vital for harnessing solar power more effectively and integrating it into the energy grid. By considering the various factors that influence solar energy production, forecasters can provide more accurate predictions, enabling better planning and operational decisions for solar power plants. Continuous improvements in modeling techniques and the increasing availability of high-quality data are expected to further enhance the accuracy and reliability of solar energy forecasts [69].

5.4 Application of Forecasting in Renewable Energy Market Operations

The integration of renewable energy sources into the electricity market has been significantly enhanced by advancements in forecasting techniques. These advancements have direct implications on energy pricing, trading, supply and demand balance, and overall grid stability and reliability [70, 71]. Understanding these applications provides insights into how forecasting is not just a technical tool but a fundamental aspect of modern energy systems.

5.4.1 Impact on Energy Pricing

Accurate forecasting of renewable energy production impacts energy pricing by providing a clearer picture of expected supply levels. In markets where prices are determined by supply and demand, the ability to predict renewable energy output helps in anticipating periods of high supply, which typically lowers prices, or low supply, which increases prices. For energy providers and consumers participating in dynamic pricing models or demand-response programs, forecasts enable more strategic decision-making regarding energy use and production, leading to potential cost savings and more efficient market operation. Moreover, better forecasting can reduce the need for expensive ancillary services and peaker plants, contributing to lower overall energy costs [69].

5.4.2 Renewable Energy Trading

Renewable energy forecasting plays a crucial role in renewable energy trading by enabling more accurate predictions of energy availability from renewable sources. Traders rely on these forecasts to make informed decisions about buying and selling energy contracts, optimizing their positions in the market based on expected supply fluctuations [72, 73]. This can lead to more liquidity in renewable energy markets and can help in the development of more sophisticated financial instruments, such as futures, options, and other derivatives based on renewable energy production. Accurate forecasts are essential for minimizing risks associated with the inherent variability of renewable sources, thereby attracting more participants and investment into renewable energy markets [74].

5.4.3 Managing Supply and Demand Balance

Forecasting is essential for managing the balance between supply and demand in real time and on future timescales. As renewable energy sources become a larger part of the energy mix, their variability poses challenges to maintaining this balance. Accurate forecasts allow grid operators and utility companies to anticipate periods of high renewable production (e.g., a sunny or windy day) and adjust their operations accordingly [75]. This might involve ramping down controllable sources of power or storing excess energy in batteries or other storage solutions. On the demand side, forecasts can be used to encourage energy usage during periods of high renewable production through demand response programs, helping to smooth out fluctuations in supply and ensuring a more stable and efficient energy system [76].

5.4.4 Enhancing Grid Stability and Reliability

The reliability and stability of the power grid are paramount, and renewable energy forecasting significantly contributes to these aspects. By accurately predicting renewable energy outputs, grid operators can better plan for the integration of these variable energy sources, ensuring that the grid remains stable despite the fluctuations in supply. Forecasting helps in the optimal dispatch of both renewable and conventional power sources, maintaining the frequency and voltage within safe limits. In addition, forecasts inform the necessary investments in grid infrastructure and technology, such as energy storage and advanced control systems, to accommodate the growing share of renewables. This proactive management supported by forecasting enhances the resilience of the energy system against unpredictable changes, ensuring a constant, reliable supply of electricity to consumers [77].

5.4.5 Investment and Financial Planning

Accurate renewable energy forecasting is invaluable for investors and financial planners in the renewable energy sector. It provides a robust basis for evaluating the viability and profitability of renewable energy projects by predicting long-term energy generation potential. This, in turn, helps in assessing expected revenues, calculating return on investment (ROI), and managing risks associated with the variability of renewable resources [78]. Forecasting can influence decisions on where and when to build new renewable energy installations and guide policymakers in creating incentives or subsidies for renewable energy investments [79].

5.4.6 Maintenance Scheduling

Forecasting plays a crucial role in optimizing the maintenance scheduling of renewable energy assets. By predicting periods of low energy production in advance, maintenance activities can be planned during these downtimes to minimize the impact on overall energy output. This strategic scheduling ensures that turbines, solar panels, and other equipment are maintained in optimal condition without significantly disrupting the energy supply [80]. Moreover, forecasting can help identify patterns that may indicate the need for maintenance or replacement, thus preventing unexpected failures and extending the lifespan of renewable energy installations [81].

5.4.7 Energy Storage Optimization

The integration of energy storage systems with renewable energy sources is essential for mitigating the variability and ensuring a steady energy supply. Forecasting facilitates the optimal charging and discharging strategies for these storage systems. By accurately predicting periods of excess production, energy can be stored for use during times of low production or high demand. This not only helps in smoothing out the supply but also in maximizing the economic benefits of storage systems by exploiting price differentials in energy markets. As a result, forecasting enhances the overall efficiency and profitability of renewable energy systems integrated with storage solutions [81].

5.4.8 Demand Response Programs

Forecasting is a cornerstone for the effective implementation of demand response programs, which aim to adjust the power demand instead of adjusting the supply. By accurately forecasting periods of high renewable energy production, utilities can incentivize consumers to shift their energy consumption to these times, thereby aligning demand with renewable energy availability. This can include encouraging the use of energy-intensive appliances during periods of high solar irradiance or lower energy usage when renewable resources are scarce. Through such programs, forecasting enables a more dynamic and interactive approach to energy management, promoting energy efficiency and reducing the need for fossil-fuel-based peak power plants.

The application of forecasting in renewable energy market operations is multifaceted, touching on crucial aspects like pricing, trading, supply-demand management, and grid stability. As forecasting techniques continue to improve, their role in facilitating the transition to a more sustainable and efficient energy system will only grow, underscoring the importance of investment in research and development in this area [82]. Figure 5.2 shows the application of forecasting in renewable energy market operations.

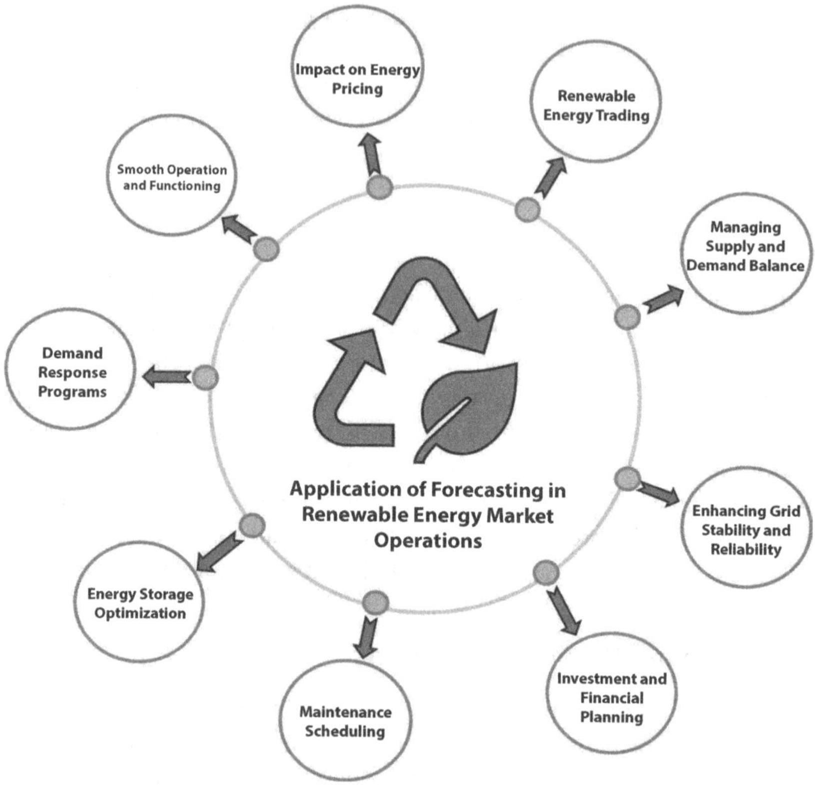

Figure 5.2 Application of forecasting in renewable energy market operations.

5.5 Advanced Topics in Renewable Energy Forecasting

The field of renewable energy forecasting is continually evolving, with new challenges and opportunities emerging as the global energy landscape shifts toward more sustainable sources. Advanced topics in this area are focusing on integrating broader environmental considerations, tapping into new sources of renewable energy, and leveraging technological advancements to enhance forecasting accuracy and utility. Below is a detailed exploration of these advanced topics [52].

5.5.1 Incorporating Climate Change Projections

Incorporating climate change projections into renewable energy forecasting represents a critical advancement in ensuring the long-term sustainability and reliability of renewable energy sources. Climate change can significantly alter weather patterns, affecting the predictability and variability of resources like wind and solar power [83]. By integrating climate models and projections into renewable energy forecasting:

- **Long-term Planning:** Forecasters can provide more accurate predictions availability of renewable resources, aiding in strategic planning and investment decisions.
- **Resilience Building:** Energy systems can be designed with greater resilience to withstand expected changes in climate, such as increased frequency and intensity of extreme weather events that could impact renewable energy production.
- **Adaptive Strategies:** Utilities and grid operators can develop adaptive strategies to manage the anticipated shifts in renewable energy generation patterns, ensuring a stable and reliable energy supply [84].

5.5.2 Forecasting for Offshore Renewable Energy Sources

Offshore renewable energy sources, particularly wind and tidal energy, offer vast potential for clean energy generation. Forecasting for these sources, however, poses unique challenges due to the complex marine environment:

- **Wind Speed and Wave Dynamics:** Offshore wind forecasting must account for higher wind speeds and more consistent patterns compared to onshore, as well as complex wave dynamics that can affect turbine performance and stability.
- **Tidal and Current Predictions:** For tidal energy, accurate forecasting relies on understanding the lunar and solar gravitational forces that drive tidal cycles, as well as local sea floor topography.
- **Access and Maintenance:** Forecasting also plays a role in planning access for maintenance and operation, considering the logistical challenges posed by the marine environment [84].

5.5.3 Role of Big Data and IoT in Forecasting

The integration of big data analytics and the Internet of Things (IoT) technologies into renewable energy forecasting marks a significant leap forward in harnessing computational and data-driven approaches:

- **Real-Time Data Collection:** IoT devices and sensors deployed across renewable energy installations collect vast amounts of real-time data on weather conditions, equipment performance, and energy production.
- **Data Analytics and Machine Learning:** Big data analytics and machine learning algorithms process this wealth of data to uncover patterns, trends, and anomalies. This enables more accurate, granular, and dynamic forecasts than previously possible.
- **Predictive Maintenance and Optimization:** Beyond energy production forecasting, these technologies also enable predictive maintenance of renewable energy assets and optimization of energy distribution and storage, contributing to overall system efficiency and reliability.

The advanced topics in renewable energy forecasting are pushing the boundaries of what is possible in predicting and managing renewable energy production. By incorporating climate change projections, addressing the unique challenges of offshore renewable sources, and leveraging the latest in data analytics and IoT technologies, the field is poised to make significant contributions to the global transition toward a more sustainable and resilient energy future. These advancements not only enhance forecasting accuracy but also support strategic decision-making, operational efficiency, and innovation in renewable energy technologies [85].

5.6 Challenges and Future Directions

As renewable energy continues to constitute a larger portion of the global energy mix, the forecasting of its generation faces evolving challenges. Addressing these challenges is paramount for maximizing the potential of renewable resources and ensuring their harmonious integration into existing energy systems. Concurrently, the sector is poised to evolve, with future directions focusing on leveraging technological innovations and adapting to the policy and regulatory landscape [86].

5.6.1 Addressing Variability and Uncertainty

One of the fundamental challenges in renewable energy forecasting is the inherent variability and uncertainty of renewable resources like wind and solar power. This variability stems from the dependency of these energy sources on weather conditions, which are naturally unpredictable and can change rapidly.

- **Improving Forecasting Accuracy:** Efforts to enhance forecasting accuracy involve developing more sophisticated models that can better account for the complexities of weather systems and their impact on renewable generation. This includes refining NWP models and incorporating higher-resolution data.
- **Probabilistic Forecasting:** Transitioning from deterministic forecasts to probabilistic ones offers a way to deal with uncertainty by providing a range of possible outcomes with associated probabilities. This approach helps grid operators and energy traders manage risks more effectively.
- **Enhanced Data Analytics:** Leveraging advanced data analytics and machine learning techniques to analyze vast datasets can uncover patterns that traditional models may miss, potentially reducing forecasting errors [87].

5.6.2 Integrating Emerging Technologies

The integration of emerging technologies presents both a challenge and an opportunity for the future of renewable energy forecasting. Technologies such as artificial intelligence (AI), the Internet of Things (IoT), and blockchain have the potential to transform forecasting methodologies and applications.

- **AI and Machine Learning:** The continued development and integration of AI and machine learning into forecasting models promise improvements in accuracy and the ability to dynamically adapt to changing conditions.
- **IoT for Real-Time Data Collection:** Expanding the network of IoT devices can enhance real-time data collection capabilities, providing more granular and immediate inputs for forecasting models, which is crucial for short-term forecasts.
- **Blockchain for Data Integrity:** Implementing blockchain technology could ensure the integrity and security of the data used in forecasting, facilitating more reliable and transparent energy transactions and grid management practices [14].

5.6.3 Policy and Regulatory Considerations

Policy and regulatory frameworks play a crucial role in the adoption and effectiveness of renewable energy forecasting. As the sector evolves, so too must the policies and regulations that govern it, adapting to new technologies and the growing importance of renewable energy.

- **Support for Research and Development:** Policymakers can encourage advancements in forecasting by supporting research and development initiatives and providing funding and resources necessary for innovation.
- **Adapting Grid Regulations:** Regulatory bodies need to adapt grid operation regulations to account for the increased integration of renewable energies, ensuring that forecasting models and practices are effectively incorporated into grid management strategies.
- **Incentivizing Technological Adoption:** Incentives for the adoption of advanced forecasting technologies by utilities and renewable energy producers can accelerate the integration of these tools into the market, enhancing the overall efficiency and reliability of renewable energy generation [88].

5.7 Future Directions

Looking forward, the field of renewable energy forecasting is set to grow in both scope and sophistication. Continued advancements in computational power, data analytics, and sensor technology will drive improvements in forecasting accuracy and reliability. Concurrently, the development of more adaptable and resilient energy grids, capable of integrating diverse renewable sources, will be critical. As this integration deepens, the interplay between technological innovations, policy adaptations, and market mechanisms will shape the trajectory of renewable energy's role in global energy systems, steering toward a more sustainable and secure energy future.

References

1. Wang, H., Lei, Z., Zhang, X., Zhou, B., Peng, J., A review of deep learning for renewable energy forecasting. *Energy Convers. Manage.*, 198, 111799, 2019, doi: https://doi.org/10.1016/j.enconman.2019.111799.

2. Alessandrini, S. and Sperati, S., Characterization of forecast errors and benchmarking of renewable energy forecasts, in: *Renewable Energy Forecasting: From Models to Applications*, pp. 235–256, Elsevier Inc., Cambridgeshire, United Kingdom, 2017, doi: 10.1016/B978-0-08-100504-0.00009-3.
3. Singh, V.P., Kumar, A., Meena, C.S., Dwivedi, G., Introduction to Energy-Efficient Vehicles for Sustainable Transportion: Transitions and Challenges, in: *Energy Efficient Vehicles*, pp. 1–18, CRC Press, Boca Raton, FL, USA, 2024, doi: https://doi.org/10.1201/9781003464556-1.
4. Ahmad, T., Zhang, H., Yan, B., A review on renewable energy and electricity requirement forecasting models for smart grid and buildings. *Sustain. Cities Soc.*, 55, 102052, 2020, doi: https://doi.org/10.1016/j.scs.2020.102052.
5. Stratigakos, A., Camal, S., Michiorri, A., Kariniotakis, G., Prescriptive Trees for Integrated Forecasting and Optimization Applied in Trading of Renewable Energy. *IEEE Trans. Power Syst.*, 37, 6, 4696–4708, 2022, doi: 10.1109/TPWRS.2022.3152667.
6. Ruhnau, O., Hennig, P., Madlener, R., Economic implications of forecasting electricity generation from variable renewable energy sources. *Renew. Energy*, 161, 1318–1327, 2020, doi: https://doi.org/10.1016/j.renene.2020.06.110.
7. Singh, V.P. and Jain, S., Economic analysis of a large scale solar updraft tower power plant. *Sustain. Energy Technol. Assess.*, 58, 103325, 2023, doi: https://doi.org/10.1016/j.seta.2023.103325.
8. Singh, V.P. and Kumar, A., Techno-Economic and Future Aspects of the HEV-EV-FCV: Decarbonization, Digitalization, and Sustainability, in: *Energy Efficient Vehicles*, 1st ed., vol. 1, pp. 212–231, CRC Press, Boca Raton, FL, USA, 2024, doi: https://doi.org/10.1201/9781003464556-11.
9. Dagoumas, A.S. and Koltsaklis, N.E., Review of models for integrating renewable energy in the generation expansion planning. *Appl. Energy*, 242, 1573–1587, 2019, doi: https://doi.org/10.1016/j.apenergy.2019.03.194.
10. Dash, S., *et al.*, A Novel Algorithmic Forex Trade and Trend Analysis Framework Based on Deep Predictive Coding Network Optimized with Reptile Search Algorithm. *Axioms*, 11, 8, 1–29, Aug. 2022, doi: 10.3390/axioms11080396.
11. Morstyn, T., Farrell, N., Darby, S.J., McCulloch, M.D., Using peer-to-peer energy-trading platforms to incentivize prosumers to form federated power plants. *Nat. Energy*, 3, 2, 94–101, 2018, doi: 10.1038/s41560-017-0075-y.
12. Masini, A. and Menichetti, E., Investment decisions in the renewable energy sector: An analysis of non-financial drivers. *Technol. Forecast. Soc. Chang.*, 80, 3, 510–524, 2013, doi: https://doi.org/10.1016/j.techfore.2012.08.003.
13. Lee, S.-C. and Shih, L.-H., Renewable energy policy evaluation using real option model — The case of Taiwan. *Energy Econ.*, 32, S67–S78, 2010, doi: https://doi.org/10.1016/j.eneco.2010.04.010.
14. Tsai, S.-B., *et al.*, Models for forecasting growth trends in renewable energy. *Renew. Sustain. Energy Rev.*, 77, 1169–1178, 2017, doi: https://doi.org/10.1016/j.rser.2016.06.001.

15. Taik, A., Nour, B., Cherkaoui, S., Empowering Prosumer Communities in Smart Grid with Wireless Communications and Federated Edge Learning. *IEEE Wireless Commun.*, 28, 6, 26–33, 2021, doi: 10.1109/MWC.017.2100187.
16. Al Irsyad, M.I., Halog, A., Nepal, R., Renewable energy projections for climate change mitigation: An analysis of uncertainty and errors. *Renew. Energy*, 130, 536–546, 2019, doi: https://doi.org/10.1016/j.renene.2018.06.082.
17. Upadhyay, R.K., Singh, V.P., Kumar, A., Sustainable Transportation: Policy, Planning, and Implementation, in: *Energy Efficient Vehicles*, 1st ed., pp. 74–95, CRC Press, Boca Raton, FL, USA, 2024, doi: https://doi.org/10.1201/9781003464556-5.
18. Verma, M. and Singh, V.P., Design optimization and sensitivity analysis in aerospace engineering, in: *Computational Fluid Flow and Heat Transfer*, pp. 112–132, CRC Press, Boca Raton, FL, USA, 2024, doi: https://doi.org/10.1201/9781003465171-7.
19. Yang, G., Du, S., Duan, Q., Su, J., Short-Term Demand Forecasting Method in Power Markets Based on the KSVM-TCN-GBRT. *Comput. Intell. Neurosci.*, 2022, 1–13, 2022, doi: 10.1155/2022/6909558.
20. Singh, V.P., Kumar, A., Meena, C.S., Dwivedi, G., *Energy Efficient Vehicles*, 1st ed., CRC Press, Boca Raton, 2024, doi: 10.1201/9781003464556.
21. Zhou, K. and Yang, S., Smart Energy Management, in: *Comprehensive Energy Systems: Volumes 1-5*, vol. 5, pp. V5-423–V5-456, Elsevier, Amsterdam, The Netherlands, 2018, doi: 10.1016/B978-0-12-809597-3.00525-3.
22. Sims, R.E.H., Rogner, H.-H., Gregory, K., Carbon emission and mitigation cost comparisons between fossil fuel, nuclear and renewable energy resources for electricity generation, *Energy Policy*, 31, 13, 1315–1326, 2003, ISSN: 0301-4215, https://doi.org/10.1016/S0301-4215(02)00192-1.
23. Corizzo, R., Ceci, M., Fanaee-T, H., Gama, J., Multi-aspect renewable energy forecasting. *Inf. Sci. (N Y)*, 546, 701–722, 2021, doi: https://doi.org/10.1016/j.ins.2020.08.003.
24. Oree, V., Sayed Hassen, S.Z., Fleming, P.J., Generation expansion planning optimisation with renewable energy integration: A review. *Renew. Sustain. Energy Rev.*, 69, 790–803, 2017, doi: https://doi.org/10.1016/j.rser.2016.11.120.
25. Kumar, A., Gori, Y., Meena, C.S., Singh, V.P., Sharma, V.K., Design and analysis of heavy vehicle medium duty transmission gearbox system, in: *Additive Manufacturing in Industry 4.0: Methods, Techniques, Modeling, and Nano Aspects*, pp. 67–86, 2022, doi: 10.1201/9781003360001-4.
26. Bawankar, S., *et al.*, Environmental impact assessment of lithium ion battery employing cradle to grave. *Sustain. Energy Technol. Assess.*, 60, 1–10, May, 2023, doi: 10.1016/j.seta.2023.103530.
27. Kumar, A., Singh, V.P., Meena, C.S., Dutt, N., Preface of Thermal Energy Systems, in: *Thermal Energy Systems: Design, Computational Techniques, and Applications*, pp. x–xiii, CRC Press, Technical Education Department, Kanpur, Uttar Pradesh, India, 2023, [Online]. Available: https://www.scopus.

com/inward/record.uri?eid=2-s2.0-85161660089&partnerID=40&md5=056f377c943550d48151824d45dcb030.

28. Kumar, A., Singh, V.P., Meena, C.S., Dutt, N., Aim and Scope of Thermal Energy Systems: Design, Computational Techniques, and Applications, in: *Thermal Energy Systems: Design, Computational Techniques, and Applications*, A. Kumar, V.P. Singh, C.S. Meena, N. Dutt (Eds.), pp. 9–10, CRC Press, Boca Raton, FL, USA, 2022.
29. Sharma, B., Kumar, S., Tiwari, P., Yadav, P., Nezhurina, M.I., ANN based short-term traffic flow forecasting in undivided two lane highway. *J. Big Data*, 5, 1, 1–16, Dec. 2018, doi: 10.1186/s40537-018-0157-0.
30. Nayal, M., Sharma, A.K., Jain, S., Singh, V.P., Chapter 11 Green Hydrogen Production Methods, Designs and Applications, in: *Highly Efficient Thermal Renewable Energy Systems Design, Optimization and Applications*, 1st ed., R.W., V. Verma, S. Thangavel, N. Dutt, A. Kumar (Eds.), pp. 178–193, CRC Press, Boca Raton, FL, USA, 2024, [Online]. Available: https://www.routledge.com/Highly-Efficient-Thermal-Renewable-Energy-Systems-Design-Optimization/Verma-Thangavel-Dutt-Kumar-Weerasinghe/p/book/9781032595641.
31. Arroyo M., F.R. and Miguel, L.J., The Role of Renewable Energies for the Sustainable Energy Governance and Environmental Policies for the Mitigation of Climate Change in Ecuador. *Energies (Basel)*, 13, 15, 1–18, 2020, doi: 10.3390/en13153883.
32. Natarajan, V.A. and Karatampati, P., Survey on renewable energy forecasting using different techniques, in: *2019 2nd International Conference on Power and Embedded Drive Control (ICPEDC)*, pp. 349–354, 2019, doi: 10.1109/ICPEDC47771.2019.9036569.
33. Sharma, B., Katiyar, V.K., Kumar, K., Traffic Accident Prediction Model Using Support Vector Machines with Gaussian Kernel, in: *Proceedings of Fifth International Conference on Soft Computing for Problem Solving*, M. Pant, K. Deep, J.C. Bansal, A. Nagar, K.N. Das (Eds.), pp. 1–10, Springer Singapore, Singapore, 2016.
34. Sharma, A.K., Nayal, M., Jain, S., Singh, V.P., Chapter 07 Optimization Techniques of Solar Thermal and Hybrid Energy Systems, in: *Highly Efficient Thermal Renewable Energy Systems Design, Optimization and Applications*, 1st ed., R.W., V. Verma, S. Thangavel, N. Dutt, A. Kumar (Eds.), pp. 1–14, CRC Press, Boca Raton, FL, USA, 2024, [Online]. Available: https://www.routledge.com/Highly-Efficient-Thermal-Renewable-Energy-Systems-Design-Optimization/Verma-Thangavel-Dutt-Kumar-Weerasinghe/p/book/9781032595641.
35. Kumar, V., Jain, V., Sharma, B., Chatterjee, J.M., Shrestha, R., *Smart City Infrastructure: The Blockchain Perspective*, Wiley, United States, 2022, ISBN:9781119785545, [Online]. Available: https://books.google.co.in/books?id=iM1gEAAAQBAJ.

36. Cros, S., Badosa, J., Szantaï, A., Haeffelin, M., Reliability Predictors for Solar Irradiance Satellite-Based Forecast. *Energies (Basel)*, 13, 21, 1–21, 2020, doi: 10.3390/en13215566.
37. Nayal, M., Sharma, A.K., Jain, S., Singh, V.P., Chapter 06 Design and Modelling of Solar, Geothermal and Hybrid Energy Systems.pdf, in: *Highly Efficient Thermal Renewable Energy Systems Design, Optimization and Applications*, 1st ed., R.W., V. Verma, S. Thangavel, N. Dutt, A. Kumar (Eds.), pp. 1–15, CRC Press, Boca Raton, FL, USA, 2024, [Online]. Available: https://www.routledge.com/Highly-Efficient-Thermal-Renewable-Energy-Systems-Design-Optimization/Verma-Thangavel-Dutt-Kumar-Weerasinghe/p/book/9781032595641.
38. Rathore, G.S., S.R., A.N., Karn, A., Singh, V.P., Recent Progress in Solar Dryers using Phase Changing Material: A Review. *Int. J. Energy Resour. Appl. (IJERA)*, 2, 1, 57–77, 2023, doi: 10.56896/IJERA.2023.2.1.005.
39. Pant, G., Meena, C.S., Saxena, A., Kumar, A., Singh, V.P., Dutt, N., Study the Temperature Variation in Alternate Coils of Insulated Condenser Cum Storage Tank: Experimental Study, in: *Lecture Notes in Mechanical Engineering*, B.S. Sikarwar, S.K. Sharma, A. Jain, K.M. Singh (Eds.), pp. 627–638, Springer Nature Singapore, Singapore, 2023, doi: https://doi.org/10.1007/978-981-99-2382-3_52.
40. Dutt, N., Jageshwar, A., Ashwani, H., Mukesh, K., Awasthi, K., Pratap, V., Thermo-hydraulic performance of solar air heater having discrete D-shaped ribs as artificial roughness. *Environ. Sci. Pollut. Res.*, 30, 28, 1–22, 2023, doi: 10.1007/s11356-023-28247-9.
41. Agarwal, V. and Sharma, S., Q Learning Algorithm for Network Resource Management in Vehicular Communication Network, in: *Autonomous Vehicles Volume 2*, pp. 259–274, 2022, doi: https://doi.org/10.1002/9781394152636.ch13.
42. Noori, M.A.R., Sharma, B., Mehra, R., Feature selection from disaster tweets using Spark-based parallel meta-heuristic optimizers. *Soc. Netw. Anal. Min.*, 12, 1, 93, 2022, doi: 10.1007/s13278-022-00930-8.
43. Azadeh, A., Babazadeh, R., Asadzadeh, S.M., Optimum estimation and forecasting of renewable energy consumption by artificial neural networks. *Renew. Sustain. Energy Rev.*, 27, 605–612, 2013, doi: https://doi.org/10.1016/j.rser.2013.07.007.
44. Rahman, M.M., *et al.*, Prospective Methodologies in Hybrid Renewable Energy Systems for Energy Prediction Using Artificial Neural Networks. *Sustainability*, 13, 4, 1–21, 2021, doi: 10.3390/su13042393.
45. Singh, V.P. and Dwivedi, G., Technical Analysis of a Large-Scale Solar Updraft Tower Power Plant. *Energies (Basel)*, 16, 1, 103325, 2023, doi: 10.3390/en16010494.
46. Ahmad, M.W., Reynolds, J., Rezgui, Y., Predictive modelling for solar thermal energy systems: A comparison of support vector regression, random

forest, extra trees and regression trees. *J. Clean. Prod.*, 203, 810–821, 2018, doi: https://doi.org/10.1016/j.jclepro.2018.08.207.
47. Kundu, A., Kumar, A., Singh, V.P., Meena, C.S., Dutt, N., Chapter 1: Introduction to Thermal Energy Resources and Their Smart Applications, in: *Thermal Energy Systems: Design, Computational Techniques, and Applications*, 1st Edition, V. P. S. C. S. M. N. D. A. Kumar (Ed.), pp. 1–15, CRC Press, Boca Raton, 2023, doi: 10.1201/9781003395768-1.
48. Sharma, B., Katiyar, V.K., Gupta, A.K., Singh, A., The Automated Vehicle Detection of Highway Traffic Images by Differential Morphological Profile. *J. Transp. Technol.*, 04, 02, 150–156, 2014, doi: 10.4236/jtts.2014.42015.
49. Zendehboudi, A., Baseer, M.A., Saidur, R., Application of support vector machine models for forecasting solar and wind energy resources: A review. *J. Clean. Prod.*, 199, 272–285, 2018, doi: https://doi.org/10.1016/j.jclepro.2018.07.164.
50. Kumar, A., Singh, V.P., Meena, C.S., Dutt, N., *Thermal Energy Systems: Design, Computational Techniques, and Applications*, 1st ed., CRC Press, Boca Raton, FL, USA, 2023, doi: https://doi.org/10.1201/9781003395768.
51. Sivasakthivel, T., Verma, V., Tarodiya, R., Meena, C.S., Singh, V.P., Kumar, R., Chapter 11: Analysis of Optimum Operating Parameters for Ground Source Heat Pump System for Different Cases of Building Heating and Cooling Mode Operations, in: *Thermal Energy Systems: Design, Computational Techniques, and Applications*, 1st ed., A. Kumar, N. Dutt, V. Singh, C. Meena (Eds.), pp. 183–207, CRC Press, Boca Raton, FL, USA, 2023, doi: 10.1201/9781003395768-11.
52. Zheng, J., Du, J., Wang, B., Klemeš, J.J., Liao, Q., Liang, Y., A hybrid framework for forecasting power generation of multiple renewable energy sources. *Renew. Sustain. Energy Rev.*, 172, 113046, 2023, doi: https://doi.org/10.1016/j.rser.2022.113046.
53. Kundu, A., Kumar, A., Dutt, N., Singh, V.P., Meena, C.S., Chapter 7: Modelling and Simulation of Thermal Energy System for Design Optimization, in: *Thermal Energy Systems: Design, Computational Techniques, and Applications*, A. Kumar, N. Dutt, V. Singh, C. Meena (Eds.), pp. 103–137, CRC Press, Boca Raton, FL, USA, 2023, doi: 10.1201/9781003395768-07.
54. Chauhan, K. and Singh, V.P., Prospect of biomass to bioenergy in India : An overview. *Mater. Today Proc.*, 72, 5, 941–953, 2023, doi: 10.1016/j.matpr.2023.01.419.
55. Alghamdi, A.A., Ibrahim, A., El-Kenawy, E.-S.M., Abdelhamid, A.A., Renewable Energy Forecasting Based on Stacking Ensemble Model and Al-Biruni Earth Radius Optimization Algorithm. *Energies (Basel)*, 16, 3, 1–21, 2023, doi: 10.3390/en16031370.
56. Singh, A.R., Singh, S.K., Singh, V.P., Process parameters optimization of carbon nano tube based catalytic transesterification of algal oil. *Mater. Today Proc.*, 72, 5, 684–691, 2023, doi: 10.1016/j.matpr.2023.01.418.

57. Vaccaro, A., Mercogliano, P., Schiano, P., Villacci, D., An adaptive framework based on multi-model data fusion for one-day-ahead wind power forecasting. *Electr. Power Syst. Res.*, 81, 3, 775–782, 2011, doi: https://doi.org/10.1016/j.epsr.2010.11.009.
58. Sharma, B. and Kumar, S., Delay Optimization Using Genetic Algorithm at the Road Intersection. *Int. J. Inf. Retr. Res.*, 9, 2, 1–10, Apr. 2019, doi: 10.4018/IJIRR.2019040101.
59. Singh, V.P., Meena, C.S., Kumar, A., Dutt, N., Double Pass Solar Air Heater : A Review. *Int. J. Energy Resour. Appl. (IJERA)*, 1, 2, 22–43, 2022, doi: 10.56896/IJERA.2022.1.2.009.
60. Jain, S., Kumar, N., Singh, V.P., Mishra, S., Sharma, N.K., Transesterification of Algae Oil and Little Amount of Waste Cooking Oil Blend at Low Temperature in the Presence of NaOH. *Energies (Basel)*, 16, 1, 1–12, 2023, doi: https://doi.org/10.3390/ en16031293.
61. Shi, J. and Teh, J., Load forecasting for regional integrated energy system based on complementary ensemble empirical mode decomposition and multi-model fusion. *Appl. Energy*, 353, 122146, 2024, doi: https://doi.org/10.1016/j.apenergy.2023.122146.
62. Kumar, R., Kumar, A., Laxmikant, Dutt, N., Singh, V.P., Cutting Parameters Optimization in Turning Operation using Taguchi Method. *Int. J. Mater. Manuf. Sustain. Technol. (IJMMST)*, 1, 2, 22–30, 2022, doi: https://doi.org/10.56896/IJMMST.2022.1.2.009.
63. Alkhayat, G. and Mehmood, R., A review and taxonomy of wind and solar energy forecasting methods based on deep learning. *Energy AI*, 4, 100060, 2021, doi: https://doi.org/10.1016/j.egyai.2021.100060.
64. Khosravi, A., Koury, R.N.N., Machado, L., Pabon, J.J.G., Prediction of wind speed and wind direction using artificial neural network, support vector regression and adaptive neuro-fuzzy inference system. *Sustain. Energy Technol. Assess.*, 25, 146–160, 2018, doi: https://doi.org/10.1016/j.seta.2018.01.001.
65. Kumar, R., *et al.*, Experimental and RSM-Based Process-Parameters Optimisation for Turning Operation of EN36B Steel. *Materials*, 16, 1, 1–18, 2023, doi: 10.3390/ma16010339.
66. Singh, V.P., Karn, A., Dwivedi, G., Alam, T., Kumar, A., Experimental Assessment of Variation in Open Area Ratio on Thermohydraulic Performance of Parallel Flow Solar Air Heater. *Arabian J. Sci. Eng.*, 41, 12, 1–17, 2022, doi: 10.1007/s13369-022-07525-7.
67. Singh, V.P., *et al.*, Heat transfer and friction factor correlations development for double pass solar air heater artificially roughened with perforated multi-V ribs. *Case Stud. Therm. Eng.*, 39, September, 102461, 2022, doi: https://doi.org/10.1016/j.csite.2022.102461.
68. Singh, V.P., Jain, S., Gupta, J.M.L., Analysis of the effect of perforation in multi-v rib artificial roughened single pass solar air heater: - Part A. *Exp. Heat Transfer*, 36, 2, 1–20, Oct. 2021, doi: 10.1080/08916152.2021.1988761.

69. Aslam, S., Herodotou, H., Mohsin, S.M., Javaid, N., Ashraf, N., Aslam, S., A survey on deep learning methods for power load and renewable energy forecasting in smart microgrids. *Renew. Sustain. Energy Rev.*, 144, 110992, 2021, doi: https://doi.org/10.1016/j.rser.2021.110992.
70. Singh, V.P., *et al.*, Nanomanufacturing and Design of High-Performance Piezoelectric Nanogenerator for Energy Harvesting, in: *Nanomanufacturing and Nanomaterials Design: Principles and Applications*, 1st ed., S. K. B. A. K. K. V. S. Singh (Ed.), pp. 241–272, CRC Press, Boca Raton, 2022, doi: 10.1201/9781003220602-15.
71. Ashwani Kumar, Y.G., Gangwar, A.K.S., Kumar, A., Meena, C.S., Singh, V.P., Dutt, N., Prasad, A., Biomedical study of femur bone fracture and healing, in: *Advanced Materials for Biomedical Applications*, 1st edition, Y. G. A. K. C. S. M. N. D. A. Kumar (Ed.), pp. 283–298, CRC Press, Boca Raton, 2022, doi: 10.1201/9781003344810-14.
72. Datta, A., Kumar, A., Kumar, A., Kumar, A., Singh, V.P., Advanced materials in biological implants and surgical tools, in: *Advanced Materials for Biomedical Applications*, 1st ed., Ashwani Kumar, N.D., Gori, Y., Kumar, A., Meena, C.S. (Eds.), pp. 283–298, 2022, doi: 10.1201/9781003344810-2.
73. Shubham Srivastava, R.K., Verma, D., Thusoo, S., Kumar, A., Singh, V.P., Nanomanufacturing for Energy Conversion and Storage Devices, in: *Nanomanufacturing and Nanomaterials Design: Principles and Applications*, 1st ed., S. K. B. A. K. K. V. S. Singh (Ed.), pp. 165–174, CRC Press, Boca Raton, 2022, doi: 10.1201/9781003220602-10.
74. Carriere, T. and Kariniotakis, G., An Integrated Approach for Value-Oriented Energy Forecasting and Data-Driven Decision-Making Application to Renewable Energy Trading. *IEEE Trans. Smart Grid*, 10, 6, 6933–6944, 2019, doi: 10.1109/TSG.2019.2914379.
75. Dutt, N., Binjola, A., Hedau, A.J., Kumar, A., Singh, V.P., Meena, C.S., Comparison of CFD Results of Smooth Air Duct with Experimental and Available Equations in Literature. *Int. J. Energy Resour. Appl. (IJERA)*, 1, 1, 40–47, 2022, doi: https://doi.org/10.56896/IJERA.2022.1.1.006.
76. Minhas, D.M., Khalid, R.R., Frey, G., Load control for supply-demand balancing under Renewable Energy forecasting, in: *2017 IEEE Second International Conference on DC Microgrids (ICDCM)*, pp. 365–370, 2017, doi: 10.1109/ICDCM.2017.8001071.
77. Sarita, K., *et al.*, Power Enhancement With Grid Stabilization of Renewable Energy-Based Generation System Using UPQC-FLC-EVA Technique. *IEEE Access*, 8, 207443–207464, 2020, doi: 10.1109/ACCESS.2020.3038313.
78. Yadav, A.K., Singh, V.P., Kumar, A., Gori, Y., Kumar, A., Study the Properties of High-Density Polyethylene with Nano Particles Silica Composite. *Int. J. Mater. Manuf. Sustain. Technol. (IJMMST)*, 1, 1, 29–36, 2022, doi: https://doi.org/10.56896/IJMMST.2022.1.1.005.

79. Mohseni, S., Brent, A.C., Kelly, S., Browne, W.N., Demand response-integrated investment and operational planning of renewable and sustainable energy systems considering forecast uncertainties: A systematic review. *Renew. Sustain. Energy Rev.*, 158, 112095, 2022, doi: https://doi.org/10.1016/j.rser.2022.112095.
80. Singh, V.P., Jain, S., Kumar, A., Establishment of correlations for the Thermo-Hydraulic parameters due to perforation in a multi-V rib roughened single pass solar air heater. *Exp. Heat Transfer*, 35, 5, 1–20, 2022, doi: 10.1080/08916152.2022.2064940.
81. Sharifi, V., Abdollahi, A., Rashidinejad, M., Flexibility-based generation maintenance scheduling in presence of uncertain wind power plants forecasted by deep learning considering demand response programs portfolio. *Int. J. Electr. Power Energy Syst.*, 141, 108225, pp. 1–17, 2022, doi: https://doi.org/10.1016/j.ijepes.2022.108225.
82. Ullah, K., *et al.*, An Optimal Energy Optimization Strategy for Smart Grid Integrated with Renewable Energy Sources and Demand Response Programs. *Energies (Basel)*, 13, 21, 1–17, 2020, doi: 10.3390/en13215718.
83. Singh, V.P., Jain, S., Karn, A., Kumar, A., Dwivedi, G., Mathematical Modeling of Efficiency Evaluation of Double Pass Parallel Flow Solar Air Heater. *Sustainability*, 14, 17, 1–22, 2022, doi: https://doi.org/10.3390/su141710535.
84. Russo, M.A., Carvalho, D., Martins, N., Monteiro, A., Forecasting the inevitable: A review on the impacts of climate change on renewable energy resources. *Sustain. Energy Technol. Assess.*, 52, 102283, 2022, doi: https://doi.org/10.1016/j.seta.2022.102283.
85. Li, J., Herdem, M.S., Nathwani, J., Wen, J.Z., Methods and applications for Artificial Intelligence, Big Data, Internet of Things, and Blockchain in smart energy management. *Energy AI*, 11, 100208, 2023, doi: https://doi.org/10.1016/j.egyai.2022.100208.
86. Kashif, M., *et al.*, Untapped renewable energy potential of crop residues in Pakistan: Challenges and future directions. *J. Environ. Manage.*, 256, 109924, 2020, doi: https://doi.org/10.1016/j.jenvman.2019.109924.
87. Dobschinski, J., *et al.*, Uncertainty Forecasting in a Nutshell: Prediction Models Designed to Prevent Significant Errors. *IEEE Power Energy Mag.*, 15, 6, 40–49, 2017, doi: 10.1109/MPE.2017.2729100.
88. Nam, K., Hwangbo, S., Yoo, C., A deep learning-based forecasting model for renewable energy scenarios to guide sustainable energy policy: A case study of Korea. *Renew. Sustain. Energy Rev.*, 122, 109725, 2020, doi: https://doi.org/10.1016/j.rser.2020.109725.

6

Machine Learning Techniques for Demand Forecasting in the Electricity Sector

Firuz Ahamed Nahid[1,2*], Hussain Mahmud Chowdhury[2] and Mohammad Nayeem Jahangir[2]

[1]*Department of Energy, Environment and Climate Change, Asian Institute of Technology, Pathum Thani, Thailand*
[2]*Department of Electrical and Electronic Engineering, International University of Business Agriculture and Technology – IUBAT, Dhaka, Bangladesh*

Abstract
This chapter presents a comprehensive review of machine learning (ML) techniques as applied to forecasting demand within the electricity sector—a pivotal factor in the optimization of energy systems. The recent influx of renewable energy sources and the advent of smart grid technologies have underscored the limitations of traditional forecasting methods. In response, ML offers an array of advanced, data-centric algorithms that adeptly capture complex patterns, enabling more precise forecasts of future electricity demand. The chapter progresses through an evaluation of fundamental ML methodologies pertinent to demand forecasting, spanning time-series models, regression analyses, ensemble learning, and state-of-the-art deep learning and hybrid algorithms. It assesses each methodology's inherent advantages and constraints and furnishes an empirical perspective on their operational efficacy in the energy domain. This chapter showcases the application and analytical benefits of ML in actual forecasting scenarios, underscoring the enhanced accuracy and adaptability these methods bring to the fore. Future challenges, including data integrity, system integration, and scalability, are addressed, with a forward-looking analysis on the evolution of ML in tandem with the continuous transformation of energy networks. In addition, this chapter presents a case study where hybrid convolutional long-short term memory recurrent neural network has been applied to forecast electricity demand of Thailand's metropolitan region showcasing the application of advanced ML algorithms in electricity sector's demand forecasting area.

**Corresponding author*: firuz@ait.asia

Jai Govind Singh, Rupendra Kumar Pachauri and Sasidharan Sreedharan (eds.) Forecasting Methods for Renewable Power Generation, (131–172) © 2025 Scrivener Publishing LLC

Keywords: Machine learning, demand forecasting, electricity sector, renewable energy, microgrid, short-term forecasting

6.1 Introduction

The electricity sector is experiencing a pivotal transformation, propelled by the rapid proliferation of renewable energy sources, the advent of smart grids, and the escalating demands for sustainable and efficient energy management [1]. Achieving a harmonious balance between power generation and consumption remains an intricate challenge within this dynamic environment. Accurate demand forecasting emerges as a critical factor for enhancing resource utilization, grid reliability, and the integration of a diverse array of energy resources. The expanding network of smart grids, smart meters, and data analytics tools presents novel opportunities for leveraging the wealth of data available to minimize the disparities between electricity supply and demand and mitigate associated costs [2].

In the midst of these advancements, machine learning (ML) emerges as a compelling subset of artificial intelligence, providing sophisticated, data-driven methodologies that cut through the complexities of electric demand forecasting [3]. Unlike traditional statistical approaches that often stumble in their efforts to encapsulate the sector's complexity and variability, ML algorithms proficiently synthesize vast datasets encompassing historical consumption patterns, climatic conditions, economic factors, and consumer behaviors [4]. This suite of algorithms—spanning from ensemble methods like random forests and non-linear techniques such as support vector machines (SVMs) to advanced deep learning architectures like neural networks—exceptionally identifies intricate correlations within data, leading to enhanced forecasting accuracy [5–7].

These ML approaches, when applied to electric demand forecasting, hold several distinct advantages. The models are adept at processing multi-dimensional data arrays, thereby enabling a multifaceted look into the myriad factors that drive electricity consumption [8]. Moreover, ML algorithms are innately designed to evolve with incoming real-time data, yielding a more agile and precise forecasting system that is essential for navigating swiftly changing demand patterns [9]. Over time, the iterative refinement of ML models perpetually sharpens their predictive abilities, which is indispensable for informed planning, efficient resource distribution, and maintaining grid equilibrium [10].

Exploring the vast landscape of demand and load forecasting, scholars have made considerable headway, unlocking robust insights into the operational efficacy of different ML techniques tailored to various forecasting time frames. The granularity of forecasting ranges from short-term [11], which holds critical value for operational management in power markets from hours to a week ahead, to medium-term [3], which steers fuel and maintenance decisions over a monthly projection, and, finally, to long-term load forecasting [12] that guides overarching strategies for generation, transmission, and distribution planning on a multi-year horizon. The specific forecasting objective dictates the temporal scope and thus catalyzes the development of complex, context-sensitive models that can adapt to data variability and an assortment of influential factors, from meteorological to socio-economic dimensions. Detailed models are thus constructed to effectively serve various load forecasting needs, from individual infrastructure management to addressing the broader demands of national energy requirements.

6.1.1 Motivation and Contribution

This chapter is motivated by the urgent need to harmonize state-of-the-art ML techniques with the evolving demands of the energy sector. It aims to demystify the advanced realm of ML algorithms, contextualizing their relevance and application within the specialized field of electric load forecasting. A robust assessment of these methodologies is presented, scrutinizing fundamental time-series models, comprehensive regression analyses, innovative ensemble approaches, and cutting-edge deep learning and hybrid models. Contributions of this work are multi-fold:

- It offers a holistic perspective on how ML models successfully interpret and utilize the deluge of data available in modern energy systems.
- This chapter provides a critical evaluation of various ML techniques, distinguishing their unique advantages and limitations in the context of electric load forecasting.
- It delivers empirical insights by presenting case studies where ML methods have been applied, illustrating their practical benefits and forecasting excellence.

In combining theoretical exposition with applied research, this chapter aims to underscore the transformative potential of ML technologies in empowering energy stakeholders to make data-driven decisions and effectively manage the intricate dynamics of today's electric grid.

6.2 Overview of Demand Forecasting

Demand forecasting stands as a cornerstone of strategic planning for the electric power sector, bridging the gap between energy supply chain management and consumption patterns [1]. It encompasses the confluence of data analytics, statistical acumen, and domain-specific insights to anticipate future electricity demand over established time horizons, which can range from hours to decades [13]. This critical process informs not only the operational scheduling of power generation but also the infrastructural and economic strategies of utilities and policymakers alike.

Electricity demand forecasting is intrinsically complex, given its dependency on a constellation of variables both exogenous and endogenous in nature [14]. Weather conditions, such as temperature, humidity, and sunlight, have profound impacts on demand, driving the need for heating or cooling and thus directly influencing consumption patterns. Economic indicators often correlate with energy usage, where periods of prosperity see heightened industrial activity and increased consumption [15]. Concurrently, socio-demographic trends, technological advances, and the evolution of consumer behaviors all dictate shifting patterns of electricity usage. The interplay of these variables necessitates a forecasting model that is not only responsive to past consumption trends but is also sufficiently sophisticated to incorporate a myriad of predictive factors.

In the background of the electric power sector, the application of ML and advanced statistical methods to demand forecasting represents a seismic shift in the ability to process large datasets and glean actionable insights [3]. These ML methods excel at identifying underlying patterns and adapting to new data, often uncovering relationships among variables that might elude conventional statistical models. Predictive analytics powered by ML introduce a higher degree of granularity and precision in forecasts, enabling more nuanced demand predictions that account for the multifaceted nature of electricity consumption [16]. Demand forecasting, therefore, is not merely an exercise in statistical extrapolation but a nuanced field that sits at the nexus of technology, economics, and environmental sciences. By embracing the complexity of the task and the breadth of influencing factors, energy stakeholders are better equipped to design flexible, resilient.

6.2.1 Classification of Demand Forecasting

Demand forecasting serves as a navigation tool for the electric power sector, facilitating informed decision-making across various temporal

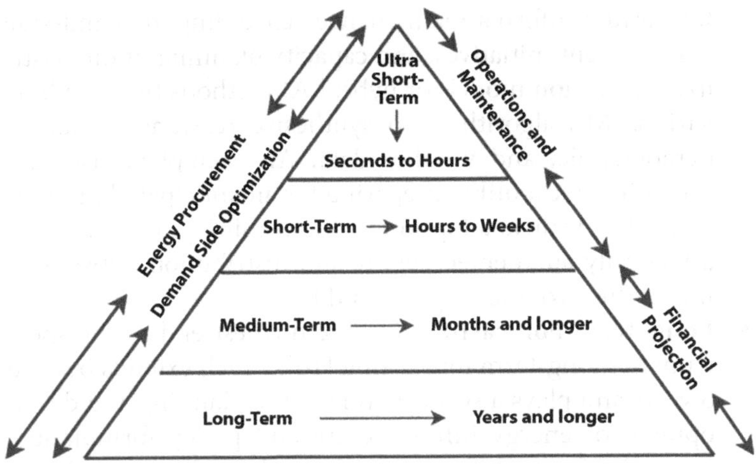

Figure 6.1 Electric demand forecasting classification.

dimensions [17]. To address the unique challenges and opportunities that accompany different planning horizons, demand forecasting is categorically segregated into short-term, medium-term, and long-term timelines as presented in Figure 6.1.

- **Short-Term Forecasting:** Short-term forecasting [11], often spanning from a few hours to a few weeks, is crucial for the operational management of electricity generation and distribution. It is characterized by its granular focus on immediate demand, enabling utility operators to respond to real-time changes swiftly. The associated models, such as autoregressive integrated moving average, exponential smoothing state space model, and ML techniques like neural networks, are optimized for precision and responsiveness [18]. Such models exploit finely detailed historical data along with immediate indicators such as current weather conditions and specific event schedules. For instance, the use of short-term forecasting during heatwaves allows for the precise allocation of additional resources to accommodate the spike in cooling demand, thereby preventing potential blackouts [19].
- **Medium-Term Forecasting**: Medium-term forecasting navigates the intermediate timeframe [3], typically spanning from weeks to up to 5 years. This tactical level of

forecasting informs maintenance scheduling, demand-side management initiatives, and capacity planning. It often utilizes regression models or hybrid AI methods that combine various ML algorithms to synthesize relevant economic, demographic, and weather data. An exemplary use case might include a utility preparing for an anticipated increase in electric vehicle adoption in the coming years, requiring a tactically augmented supply and distribution network to match this evolutionary demand [20].

- **Long-Term Forecasting**: On the furthest end of the spectrum lies long-term forecasting [16], which extends beyond 5 years and plays a strategic role in the planning and development of energy infrastructure and policy formulation. Here, the focus shifts to broad trends such as population growth, urbanization, technological advancements, and policy impacts. Models for long-term forecasting may incorporate scenario analysis and simulations that extend decades into the future. This foresight is pivotal when planning for large-scale infrastructure projects or the integration of renewable energy sources that have a considerable lead time on investment and development. For instance, a country's commitment to reducing carbon emissions by a certain year will significantly influence the long-term forecasting of energy demands and the corresponding adaptations in power generation portfolios [21].

6.2.2 Benefits of Load Forecasting

Load forecasting holds paramount importance in the electricity sector due to its pivotal role in ensuring the efficient utilization of resources and maintaining the stability of the power grid. By accurately predicting electricity demand, load forecasting enables stakeholders to make informed decisions and implement strategies that yield numerous benefits across various domains [22]. Key reasons why load forecasting is indispensable include the following:

6.2.2.1 *Efficient Resource Utilization [1]*

- Facilitates the optimal matching of electricity supply with demand in real time.

- Prevents the wastage of energy and minimizes strain on the power grid by ensuring that generation meets consumption requirements effectively.

6.2.2.2 *Operational Efficiency for Energy Producers [23]*

- Enables energy producers to adjust their output efficiently based on forecasted demand.
- Enhances operational efficiency, conserves natural resources, reduces costs, and aligns with environmental goals by minimizing the reliance on costly peak power plants.

6.2.2.3 *Enhanced Grid Operations and Reliability [24]*

- Provides grid operators with precise predictions to manage the grid effectively.
- Reduces the risks of grid instability and potential blackouts by preemptively addressing supply-demand imbalances and implementing proactive measures to stabilize the grid.

6.2.2.4 *Consumer Benefits and Renewable Energy Integration [25]*

- Leads to lower energy rates for consumers by reducing operational costs through better demand-supply alignment.
- Ensures a stable and reliable grid, resulting in fewer disruptions in service and increased consumer satisfaction.
- Facilitates the integration of renewable energy sources by accurately predicting their output and planning for their optimal utilization.

6.2.2.5 *Strategic Planning and Policy Support [12]*

- Informs critical decisions related to the development and sustainability of power infrastructure.
- Enables utilities and policymakers to plan for future demand patterns, invest in infrastructure upgrades, and allocate resources to projects that promote sustainability and resilience.

In essence, load forecasting serves as a cornerstone of the electricity sector, offering multifaceted benefits that extend beyond operational efficiency to encompass economic, environmental, and strategic dimensions. Its significance is expected to grow as the energy landscape evolves, making it an indispensable tool for shaping a resilient and sustainable energy future.

6.2.3 Factors Affecting Electricity Demand Forecasting

Electricity demand forecasting is a complex endeavor shaped by various factors, each contributing to the intricate patterns of energy usage. Recognizing and understanding these factors is essential for the development of accurate forecasting models, crucial for effective energy management and planning.

6.2.3.1 Temporal Factors [26]

- Time-of-day and day-of-week patterns influence demand curves, with peak hours typically coinciding with morning and evening periods.
- Seasonal variations, including changes in heating and cooling needs, as well as holidays and special events, further impact demand fluctuations.

6.2.3.2 Meteorological Effects [27]

- Weather conditions, particularly temperature and humidity, directly affect energy demand, with extreme weather events leading to sudden fluctuations.
- Heatwaves, cold snaps, and storms can result in significant spikes or dips in demand, necessitating responsive forecasting models.

6.2.3.3 Economic Indicators [28]

- GDP growth rates and industrial activity serve as predictors of energy consumption, while energy prices influence consumer behavior and demand patterns.
- Economic expansion typically correlates with increased energy usage, particularly in industrial sectors.

6.2.3.4 *Societal Changes [29]*

- Urbanization and population growth drive electricity demand, although adoption of energy-efficient technologies and electric vehicles can mitigate growth.
- Shifts in consumer behavior toward energy conservation and renewable energy adoption impact overall demand dynamics.

6.2.3.5 *Regulatory and Policy Influences [30]*

- Government policies and incentives, such as subsidies for renewable energy and energy conservation regulations, shape demand patterns.
- Initiatives focused on decarbonization and climate change mitigation influence energy consumption by incentivizing cleaner energy sources and efficient usage practices.

6.2.3.6 *Complex Interactions [31]*

- Regulatory measures aimed at reducing carbon emissions and promoting energy efficiency can have complex, interconnected effects on demand dynamics.
- The interaction between different policies may lead to varied outcomes, highlighting the need for nuanced forecasting techniques.

6.2.4 Challenges of Demand Forecasting in the Electricity Sector

The quest for precise load forecasting models is critical for enhancing the operational efficiency and profitability of electric power generation and distribution companies. Despite the development of sophisticated techniques, several ongoing challenges hinder the achievement of the most accurate models. These challenges constitute the key obstacles in perfecting load forecasting.

- One significant hurdle is the sensitivity of electrical load forecasting to climatic conditions [32, 33]. Weather unpredictability can induce significant errors in load predictions, particularly when unforeseen weather changes occur. Furthermore, a vast electrical network may encompass regions with diverse weather patterns, complicating the forecast and potentially impacting financial outcomes.

- Consumer heterogeneity presents another set of difficulties [34]. The use of varying types of metering systems, such as smart and conventional meters with differing tariff structures, along with diverse consumer usage behaviors, necessitates a nuanced understanding of system utilization. Forecasts must be uniquely tailored for each metering arrangement and consolidated to achieve a comprehensive forecast. Neglecting these distinctions can lead to substantial errors.
- Demand forecasting is complicated further by the volatility in consumption behaviors prompted by sudden shifts in variables like pricing [4, 8]. Price elasticity of demand can alter consumption patterns, making it challenging to capture and predict these changes accurately in forecasting models.
- Power system disturbances also disrupt typical load patterns. Anomalies such as faults and transients yield exceptional data points that must be carefully factored into model design [35]. If these outliers are not appropriately addressed, then they can skew the model, leading to unreliable forecasting outcomes.
- Environmental parameters including temperature and humidity variations also play a pivotal role in electricity consumption, yet accurately integrating these into demand forecasting poses its own set of challenges [15, 28].
- Lastly, distribution companies must grapple with establishing an acceptable margin of error for short-term load forecasts [6]. The balance between precision and practicality is delicate; too fine a margin may entail untenable costs, whereas too broad a margin might result in substantial financial losses.

These diverse challenges drive practitioners to concentrate on different facets of model development, from capturing incidental data to fine-tuning algorithms, all in the pursuit of creating a load forecasting system that is as dependable as possible.

6.2.5 Demand Forecasting Model Generation Framework

Developing and validating forecasting models involves a structured series of steps to ensure accuracy and efficacy. This process, synthesized from scholarly literature [6, 11, 25, 27, 32, 35–38], unfolds as follows:

(a) Data Collection and Processing:
- Collect load data, essential for forecasting.
- Process data, enriching it with historical meteorological records.

(b) Data Quality Enhancement:
- Prune redundant or irrelevant information.
- Ensure data integrity through cleaning and evaluation.

(c) Feature Set Construction:
- Select relevant attributes crucial for predictive power.

(d) Forecasting Method Selection:
- Choose a method aligned with data nature and forecasting objectives.

(e) Model Training and Validation:
- Divide data into training, validation, and test sets.
- Train the model on the training set and fine-tune as necessary.
- Validate the model on the validation set and adjust as required.

(f) Model Testing:
- Test the final model against the test set to quantify its performance using error metrics.

(g) Results Visualization and Analysis:
- Utilize graphical representations to convey model performance.
- Apply data visualization techniques to elucidate data attributes and patterns.

(h) Innovation and Methodological Enhancements:
- Encourage innovations in load data categorization, algorithm selection, and performance comparison methods.

(i) Monitoring and Updating:
- Continuously monitor the model for performance drift.
- Update the model as needed to maintain forecasting reliability.

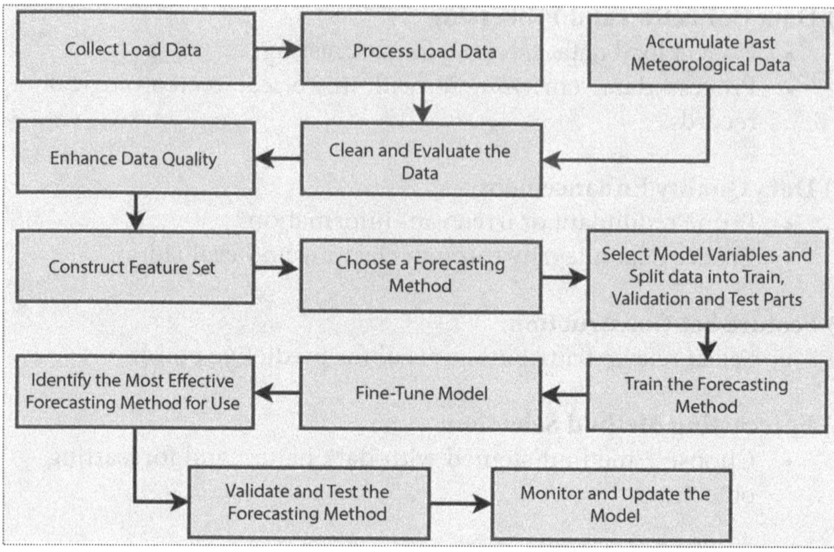

Figure 6.2 Demand forecasting modeling approach.

This structured approach ensures the development of robust forecasting models essential for demand prediction in the electricity sector. The framework has been presented in Figure 6.2.

6.3 Overview of Machine Learning in Demand Forecasting

The advent of ML has marked a paradigm shift in forecasting methodologies. By introducing self-improving algorithms capable of deciphering complex patterns from large datasets, ML has played an instrumental role in enhancing the precision and reliability of demand forecasts [39, 40].

6.3.1 Defining Machine Learning

Machine learning is a branch of artificial intelligence focused on building systems that learn from data, identifying patterns, and making decisions with minimal human intervention [3]. Unlike traditional statistical models that are explicitly programmed with rigid assumptions and rules, ML models can accommodate and adapt to non-linear relationships and high-dimensional data without being explicitly programmed for the specific task [9, 16].

6.3.2 Categorizing Machine Learning Methods

Machine learning methods pertinent to demand forecasting include the following:

- **Supervised Learning** [2]: This involves models that are "trained" using labeled datasets. They predict the target output (e.g., electricity demand) by learning from input-output pairs. Techniques like regression, neural networks, and SVMs fall under this category.
- **Unsupervised Learning** [41]: These algorithms discover hidden patterns or intrinsic structures in input data that is not labeled. Clustering and dimensionality reduction are common unsupervised learning methods beneficial for segmenting customers based on usage patterns or reducing the complexity of data.
- **Reinforcement Learning** [1]: In this approach, the model learns to make decisions by performing actions and observing the results, focusing on maximizing a notion of cumulative reward. It is less commonly applied in demand forecasting but might be utilized for optimizing energy distribution strategies.
- **Hybrid Models** [28, 42]: These combine various ML models to capitalize on their relative strengths. For example, a hybrid model might use a combination of clustering to categorize similar demand patterns and time-series forecasting for making predictions within each category.

6.3.3 Traditional vs. Machine Learning Approaches

Traditional forecasting methods and contemporary ML techniques both have their place in demand forecasting. However, their capabilities and limitations differ significantly. Table 6.1 presents the aspects of these models and shows the comparison [3, 6, 9, 23, 31, 36, 42].

6.3.4 Machine Learning Techniques in Demand Forecasting

In the rapidly evolving field of demand forecasting, ML techniques have taken a central role due to their ability to make sense of complex data and reveal nuanced patterns. This section provides an overview of these techniques, with a focus on their strengths and limitations as presented in Table 6.2.

Table 6.1 Comparison of traditional and machine learning models in demand forecast.

Aspect	Traditional methods	Machine learning techniques
Scope	Focus on simple moving averages, linear regression	Utilize decision trees, neural networks, support vector machines
Flexibility	Limited to linear relationships, lack complexity capture	Able to handle non-linear relationships, high-dimensional data
Performance	Intuitive but less accurate, struggle with complex data	Outperform traditional methods, excel with non-linear, high-dimensional data
Interpretability	Provides clear coefficients for relationships, easy to interpret	Less interpretable, often considered as "black boxes"
Computational Requirements	Low computational requirements	Resource-intensive, require powerful processors, GPUs, or TPUs
Data Preprocessing	Simple data preprocessing, less feature engineering	Extensive data preprocessing, require advanced feature engineering

6.3.5 Summary of the Reviewed Papers

The papers reviewed in the area of ML models and its application in electric demand forecasting are summarized in Table 6.3, which shows the modeling category in terms of short-, medium-, and long-term forecasting, the model utilized with their strength and limitations.

The reviewed literature on ML models for electric load forecasting reveals a diverse array of approaches and methodologies, each tailored to address specific challenges inherent in predicting energy consumption patterns. Short-term forecasting models, essential for day-to-day operational decision-making, exhibit a range of strengths and limitations. EGPR and SVM, for instance, offer expedited computational times and adept global optimization capabilities, respectively, contributing to their utility in real-time load prediction scenarios. However, the omission of crucial weather-related variables in some models,

Table 6.2 Strength and limitations of machine learning models in electric demand forecasting.

Machine learning techniques	Definition	Strength	Limitations
Time-Series Forecasting Models [43, 44]	Are a category of statistical techniques that predict future values based on previously observed values arranged in time order. These models are specifically designed to capture temporal dependencies, trends, seasonality, and cycles in time-series data.	Proficient at identifying and extrapolating time-based patterns and trends. Can incorporate previous values and forecast error information to improve future estimates. Typically requires less computational power compared to more complex models.	Can be overly sensitive to outliers or missing data, which can distort forecasts. Struggles with sudden changes in trends or seasonality not present in historical data. Limited in capturing relationships between multiple input variables and time-series data.

(*Continued*)

Table 6.2 Strength and limitations of machine learning models in electric demand forecasting. (*Continued*)

Machine learning techniques	Definition	Strength	Limitations
Regression Models [45–47]	Are a set of statistical methods for estimating the relationships among variables. They typically predict a continuous outcome variable from one or more predictor variables. These models can range from simple linear relationships to complex non-linear interactions.	Transparent and interpretable, offering a clear understanding of how input variables affect the output. Simpler and faster to train, which can be advantageous in frameworks with limited resources. Versatile and well-established, with a solid theoretical foundation underpinning their use.	Linearity assumption can misrepresent real-world complexities. Susceptible to overfitting, especially with many features. Can be heavily influenced by non-representative training data, leading to skewed predictions.
Ensemble Methods [48–50]	Involve the combination of several machine learning models to improve the robustness and accuracy of predictions. By aggregating the predictions from multiple models, ensembles reduce the risk of error from any single model and often achieve higher performance.	Aggregate predictions from multiple models, leading to improved accuracy and robustness. Effective in reducing the risk of model overfitting, a common problem in predictive modeling. Can handle a diverse set of data distributions and relationships between variables.	Computational complexity and cost are significantly higher than simpler models. Can lack transparency, making it difficult to interpret which features drive predictions. The aggregation process can dilute the contribution of any single model that may excel in certain conditions.

(*Continued*)

Table 6.2 Strength and limitations of machine learning models in electric demand forecasting. (*Continued*)

Machine learning techniques	Definition	Strength	Limitations
Deep Learning Models [51–54]	Are advanced neural network architectures with multiple layers that can learn high-level abstractions in data. These models can automatically discover the representations needed for detection or classification directly from raw data, with minimal human intervention.	Potentially unmatched in extracting complex, non-linear patterns from large datasets. Highly scalable, usually improving performance with more data and computing power. Can automatically extract and make use of features from raw data without much manual intervention.	They require vast amounts of labeled training data to perform effectively and avoid overfitting. Deep learning models are computationally expensive, requiring significant processing power and time, particularly for training. Interpretability is a significant challenge, as the layered structure of neural networks creates a "black box" phenomenon, making it difficult to understand the decision-making process.

(*Continued*)

Table 6.2 Strength and limitations of machine learning models in electric demand forecasting. (*Continued*)

Machine learning techniques	**Definition**	**Strength**	**Limitations**
Hybrid Models [55–57]	Integrate various modeling techniques into a cohesive system. They aim to leverage the distinctive strengths of different methods to achieve more accurate and reliable predictions than could be obtained from any single model alone. Hybrid models often combine traditional statistical methods with advanced machine learning or deep learning techniques	Hybrid models combine the strengths of different techniques, potentially offsetting the weaknesses of individual methods. They provide flexibility, allowing for customization to specific forecasting scenarios and data conditions. Often result in improved accuracy and reliability of forecasts, as they can capture a wider range of patterns and relationships in the data.	The complexity of these models can make them difficult to understand and interpret, similar to ensemble methods and deep learning. Optimization of hybrid models can be challenging due to the increased number of parameters and architectural decisions required. The development and tuning of hybrid models can be more resource-intensive in terms of data, computation, and expertise needed.

Table 6.3 Summary of the reviewed literatures in machine learning approaches in electric demand forecasting.

Ref	Forecast category	ML model used	Modeling strength	Modeling limitation
[23]	Short term	mRMR-WGRP - SecRPSO-SVM	Smaller MAPE compared to other models; Superior performance of SecRPSO over GA for parameter optimization; Better feature selection with mRMR compared to BPNN and mRMR-BPNN; Enhanced prediction accuracy and sensitivity to proportionality changes.	Increased complexity due to multiple feature selection algorithms; Extensive computational requirements; Limited accuracy due to specific historical load sequences selected by mRMR.
[3]	Short term	EGPR	Low computational time after training. Low RMSE, MSE, and MAE.	Slow training process Requires extensive computational resources.
[6]	Short term	SVM	Capable of rejecting local minima and achieving global optimal solutions.	Weather-related factors like humidity, windspeed, and cloud cover are not considered in feature selection.

(Continued)

Table 6.3 Summary of the reviewed literatures in machine learning approaches in electric demand forecasting. (*Continued*)

Ref	Forecast category	ML model used	Modeling strength	Modeling limitation
[58]	Short term	EEMD-NB	Can normalize input data variances; Works with single dataset input. Processes complex hourly load data into single sub-series.	Limited to day-ahead short-term forecasting; Not applicable for small-scale residential, commercial, or industrial establishments, nor nationwide forecasting; Limited by the acquired historical data.
[7]	Long term	LSTM	Applicable to nationwide electricity agencies; Avoids long-term dependency issues; Memorizes long-term information.	Requires larger-scale data collection for accuracy.
[59]	Short term and mid term	WT-GALSTM	Historical electricity demand data is subdivided using the WT method. GA generates the primary neural network parameters, minimizing forecasting errors iteratively.	GA-LSTMs have long forecasting steps, leading to potential inaccuracies in late negative forecasting. GA-LSTMs analyze only data with linear relationships, limiting their capacity to capture non-linear patterns.

(*Continued*)

Table 6.3 Summary of the reviewed literatures in machine learning approaches in electric demand forecasting. (*Continued*)

Ref	Forecast category	ML model used	Modeling strength	Modeling limitation
[36]	Long term	CST-LSTM (black box)	Combines multiple regression models and LSTM for improved accuracy; Effective for hourly resolution long-term forecasts; Captures both short-term and long-term consumption trends.	Combines regression models and LSTM, potentially increasing complexity and computational demands; Uncertainty regarding model performance for longer-term forecasts or different regions.
[9]	Long term	MNN-SVM-WOA (feedforward MLP)	Incorporates diverse input data; Utilizes correlation matrix to demonstrate relationships; Compares effectiveness using p-value and confidence intervals.	Limited generalizability to other regions/countries; No discussion on external factors impacting future demand; Lack of interpretability/explainability in predictions.

(*Continued*)

Table 6.3 Summary of the reviewed literatures in machine learning approaches in electric demand forecasting. (*Continued*)

Ref	Forecast category	ML model used	Modeling strength	Modeling limitation
[46]	Short term	E-GPR	Rational quadratic GPR and exponential GPR exhibit superior load forecasting performance. Model accuracy validated using mean absolute percentage error and R squared; Selection based on nonparametric kernel-based probabilistic nature.	Weather factors like temperature, humidity, cloud cover, and brightness are not considered in the models despite their known correlation with load characteristics.
[60]	Short term	MLR	Provides good interpretability for understanding electricity consumption behavior; Includes interactions between variables for comprehensive load fluctuation analysis; Successfully forecasts 3-year hourly energy demand for a US utility.	Assumes linear relationship may not capture non-linear data patterns; Does not account for non-stationary or uncertain load characteristics.

(*Continued*)

Table 6.3 Summary of the reviewed literatures in machine learning approaches in electric demand forecasting. (*Continued*)

Ref	Forecast category	ML model used	Modeling strength	Modeling limitation
[61]	Short term	FFNN	FFNN with error correction improves load forecasts; Processes information quickly for accurate forecasts; Allows adjustments based on consumer type.	Increased complexity and execution time compared to LSTM.
[39]	Long term	Quantile regression (QR)	Enables forecasting of electricity demand at all percentiles, facilitating variability assessment; Non-parametric estimation without distribution assumptions enhances flexibility; Provides full demand distribution per hour for uncertainty assessment and exceedance probability calculation.	Assumes linear relationship between covariates and demand quantiles, possibly missing non-linearities; Assumes forecasted demand distribution remains accurate until 2023, potentially ignoring market changes; Does not consider uncertainties or errors in input data, affecting forecast accuracy.

(*Continued*)

Table 6.3 Summary of the reviewed literatures in machine learning approaches in electric demand forecasting. (*Continued*)

Ref	Forecast category	ML model used	Modeling strength	Modeling limitation
[62]	Short term	LSTM-DFNN	Simple and easy implementation of proposed model; Performs online, enhancing accuracy in near real-time applications; Outperforms single engines in MAE, MAPE, and RMSE when applied to day-ahead load forecasting in Toronto.	LSTMs require significant computational resources and powerful hardware due to their intensive computation, especially when dealing with large datasets.
[8]	Long terms and short term	DBiLSTM	Integrates Copula correlation analysis with DBiLSTM for multi-load forecasting. Copula analysis identifies influential factors, enhancing input accuracy. DBiLSTM captures bidirectional temporal dependencies for accurate forecasts.	Real-time prices impact system: considering them in input features could enhance suitability. Determining hidden layers and neurons involves time-consuming experimentation. Abnormal or missing data from intelligent terminal equipment requires manual identification and supplementation before model training.

(*Continued*)

Table 6.3 Summary of the reviewed literatures in machine learning approaches in electric demand forecasting. (*Continued*)

Ref	Forecast category	ML model used	Modeling strength	Modeling limitation
[63]	Short term	CNN-BiLSTM-BiGRU	CNNs extract load and temperature features, enhancing recurrent neural networks' forecasting. Autocorrelation coefficients determine CNN kernel size, aiding accurate feature extraction. 2-D CNN features inputted to bidirectional GRU and LSTM units for hourly load forecasting.	Multiple models heighten complexity in feature extraction, selection, and training time.
[15]	Short term	AdaBoost(AR2)	Ensemble methods enhance generality and robustness by combining predictions from multiple base estimators. AdaBoost minimizes the loss function for the current model at each stage.	Lagged environmental temperature and humidity features yield higher errors. Not incorporating exogenous variables like direct solar irradiation affects forecasting accuracy

(*Continued*)

Table 6.3 Summary of the reviewed literatures in machine learning approaches in electric demand forecasting. (*Continued*)

Ref	Forecast category	ML model used	Modeling strength	Modeling limitation
[18]	Short term	SVR-DMD	DMD captures spatiotemporal error dynamics for accurate forecasting. Simple identification and processing precede prediction. DMD mode captures changing trends with single frequency and growth rate. Generalizes well for different date types; simple calculation and implementation without additional parameter selection or feature extraction.	EVCM enhances forecasting by limiting DMD-corrected load within a reasonable range, necessitating higher historical data accuracy. Prediction accuracy diminishes for smaller areas due to their load demand curve's increased randomness. Tracking error fluctuations with DMD reduces error accumulation when correcting load values with predicted errors. DMD's applicability is somewhat limited by sample dimension characteristics.

(*Continued*)

Table 6.3 Summary of the reviewed literatures in machine learning approaches in electric demand forecasting. (*Continued*)

Ref	Forecast category	ML model used	Modeling strength	Modeling limitation
[37]	Short term	SVR-LSTM	Enhanced accuracy: integrates SVR and LSTM strengths for improved accuracy. Handling nonlinearity: addresses complex nonlinear patterns in microgrid load data effectively. Capturing temporal patterns: utilizes LSTM's time-series capabilities for precise analysis. Flexibility: easily adjustable parameters adapt to various microgrid scenarios. Reduced overfitting: combining algorithms mitigates overfitting, enhancing model balance.	Integration and fine-tuning of SVR and LSTM components increase computational resources and training time. Challenges in selecting hyperparameters for SVR and LSTM components. Effective data preprocessing is crucial for successful hybrid model training.

(*Continued*)

Table 6.3 Summary of the reviewed literatures in machine learning approaches in electric demand forecasting. (*Continued*)

Ref	Forecast category	ML model used	Modeling strength	Modeling limitation
[25]	Short term	HALOFMI	Intelligent feature selection: HALOFMI methodology improves model performance by selecting relevant features, reducing noise. Generalization capability: MLP component exhibits superior performance on unseen data. Performance metrics and validation: validation ensures reliability and accuracy of the forecast model. Adaptability to microgrid context: methodology addresses challenges of intermittent renewable energy and flexible loads in microgrid forecasting.	HALOFMI methodology, with its hybrid approach and intelligent feature selection algorithm coupled with MLP, may pose challenges in implementation and fine-tuning compared to simpler forecasting methods.

(*Continued*)

Table 6.3 Summary of the reviewed literatures in machine learning approaches in electric demand forecasting. (*Continued*)

Ref	Forecast category	ML model used	Modeling strength	Modeling limitation
[4]	Short term	SVR, LSTM, random forest	The model's combination of LSTM, random forest, and SVR enhances robustness and adaptability to varied energy consumption patterns. A significant reduction in forecasting errors leads to improved load demand prediction accuracy. The model demonstrates efficacy in handling volatile and uncertain load patterns, crucial for microgrid energy management challenges. It excels in forecasting individual consumer loads, essential for managing diverse consumption patterns in microgrid energy management. Incorporation of LSTM facilitates effective capture of temporal correlations in load patterns, addressing inconsistencies.	The utilization of multiple ML techniques and dynamic weighting in the proposed model may increase complexity and necessitate careful parameter tuning. Historical data is required for the efficacy-based dynamic weighting method, posing challenges in some cases for obtaining predictors' previous errors and efficacies.

such as SVM, underscores the potential limitations in capturing the nuanced interplay between environmental factors and energy demand fluctuations.

In contrast, long-term forecasting models like LSTM showcase their applicability to broader temporal horizons, catering to the strategic planning needs of nationwide electricity agencies. By circumventing long-term dependency issues, LSTM models provide insights into future energy demand trends, facilitating informed resource allocation and infrastructure development. Yet, challenges persist in acquiring extensive datasets to ensure the accuracy and generalizability of LSTM-based forecasts, highlighting the importance of robust data collection mechanisms in model development and validation processes.

Hybrid approaches, which integrate multiple ML techniques like SVR and LSTM, emerge as a promising avenue for enhancing forecasting precision. By leveraging the complementary strengths of individual algorithms, hybrid models exhibit superior accuracy and flexibility in capturing complex load patterns. However, the increased complexity associated with integrating diverse methodologies necessitates meticulous parameter tuning and rigorous data preprocessing, adding layers of intricacy to model development and deployment. Moreover, the ensemble methodology, exemplified by AdaBoost, further augments model robustness by aggregating predictions from diverse base estimators yet grapples with challenges in integrating lagged environmental variables and ensuring adaptability to evolving load characteristics.

6.4 Machine Learning–Based Demand Forecasting in Thailand's Metropolitan Areas: An In-Depth Case Study

6.4.1 Overview

This case study delves into the application of an advanced ML technique, convolutional long short-term memory recurrent neural network, for electric demand forecasting in Thailand's central metropolitan region. The spotlight is on three major provinces—Bangkok, Samut Prakan, and Nontha Buri—under the purview of the Metropolitan Electricity Authority of Thailand, from where the pertinent load data was meticulously sourced [64]. A detailed view of the dataset's characteristics, enumerating the statistical parameters, is presented in Table 6.4, whereas the corresponding average load curve is depicted in Figure 6.3 for the different months of the year.

Table 6.4 Dataset statistics.

Data type	Data source	Data length	Data interval	Train	Validation	Test
Load data	Metropolitan Electricity Authority	Four years data of Metropolitan Region, Thailand	Half hourly	75%	15%	15%

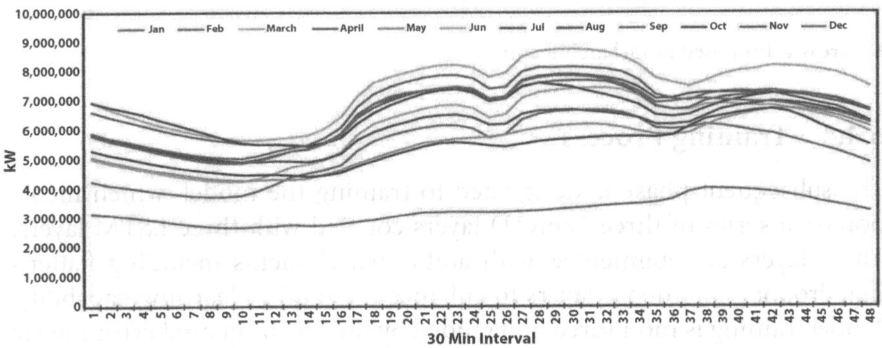

Figure 6.3 Monthly average load curve of metropolitan area, Thailand.

6.4.2 Model Design and Validation

The crux of the model proposed in this case study is captured in Figure 6.4, outlining the architectural design. An extensive description of the model's methodology and underlying mathematical framework has been thoroughly articulated in References [65, 66]; thus, those specifics are conspicuously absent from this chapter to avoid redundancy.

6.4.3 Data Management

A meticulous data cleansing process is employed as the first strategic step in the demand forecasting model. This process includes the identification and purging of anomalies, as well as imputing missing or erroneous entries through interpolation techniques.

In preparation for model training, the dataset is partitioned into three distinct subsets: training, validation, and testing. The division adheres to the widely accepted proportions of 70%, 15%, and 15% respectively, which is considered a best practice based on prevailing research norms [6, 40, 50, 51], as indicated in Table 6.4.

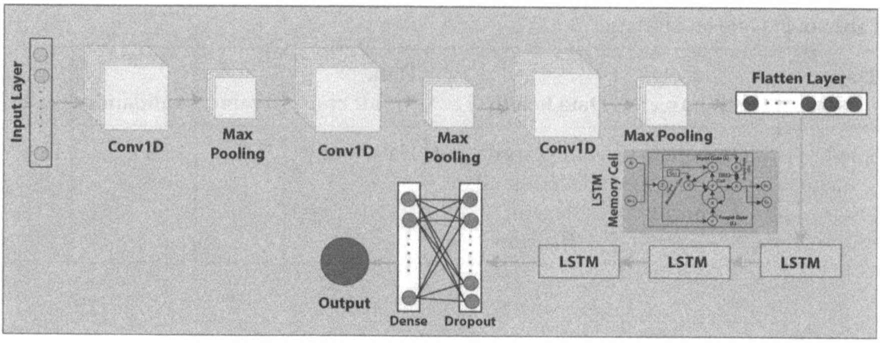

Figure 6.4 Proposed model architecture.

6.4.4 Training Process

The subsequent phase is designated to training the model, which incorporates a series of three Conv1D layers coupled with three LSTM layers. These layers are augmented with additional elements including flattening, dropout, and dense layers to enhance the model's learning capability. Model training is monitored and guided by the mean squared error metric to circumvent potential pitfalls like underfitting or overfitting. The training itinerary spans approximately 47 min, with the optimal configuration of hidden nodes derived from a grid-search procedure encapsulated in the search space detailed in Reference [67].

6.4.5 Model Performance Evaluation

Rigorous evaluations gauge the demand forecasting performance of the proposed model against three pivotal error metrics: mean absolute error, root mean squared error, and mean absolute percentage error. Precise formulations of these indices are disclosed in reference [65]. The CLSTM-RNN model's performance is subject to a comparative assessment alongside three other prominent deep learning models—convolutional neural networks, long short-term memory networks, and traditional recurrent neural networks. The comparative analysis is visually represented in Figure 6.5, which provides a clear juxtaposition of actual vs. predicted load for a select timeframe—specifically, the data from 30 December 2023, as featured in the Test dataset section. This illustration is deliberately constrained to a 1-day forecast to facilitate simplicity and clarity in the evaluation.

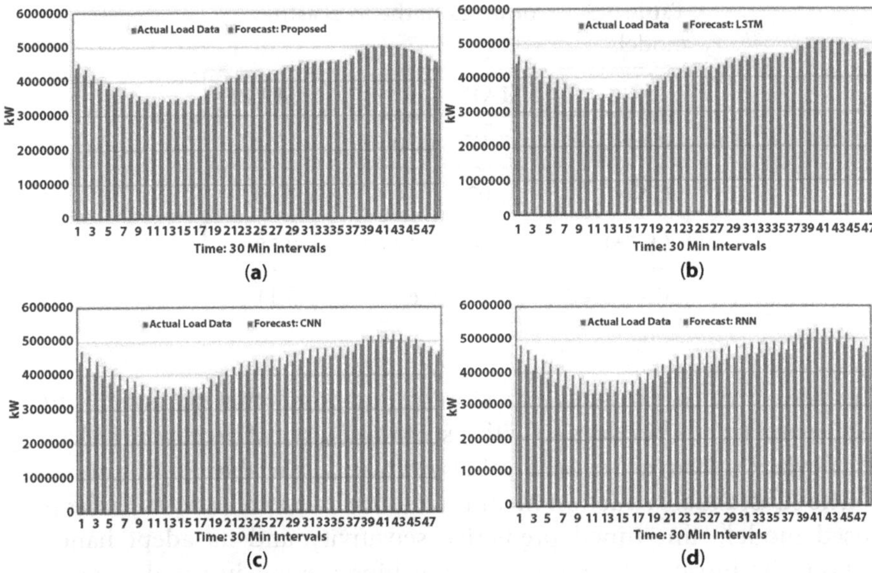

Figure 6.5 Forecasting performance of the models: (a) proposed model, (b) LSTM model, (c) CNN model, and (d) RNN model.

6.4.6 Model Performance Evaluation

The comparative performance of each model is numerically articulated in Table 6.5, where the errors are rigorously dissected across three primary metrics: mean absolute error (MAE), root mean squared error (RMSE), and mean absolute percentage error (MAPE).

The traditional recurrent neural network model exhibits a MAE of 17.17, an RMSE of 20.72, and a MAPE of 13.10, suggesting moderate forecasting proficiency but with notable room for improvement. The convolutional neural network stands as a significant stride forward, reducing MAE to 12.03, RMSE to 14.67, and MAPE to 10.57, indicating a sharper ability to track and predict energy demand trends in the dataset.

The long short-term memory network, renowned for its temporal pattern recognition, further refines forecast precision with an MAE of 10.56, an RMSE of 12.05, and a MAPE of 8.58. These figures underscore LSTM's advancement over its predecessors when applied to electricity demand forecasting scenarios.

However, the proposed convolutional long-short term memory recurrent neural network model outshines all with its remarkably low error rates: MAE plummets to 5.87, RMSE to 6.73, and MAPE to a mere 3.11.

Table 6.5 Error index of the forecasting models.

	MAE	RMSE	MAPE
RNN	17.17	20.72	13.10
CNN	12.03	14.67	10.57
LSTM	10.56	12.05	8.58
Proposed	5.87	6.73	3.11

This performance translates into a forecasting model that not only captures the general trends but does so with a striking degree of precision, reflecting less deviation from actual load values even amid variable conditions.

The substantial reduction in all three error measures attests to the proposed model's fine-tuned predictive sensitivity and its adept handling of both the intricate nonlinear relationships present in the data and the sequential time-dependent patterns that characterize electrical demand. The progressive descent in error rates from RNN to CNN to LSTM and culminating in the proposed CLSTM-RNN model, as demonstrated in Table 6.5, highlights the evolution of ML capabilities and the remarkable benefits harnessed through this advanced hybrid modeling approach.

In summary, the assessment of Table 6.5 underlines the CLSTM-RNN's distinct advantage in forecasting accuracy within the metropolitan context, promising enhanced decision-making support for power generation and distribution. The efficacy reflected by the low error margins intensifies the case for adopting such sophisticated models in utility operations, potentially leading to optimized resource allocation and heightened.

6.4.7 Discussion

- The case study demonstrates a significant evolution in electric demand forecasting, moving from rudimentary models to sophisticated convolutional long-short term memory recurrent neural networks, highlighting the growing reliance on ML to analyze and predict complex patterns in electricity usage effectively.
- With its geographic focus on Thailand's metropolitan areas, the case highlights not only the urban-specific consumption patterns but also the potential for such advanced forecasting

models to be tailored to regional nuances, thereby improving the accuracy and reliability of predictions in diverse settings.

- The study underscores the importance of accurate demand forecasting in electricity as a cornerstone for optimizing resource distribution, mitigating potential inefficiency, and contributing to the overall stability and sustainability of the power supply.
- The methodology of the study, which involves substantial data processing and validation techniques such as data cleansing and train-test splits, reflects the detailed and meticulous approach necessary for the effective application of ML in real-world scenarios.
- By quantitatively comparing the proposed CLSTM-RNN model's performance with other models, the case study provides concrete evidence of its enhanced forecasting accuracy, offering valuable insights for future research and development in the field.
- The case study encapsulates the challenges and opportunities of integrating renewable energy sources and achieving real-time processing scalability, setting a blueprint for future advancements in the domain of electricity demand forecasting using ML technologies.

6.5 Conclusion

Throughout this chapter, we delved into the wide-ranging ML approaches that have been harnessed to predict demand in energy systems. Progression from classical regression models to cutting-edge deep learning approaches has characterized the trajectory of load forecasting methods, showcasing a field in flux, continually adapting to the intricate demands of modern energy systems.

The efficacy of ensemble learning techniques is a standout revelation from our discourse, affirming the power of aggregating predictions from diverse models. Methods such as stacking, bagging, and boosting have outshined their individual counterparts, enhancing forecast robustness and precision. Equally notable is the ascent of deep learning, which has carved out a niche for itself in load forecasting with an innate aptitude for distilling complex patterns from voluminous datasets. Recurrent neural networks, convolutional neural networks, and their derivatives, in particular, have proven adept in parsing the temporal and nonlinear complexities typical of energy consumption data.

The development of hybrid models represents another stride forward, merging diverse predictive strategies to capitalize on the collective analytical strength they offer. This confluence of statistical, algorithmic, and experiential insights into hybrid models has fortified their predictive power, proving especially beneficial in heterogenous and fluctuating energy scenarios.

Furthermore, the nuances of data preparation, feature construction, and model refinement have emerged as pivotal elements in the optimization of forecasting models. These steps, which encompass everything from data normalization to the nuanced adjustment of model parameters and the regulation of ensemble diversity, have been instrumental in the development and fine-tuning of forecasting models.

Anticipating the path ahead, the advancement of load forecasting is set to entwine closely with the unfolding narrative of smart grids, the Internet of Things, and the renewable energy panorama. As real-time data, adaptive algorithms, and autonomous decision-making become increasingly integrated, forecasting systems are poised to transform into proactive, agile entities, facilitating efficient energy management and reinforcing the reliability and ecological balance of the grid.

In essence, the evolution from basic statistical forecasting to advanced ML methodologies represents more than mere technical refinement; it signifies the convergence of energy sciences, data analytics, and artificial intelligence. This confluence is shaping a complex, multidisciplinary domain charged with the pivotal goal of developing scalable, precise load forecasting mechanisms—a goal central to fueling innovation and steering us toward a smarter and environmentally conscious energy landscape.

References

1. Wazirali, R., Yaghoubi, E., Abujazar, M.S.S., Ahmad, R., Vakili, A.H., State-of-the-art review on energy and load forecasting in microgrids using artificial neural networks, machine learning, and deep learning techniques. *Electr. Power Syst. Res.*, 225, September, 109792, 2023, doi: 10.1016/j.epsr.2023.109792.
2. Román-Portabales, A., López-Nores, M., Pazos-Arias, J.J., Systematic review of electricity demand forecast using ann-based machine learning algorithms. *Sensors*, 21, 13, 1–23, 2021, doi: 10.3390/s21134544.

3. Bareth, R., Kochar, M., Yadav, A., Comparative Analysis of different Machine learning Models for Load Forecasting. *2023 IEEE IAS Global Conference on Renewable Energy and Hydrogen Technologies, GlobConHT 2023*, no. July 2022, pp. 1–5, 2023, doi: 10.1109/GlobConHT56829.2023.10087406.
4. Muzumdar, A., Modi, C., Vyjayanthi, C., Designing a robust and accurate model for consumer-centric short-term load forecasting in microgrid environment. *IEEE Syst. J.*, 16, 2448–2459, 2021, Accessed: May 11, 2024. [Online]. Available: https://ieeexplore.ieee.org/abstract/document/9426924/.
5. Gokce, M.M. and Duman, E., Performance Comparison of Simple Regression, Random Forest and XGBoost Algorithms for Forecasting Electricity Demand. *3rd International Informatics and Software Engineering Conference, IISEC 2022*, 2022, doi: 10.1109/IISEC56263.2022.9998213.
6. Chauhan, M., Gupta, S., Sandhu, M., Short-Term Electric Load Forecasting Using Support Vector Machines. *ECS Trans.*, 107, 1, 9731–9737, 2022, doi: 10.1149/10701.9731ecst.
7. Nguyen, V.B., Duong, M.T., Le, M.H., Electricity Demand Forecasting for Smart Grid Based on Deep Learning Approach. *Proceedings of 2020 5th International Conference on Green Technology and Sustainable Development, GTSD 2020*, pp. 353–357, 2020, doi: 10.1109/GTSD50082.2020.9303164.
8. Zheng, J., *et al.*, Multiple-Load Forecasting for Integrated Energy System Based on Copula-DBiLSTM. *Energies*, 14, 8, 2188, Apr. 2021, doi: 10.3390/EN14082188.
9. Saglam, M., Spataru, C., Karaman, O.A., Forecasting Electricity Demand in Turkey Using Optimization and Machine Learning Algorithms. *Energies*, 16, 11, 4499, Jun. 2023, doi: 10.3390/EN16114499.
10. Moroff, N.U., Kurt, E., Kamphues, J., Machine Learning and Statistics: A Study for assessing innovative Demand Forecasting Models. *Procedia Comput. Sci.*, 180, 40–49, Jan. 2021, doi: 10.1016/J.PROCS.2021.01.127.
11. Kong, W., Dong, Z.Y., Jia, Y., Hill, D.J., Xu, Y., Zhang, Y., Short-Term Residential Load Forecasting Based on LSTM Recurrent Neural Network. *IEEE Trans. Smart Grid*, 10, 1, 841–851, Jan. 2019, doi: 10.1109/TSG.2017.2753802.
12. Tumiran, T., Sarjiya, S., Putranto, L.M., Putra, E.N., Budi, R.F.S., Nugraha, C.F., Long-Term Electricity Demand Forecast Using Multivariate Regression and End-Use Method: A Study Case of Maluku-Papua Electricity System. *ICT-PEP 2021 - International Conference on Technology and Policy in Energy and Electric Power: Emerging Energy Sustainability, Smart Grid, and Microgrid Technologies for Future Power System, Proceedings*, pp. 258–263, 2021, doi: 10.1109/ICT-PEP53949.2021.9601144.
13. Verwiebe, P.A., Seim, S., Burges, S., Schulz, L., Müller-Kirchenbauer, J., Modeling Energy Demand—A Systematic Literature Review. *Energies*, 14, 23, 7859, Nov. 2021, doi: 10.3390/EN14237859.

14. Ahmad, T. and Zhang, D., A critical review of comparative global historical energy consumption and future demand: The story told so far. *Energy Rep.*, 6, Elsevier Ltd, 1973–1991, Nov. 01, 2020, doi: 10.1016/j.egyr.2020.07.020.
15. Silva, J., Praça, I., Pinto, T., Vale, Z., Energy consumption forecasting using ensemble learning algorithms, in: *Advances in Intelligent Systems and Computing*, vol. 1004, pp. 5–13, 2020, doi: 10.1007/978-3-030-23946-6_1.
16. Rawal, K. and Ahmad, A., A Comparative Analysis of Supervised Machine Learning Algorithms for Electricity Demand Forecasting. *ICPC2T 2022 - 2nd International Conference on Power, Control and Computing Technologies, Proceedings*, pp. 1–6, 2022, doi: 10.1109/ICPC2T53885.2022.9776960.
17. Waheed, W., A Data-driven Long-Short Term Load Forecasting with LSTM-RNN Networks A Data-driven Long-Short Term Load Forecasting, pp. 0–22, 2023.
18. Kong, X., Li, C., Wang, C., Zhang, Y., Zhang, J., Short-term electrical load forecasting based on error correction using dynamic mode decomposition. *Appl. Energy*, 261, September 2019, 114368, 2020, doi: 10.1016/j.apenergy.2019.114368.
19. Vardhan, B.V.S., Khedkar, M., Srivastava, I., Thakre, P., Bokde, N.D., A Comparative Analysis of Hyperparameter Tuned Stochastic Short Term Load Forecasting for Power System Operator. *Energies (Basel)*, 16, 3, 1243–1243, Jan. 2023, doi: 10.3390/EN16031243.
20. Eskandari, H., Imani, M., Moghaddam, M.P., Convolutional and recurrent neural network based model for short-term load forecasting. *Electr. Power Syst. Res.*, 195, February, 107173, 2021, doi: 10.1016/j.epsr.2021.107173.
21. Li, J., Lei, Y., Yang, S., Mid-long term load forecasting model based on support vector machine optimized by improved sparrow search algorithm. *Energy Rep.*, 8, 491–497, Aug. 2022, doi: 10.1016/J.EGYR.2022.02.188.
22. Setiawan, A., Arifin, Z., Sudiarto, B., Jufri, F.H., Haramaini, Q., Garniwa, I., Comparison of Medium-Term Load Forecasting Methods (Splitted Linear Regression and Artificial Neural Networks) in Electricity Systems Located in Tropical Regions, in: *2022 3rd International Conference on Clean and Green Energy Engineering, CGEE 2022*, Institute of Electrical and Electronics Engineers Inc., pp. 84–88, 2022, doi: 10.1109/CGEE55282.2022.9976521.
23. Dai, Y. and Zhao, P., A hybrid load forecasting model based on support vector machine with intelligent methods for feature selection and parameter optimization. *Appl. Energy*, 279, 115332, Dec. 2020, doi: 10.1016/J.APENERGY.2020.115332.
24. Waheed, W. and Xu, Q., A Data-driven Long-Short Term Load Forecasting with LSTM-RNN Networks, Jun. 2023, doi: 10.21203/RS.3.RS-2998305/V1.
25. Guimarães, T., Costa, L., H.L., A hybrid approach to load forecast at a micro grid level through machine learning algorithms, 2020, ieeexplore.ieee.org, Accessed: May 11, 2024. [Online]. Available: https://ieeexplore.ieee.org/abstract/document/9203308/.

26. Hernandez, L., *et al.*, A survey on electric power demand forecasting: Future trends in smart grids, microgrids and smart buildings. *IEEE Commun. Surv. Tutorials*, 16, 3, 1460–1495, 2014, doi: 10.1109/SURV.2014.032014.00094.
27. González Grandón, T., Schwenzer, J., Steens, T., Breuing, J., Electricity demand forecasting with hybrid classical statistical and machine learning algorithms: Case study of Ukraine. *Appl. Energy*, 355, 122249, 2024, doi: 10.1016/j.apenergy.2023.122249.
28. Izidio, D.M.F., de Mattos Neto, P.S.G., Barbosa, L., de Oliveira, J.F.L., da N. Marinho, M.H., Rissi, G.F., Evolutionary hybrid system for energy consumption forecasting for smart meters. *Energies (Basel)*, 14, 7, 1–19, 2021, doi: 10.3390/en14071794.
29. Abbasi, A., Ashraf, E., Pandita, R., Jamil, M., Bayesian Regularization Based Electrical Load Forecasting. *2023 International Conference on Recent Advances in Electrical, Electronics and Digital Healthcare Technologies, REEDCON 2023*, pp. 415–419, 2023, doi: 10.1109/REEDCON57544.2023.10150744.
30. Qinghe, Z., Wen, X., Boyan, H., Jong, W., Junlong, F., Optimised extreme gradient boosting model for short term electric load demand forecasting of regional grid system. *Sci. Rep.*, 12, 1, 19282, Nov. 2022, doi: 10.1038/S41598-022-22024-3.
31. Sun, H. and Bi, Z., Hybrid forecasting model of electric heating load based on CART decision tree regression. *2021 IEEE International Conference on Artificial Intelligence and Computer Applications, ICAICA 2021*, Jun. 2021, pp. 204–208, doi: 10.1109/ICAICA52286.2021.9498065.
32. Dhaval, B. and Deshpande, A., Short-term load forecasting with using multiple linear regression. *Int. J. Electr. Comput. Eng.*, 10, 4, 3911–3917, Aug. 2020, doi: 10.11591/IJECE.V10I4.PP3911-3917.
33. Hameed, J., Khalid, R., Javed, M.U., Javaid, S., Ahmed, S., Javaid, N., Enhanced Classification with Logistic Regression for Short Term Price and Load Forecasting in Smart Homes. *2020 3rd International Conference on Computing, Mathematics and Engineering Technologies: Idea to Innovation for Building the Knowledge Economy, iCoMET 2020*, Jan. 2020, doi: 10.1109/ICOMET48670.2020.9074059.
34. Xu, W., Shen, Z., Fan, X., Liu, Y., Short-term wind power prediction based on anomalous data cleaning and optimized LSTM network. *Front. Energy Res.*, 11, 1268494, Nov. 2023, doi: 10.3389/FENRG.2023.1268494/BIBTEX.
35. Xing, H., Zhao, B., Wang, Z., Short-term Power Generation Prediction of Photovoltaic Panels Based on Meteorological Parameters and Support Vector Machine. *ACM Cloud and Autonomic Computing Conference*, Nov. 2020, pp. 6018–6022, doi: 10.1109/CAC51589.2020.9326631.
36. Grandon, T.G., Schwenzer, J., Steens, T., Breuing, J., Electricity Demand Forecasting with Hybrid Statistical and Machine Learning Algorithms: Case Study of Ukraine, Apr. 2023, Accessed: Nov. 19, 2023. [Online]. Available: https://arxiv.org/abs/2304.05174v1.

37. Moradzadeh, A., Zakeri, S., Shoaran, M., Mohammadi-Ivatloo, B., Mohammadi, F., Short-term load forecasting of microgrid via hybrid support vector regression and long short-term memory algorithms. *Sustainability*, 12, 7076, 2020, •mdpi.com, Accessed: May 11, 2024. [Online]. Available: https://www.mdpi.com/2071-1050/12/17/7076.
38. Shahare, K., Mitra, A., Naware, D., Keshri, R., Suryawanshi, H.M., Performance analysis and comparison of various techniques for short-term load forecasting. *Energy Rep.*, 9, 799–808, Mar. 2023, doi: 10.1016/J.EGYR.2022.11.086.
39. Mokilane, P., Galpin, J., Sarma Yadavalli, V.S., Debba, P., Koen, R., Sibiya, S., Density forecasting for long-term electricity demand in South Africa using quantile regression. *S. Afr. J. Econ. Manag. Sci.*, 21, 1, 1–14, 2018, doi: 10.4102/sajems.v21i1.1757.
40. Gebremeskel, D.H., Ahlgren, E.O., Beyene, G.B., Long-term evolution of energy and electricity demand forecasting: The case of Ethiopia. *Energy Strategy Rev.*, 36, 100671, Jul. 2021, doi: 10.1016/j.esr.2021.100671.
41. Giamarelos, N., Papadimitrakis, M., Stogiannos, M., Zois, E.N., Livanos, N.A.I., Alexandridis, A., A Machine Learning Model Ensemble for Mixed Power Load Forecasting across Multiple Time Horizons. *Sensors*, 23, 12, 5436, 2023, doi: 10.3390/s23125436.
42. Kim, M., Choi, W., Jeon, Y., Liu, L., A hybrid neural network model for power demand forecasting. *Energies (Basel)*, 12, 5, 931, 2019, doi: 10.3390/en12050931.
43. Paparoditis, E. and Sapatinas, T., Short-term load forecasting: The similar shape functional time-series predictor. *IEEE Trans. Power Syst.*, 28, 4, 3818–3825, 2013, doi: 10.1109/TPWRS.2013.2272326.
44. Nguyen, H. and Hansen, C.K., Short-term electricity load forecasting with Time Series Analysis. *2017 IEEE International Conference on Prognostics and Health Management, ICPHM 2017*, pp. 214–221, Jul. 2017, doi: 10.1109/ICPHM.2017.7998331.
45. Lee, G.C., Regression-Based Methods for Daily Peak Load Forecasting in South Korea. *Sustainability*, 14, 7, 3984–3984, Mar. 2022, doi: 10.3390/SU14073984.
46. Gochhait, S. and Sharma, D., Regression model-based short-term load forecasting for load despatch centre. *Journal of Applied Engineering and Technological Science (JAETS)*, 4, 2, 693–710, 2023.
47. Luo, J., Hong, T., Gao, Z., Fang, S.C., A robust support vector regression model for electric load forecasting. *Int. J. Forecast.*, 39, 2, 1005–1020, Apr. 2023, doi: 10.1016/J.IJFORECAST.2022.04.001.
48. Shi, J., Li, C., Yan, X., Artificial intelligence for load forecasting: A stacking learning approach based on ensemble diversity regularization. *Energy*, 262, 125295, Jan. 2023, doi: 10.1016/J.ENERGY.2022.125295.

49. Moon, J., Jung, S., Rew, J., Rho, S., Hwang, E., Combination of short-term load forecasting models based on a stacking ensemble approach. *Energy Build.*, 216, 109921, Jun. 2020, doi: 10.1016/J.ENBUILD.2020.109921.
50. Chu, Y., *et al.*, Short-term metropolitan-scale electric load forecasting based on load decomposition and ensemble algorithms. *Energy Build.*, 225, 110343, Oct. 2020, doi: 10.1016/J.ENBUILD.2020.110343.
51. Sehovac, L. and Grolinger, K., Deep Learning for Load Forecasting: Sequence to Sequence Recurrent Neural Networks with Attention. *IEEE Access*, 8, 36411–36426, 2020, doi: 10.1109/ACCESS.2020.2975738.
52. Hafeez, G., Alimgeer, K.S., Khan, I., Electric load forecasting based on deep learning and optimized by heuristic algorithm in smart grid. *Appl. Energy*, 269, 114915, Jul. 2020, doi: 10.1016/J.APENERGY.2020.114915.
53. Aslam, S., Herodotou, H., Mohsin, S.M., Javaid, N., Ashraf, N., Aslam, S., A survey on deep learning methods for power load and renewable energy forecasting in smart microgrids. *Renew. Sustain. Energy Rev.*, 144, 110992, Jul. 2021, doi: 10.1016/J.RSER.2021.110992.
54. Shen, Y., Ma, Y., Deng, S., Huang, C.J., Kuo, P.H., An Ensemble Model based on Deep Learning and Data Preprocessing for Short-Term Electrical Load Forecasting. *Sustainability*, 13, 4, 1694, Feb. 2021, doi: 10.3390/SU13041694.
55. Alhussein, M., Aurangzeb, K., Haider, S.I., Hybrid CNN-LSTM Model for Short-Term Individual Household Load Forecasting. *IEEE Access*, 8, 180544–180557, 2020, doi: 10.1109/ACCESS.2020.3028281.
56. Lee, J. and Cho, Y., National-scale electricity peak load forecasting: Traditional, machine learning, or hybrid model? *Energy*, 239, 122366, Jan. 2022, doi: 10.1016/J.ENERGY.2021.122366.
57. Al Mamun, A., Sohel, M., Mohammad, N., Haque Sunny, M.S., Dipta, D.R., Hossain, E., A Comprehensive Review of the Load Forecasting Techniques Using Single and Hybrid Predictive Models. *IEEE Access*, 8, 134911–134939, 2020, doi: 10.1109/ACCESS.2020.3010702.
58. Angeles, E.R., Badreldin, M.D., Santos, A.J.C., Ostia, C.F., Day Ahead Load Forecasting Model Using Gaussian Naïve Bayes with Ensemble Empirical Mode Decomposition. *TENSYMP 2021 - 2021 IEEE Region 10 Symposium*, pp. 1–6, 2021, doi: 10.1109/TENSYMP52854.2021.9550820.
59. Salami, M., Niakan, M.R., Hasani, M., Electricity load forecasting: A new hybrid model combining wavelet transform, genetic algorithm and LSTM models. *International Conference on Electrical, Computer, Communications and Mechatronics Engineering, ICECCME 2021*, Oct. 2021, doi: 10.1109/ICECCME52200.2021.9591091.
60. Hong, T., Gui, M., Baran, M.E., Willis, H.L., Modeling and forecasting hourly electric load by multiple linear regression with interactions. *IEEE PES General Meeting, PES 2010*, pp. 1–8, 2010, doi: 10.1109/PES.2010.5589959.

61. Machado, E., Pinto, T., Guedes, V., Morais, H., Electrical Load Demand Forecasting Using Feed-Forward Neural Networks. *Energies*, 14, 22, 7644, Nov. 2021, doi: 10.3390/EN14227644.
62. Bashari, M. and Rahimi-Kian, A., Forecasting electric load by aggregating meteorological and history-based deep learning modules. *IEEE Power and Energy Society General Meeting*, vol. 2020-Augus, pp. 0–4, 2020, doi: 10.1109/PESGM41954.2020.9282124.
63. Eskandari H., Imani M., Moghaddam M. P., Convolutional and recurrent neural network based model for short-term load forecasting, *Electr. Power Syst. Res.*, 195, 107173, 2021, doi: https://doi.org/10.1016/j.epsr.2021.107173.
64. Metropolitan Electricity Authority, Accessed: May 11, 2024. [Online]. Available: https://www.mea.or.th/en/.
65. Ahamed Nahid, F., Ongsakul, W., Madhu Manjiparambil, N., Very Short-Term Wind Speed Forecasting Using Convolutional Long Short Term Memory Recurrent Neural Network. *Proceedings of the 2020 International Conference and Utility Exhibition on Energy, Environment and Climate Change, ICUE 2020*, Oct. 2020, doi: 10.1109/ICUE49301.2020.9307061.
66. Nahid, F.A., Ongsakul, W., Manjiparambil, N.M., Short term multi-steps wind speed forecasting for carbon neutral microgrid by decomposition-based hybrid model. *Energy Sustain. Dev.*, 73, 87–100, Apr. 2023, doi: 10.1016/j.esd.2023.01.016.
67. Nahid, F.A., Ongsakul, W., Manjiparambil, N.M., Singh, J.G., Roy, J., Mode decomposition-based short-term multi-step hybrid solar forecasting model for microgrid applications. *Electr. Eng.*, 106, 3349–3380, Dec. 2023, doi: https://doi.org/10.1007/s00202-023-02138-1.

7

Evaluation and Performance Metrics for Forecasting Renewable Power Generation, Demand, and Electricity Price

Firuz Ahamed Nahid[1,2*], Mohammad Nayeem Jahangir[2], Hussain Mahmud Chowdhury[2] and Khadiza Akter[2]

[1]*Department of Energy, Environment and Climate Change, Asian Institute of Technology, Pathum Thani, Thailand*

[2]*Department of Electrical and Electronic Engineering, International University of Business Agriculture and Technology - IUBAT, Dhaka, Bangladesh*

Abstract

The precision of renewable energy forecasts is crucial in navigating the complexities of an increasingly renewable-dominated energy landscape, where accurate predictions are pivotal for averting economic losses and ensuring power system reliability. However, a significant challenge persists: the absence of standardized and robust metrics for evaluating the accuracy of forecasts concerning renewable power generation, demand, and electricity price. This chapter undertakes a comprehensive exploration of the diverse methodologies employed to assess the performance of forecasting models within the renewable energy realm. By meticulously dissecting these evaluation techniques, it illuminates their distinct advantages and constraints, thereby highlighting the nuanced nature of assessing forecast accuracy in this domain. Through an extensive review, this chapter not only emphasizes the importance of accurate forecasting but also reveals the intricate tapestry of evaluation metrics essential for its realization. Ranging from traditional metrics like mean absolute error and root mean square error to more nuanced measures tailored to specific forecasting domains, such as mean percentage error and forecasting efficiency, this discourse offers a panoramic view of the evaluation landscape. By scrutinizing these metrics' applicability, reliability, and interpretability, this chapter provides readers with a deeper understanding of the challenges inherent in measuring forecast accuracy within the renewable energy context. Such analysis offers valuable insights for stakeholders across academia, industry, and policy

**Corresponding author*: firuz@ait.asia

Jai Govind Singh, Rupendra Kumar Pachauri and Sasidharan Sreedharan (eds.) Forecasting Methods for Renewable Power Generation, (173–218) © 2025 Scrivener Publishing LLC

domains, facilitating informed decision-making and driving continuous improvement in renewable energy forecasting practices.

Keywords: Renewable energy forecasting, forecast accuracy evaluation, power system reliability, evaluation metrics, renewable power generation, electricity price forecasting

7.1 Introduction

The transformative shift toward renewable energy not only heralds a new era for power generation but also brings to the forefront the challenge of precise forecasting amidst this transition. Accurate forecasting for renewable energy is indispensable, not just for matching generation with variable natural resources but also for predicting demand surges and anticipating electricity market pricing fluctuations [1]. Within the renewable energy sector, the accuracy of forecasting profoundly influences grid stability, investment decisions, regulatory policymaking, and the ultimate success of energy transition initiatives [2, 3].

7.1.1 Power Generation Forecasting

In the realm of renewable energy, power generation forecasting is not merely a routine procedure; it is the centerpiece of energy strategy, directly influencing the sustainability and reliability of the entire energy system [4, 5]. The episodic nature of resources such as solar irradiance and wind speed necessitates a predictive approach that can provide grid operators with foresight into anticipated energy supplies. The inherent variability of renewables, such as the ebb and flow of wind or the passage of clouds reducing solar output, poses unique challenges that forecasting seeks to mitigate [2]. By anticipating generation levels, operators can balance these fluctuations with consumer demand and storage capabilities, ultimately reducing the dependence on standby generation from fossil fuels, which carry both economic and environmental costs [6, 7].

Advanced forecasting techniques that incorporate weather prediction models, historical data, and real-time analytics are thus employed to create accurate generation forecasts [8]. These models help manage renewable resources to meet load requirements during peak and off-peak hours, ensuring that renewable integration does not compromise grid stability. Accurate forecasting also aids in system planning, facilitating the strategic expansion of renewable facilities, investing in storage technologies, and designing smart grids capable of handling high penetrations of renewable energy [9].

7.1.2 Demand Forecasting

Similarly, the importance of demand forecasting cannot be overstated. It is the linchpin for achieving an optimized balance between power generation and consumer needs [10]. With economic growth altering demand profiles and technological advances such as electric vehicles and smart appliances redefining consumption patterns, demand forecasting has become increasingly complex. Utilities and grid operators invest heavily in sophisticated forecasting techniques that can dissect myriad factors influencing demand — from weather patterns to consumer behavior and economic policies [1, 11].

Effective demand forecasts enable power systems to respond dynamically to changes in consumption. By understanding when and where energy usage spikes are likely to occur, utilities can reduce wasted capacity, focus on demand-side management, and more effectively integrate distributed energy resources [12]. This not only decreases operational costs but also provides decision-makers with the information necessary to build robust, adaptable, and efficient energy infrastructures for the future.

7.1.3 Electricity Price Forecasting

Finally, electricity price forecasting serves as a vital source of strategic intelligence for a range of stakeholders within the energy market. This form of forecasting hones in on the financial aspects of energy trade, where accurate predictions can mean the difference between profit and loss [3, 13]. Renewable energy's variable nature introduces a layer of complexity to price formation, acting as a catalyst for sharp fluctuations that can catch market participants off-guard.

Forecasts of electricity prices that can accurately account for renewable intermittency, demand patterns, and evolving market mechanisms allow for preemptive measures against market volatility [14]. These forecasts are crucial for traders to anticipate market movements and position themselves advantageously. For energy suppliers, an understanding of future price trends is instrumental in defining bidding strategies in power auctions and for contract negotiations. Likewise, consumers and procurement managers rely on price forecasts to secure the best rates for future energy needs, whereas investors use these insights to assess the viability of projects in the renewable sector [15]. In essence, electricity price forecasting transcends its immediate market utility to become a cornerstone of strategic planning across the renewable energy value chain.

The precise forecast of electricity prices must consider not only the fluctuating supply due to renewable sources but also the regulatory changes, fuel costs, grid infrastructure upgrades, and the entry of new technologies like energy storage, which can all significantly impact price dynamics [16]. Given the critical role that price forecasts play in market operations and planning, continuous enhancement of these predictive models is a primary focus within the sector. Progress in areas such as artificial intelligence [17], machine learning [18], and data analytics [19] holds promise for advancing the accuracy and timeliness of price forecasting, potentially leading to more stable and efficient energy markets.

With the foundational understanding of the essential nature of accurate renewable energy forecasts and the complexities involved in generation, demand, and price predictions, this chapter is motivated by the need to highlight the evaluation and performance metrics that form the basis of forecasting within the renewable energy sector. By bringing clarity to the strengths and limitations of current evaluation metrics, this work seeks to inform and improve the practices of researchers, industry professionals, and policymakers alike. As a contribution, this chapter offers a systematic review of forecasting models, elucidating the application of different metrics, and providing a critical analysis of evaluation techniques.

7.2 Understanding Power Generation, Demand, and Price Forecasting

Forecasting, at its core, is the process of making predictions about future events based on historical data, current trends, and specialized analytical techniques [20]. In the context of renewable energy, forecasts are used to predict various aspects such as power generation from different renewable resources, consumer demand for electricity, and fluctuations in electricity prices. These forecasts hinge on complex, dynamic models capable of assimilating a multitude of variable factors, from weather conditions affecting generation to macroeconomic indicators influencing demand and market price.

7.2.1 Challenges and Uncertainties in Forecasting Electric Power, Demand, and Price

Forecasting in the renewable energy sector is a complex task rife with uncertainties and challenges that stem from a variety of interdependent factors affecting power generation, demand, and electricity price. This detailed overview outlines the various hurdles that analysts and energy planners face as shown in Figure 7.1.

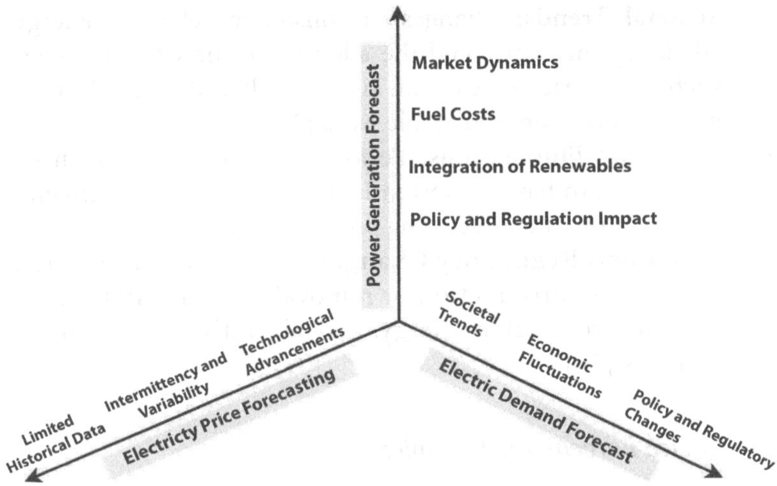

Figure 7.1 Challenges of power generation, demand, and price forecasting.

7.2.1.1 Power Generation Forecasting Challenges

Renewable power generation forecasting must contend with several challenges:

- **Intermittency and Variability:** The inherent unpredictability of weather conditions affects the availability of renewable resources [21, 22]. Wind and solar outputs can fluctuate dramatically because of changing weather patterns, making short-term and long-term forecasting challenging.
- **Limited Historical Data:** Many renewable energy technologies are relatively new, resulting in limited historical generation data which can constrain the development of accurate predictive models [23].
- **Technological Advancements:** As technology improves, the efficiency and capabilities of renewable energy systems change, which can render previous forecasting models less applicable or obsolete [23].

7.2.1.2 Demand Forecasting Uncertainties

Electricity demand forecasting also faces its unique set of challenges:

- **Societal Trends:** Changes in consumer behavior, energy efficiency measures, and the adoption of new technologies such as electric vehicles can significantly influence electricity demand in unpredictable ways [24].
- **Economic Fluctuations:** Periods of economic growth or recession can have a substantial impact on energy demand, complicating the forecasting process [25].
- **Policy and Regulatory Changes:** Governmental policies, such as the introduction or removal of tariffs and incentives for renewable energy, can abruptly shift demand patterns [26].

7.2.1.3 Price Forecasting Complexities

Electricity price forecasting is complicated by various factors:

- **Market Dynamics:** Supply and demand imbalances, often influenced by the variability of renewable generation and the operational flexibility of traditional plants, lead to price volatility [27].
- **Fuel Costs:** Fluctuations in the costs of fossil fuels can affect the overall energy mix and, in turn, impact pricing in electricity markets, particularly in systems that still rely on fossil fuel generation as a backup [28].
- **Policy and Regulation Impact:** Changes in policy, such as the implementation of carbon pricing or subsidies for renewables, can have far-reaching effects on electricity markets. Predicting these policy changes and their impacts adds another layer of complexity [15].
- **Integration of Renewables:** As the share of renewables in the energy mix grows, the influence of variable generation on market prices increases, which can result in negative pricing scenarios during periods of low demand and high renewable output, posing additional challenges to forecasters [29].

7.2.2 The Advantages of Ongoing Forecasting Evaluation in Power Generation, Demand, and Price Forecasting

Ongoing forecast evaluation is essential in the energy sector for enhancing the reliability of predictions regarding power generation, electricity demand, and electricity prices. It is a process that invariably refines forecasting methods. Below are the primary advantages of maintaining a routine in forecast evaluation:

- **Data-Driven Decisions [3, 30]:** By routinely reviewing forecasting outcomes, it is possible to uncover and rectify any systematic inaccuracies. This continuous monitoring leads to better-aligned decisions—a critical factor in fields such as energy where precision is paramount. As an example, consistently high estimates in electricity demand forecasts could indicate the need for recalibrating prediction models to avoid over generating power, thus saving costs and resources.
- **Efficient Allocation of Resources [23, 31]:** When forecasts are regularly evaluated, energy sector planners can discern emerging trends and modify resource distribution effectively. If a pattern of underestimation of peak loads is identified, then grid operators can ensure appropriate adjustments in generation capabilities or storage strategies to meet the demands while avoiding unnecessary expenses.
- **Customer Fulfillment [24]:** The goal of prediction in the energy sector, including pricing and demand, is to remain responsive to customer expectations. Through continual adjustment of forecasts based on past outcomes, utility companies can better match supply with consumer needs, ensuring satisfaction and reducing the risk of service disruptions or energy waste.
- **Operational Cost Savings [32]:** In the competitive and budget-conscious realm of energy management, iterative forecast assessment enables the optimization of both supply chain and production processes. Elimination of recurring inaccuracies in generation or demand forecasts can prevent scenarios such as surplus generation or scarcity, which, in turn, directly affects operational expenditure.
- **Collaborative Efficiency [21, 33]:** Consistent evaluation encourages synergy across different segments within an energy company, from analysts to field operators. The shared pursuit of enhancing forecast models facilitates a unified approach to strategic planning and operational execution. This united effort ensures that forecasts become more

aligned with real-world outcomes, thus improving the institution's operational performance.

In summary, a dedicated framework for continuous forecast assessment within the energy sector is key for advancing the overall effectiveness of forecasting. This entails fine-tuning existing models, adapting to real-time consumption patterns, and being responsive to market fluctuations, which together contribute to a more stable and predictive energy supply system.

7.3 Significance of Accuracy and Reliability in Forecasting Electric Power, Demand, and Price

The accuracy and reliability of forecasts in the renewable energy sector, covering electricity generation, demand, and pricing, is of paramount importance for several reasons as shown in Figure 7.2.

7.3.1 For Energy Producers

- **Grid Stability and Reliability:** Accurate generation forecasts help grid operators balance supply and demand, ensuring steady electricity supply and preventing blackouts or grid instability [34].
- **Economic Planning:** Precision in forecasting helps in making informed investment decisions, optimizing operational

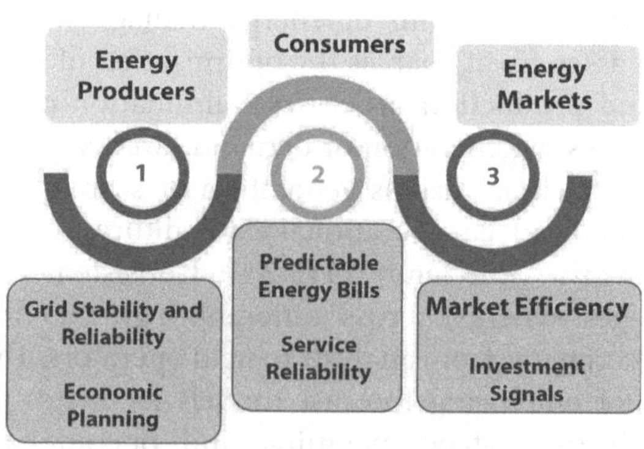

Figure 7.2 Significance of accuracy and reliability in forecasting electric power, demand, and price.

costs, and maximizing returns, especially considering the substantial capital costs associated with renewable energy projects [35].

7.3.2 For Consumers

- **Predictable Energy Bills:** Reliable demand forecasts mean that energy can be supplied efficiently, helping to keep consumer costs down and preventing the need for costly peak energy generation solutions [3].
- **Service Reliability:** Consistent and predictable electricity service is crucial for both residential consumers and industries, with the latter often relying on a stable energy supply for continuous operations [36].

7.3.3 For Energy Markets

- **Market Efficiency:** Accurate price forecasts contribute to effective market operation, enabling energy trading and risk management through future and derivative contracts, as well as informing policy on renewable energy incentives or taxes [37].
- **Investment Signals:** Clear signaling to investors is facilitated by reliable forecasts, guiding investments into the most beneficial and cost-effective energy solutions [15, 38].

7.4 Strategic Framework for Enhanced Forecast Evaluation

In the dynamic world of energy management, where the stakes are high and the variables many, the ability to predict future trends in power generation, demand, and electricity pricing cannot be overstressed. Establishing a robust forecast evaluation process forms the bedrock for operational excellence and long-term strategic success. Here are comprehensive strategies for honing this essential practice (also presented in Figure 7.3):

- **Clarifying Assessment Metrics [39]:** Initiate with explicit definition of evaluation metrics, chosen to mirror the energy organization's targeted outcomes. Metrics such as mean absolute percentage error or mean absolute deviation might be employed to ensure precision, whereas forecast lead time or bias are crucial for assessing promptness.

- **Analyzing Historical Data [40]:** Investing in the thorough analysis of historical forecasting data is essential, as it serves as the benchmark for gauging the predictiveness of current methods. Identifying trends in previous forecasts assists in course-correcting future strategies and refining the predictive algorithms used.
- **Embracing Diverse Evaluation Methods [41]:** Adopt a variety of techniques to scrutinize forecast accuracy, such as longitudinal error tracking, comparative actuals analysis, statistical validity testing, and data visualization. Such multipronged approaches provide nuanced insights into the forecasting process and unearth areas ripe for improvement.
- **Engaging Sector Stakeholders [42]:** Involve stakeholders in forecasting scrutiny, ranging from sales teams to customer service agents. Their unique perspectives on the market and customer behaviors can greatly influence the refinement of forecasts and enhance overall accuracy.

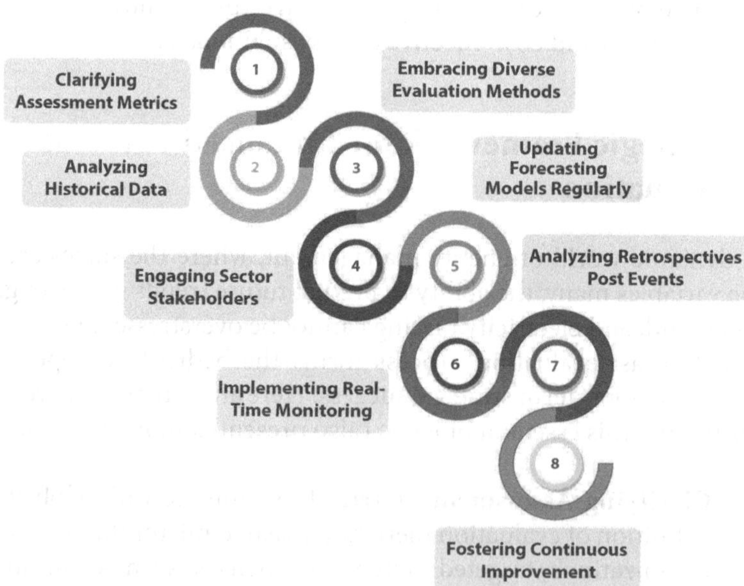

Figure 7.3 Framework for enhanced forecast evaluation.

- **Updating Forecasting Models Regularly [39]:** Adapt and update forecasting models to maintain their relevance amidst changing market conditions. Revisit assumptions, incorporate new datasets, and adjust modeling techniques to keep predictions accurate and timely.
- **Implementing Real-Time Monitoring [19]:** Set up systems for real-time performance tracking, enabling the anticipation and prompt addressing of potential forecasting anomalies. This proactive approach ensures agile response strategies and helps maintain forecast reliability.
- **Analyzing Retrospectives Post Events [43]:** After substantial forecasting episodes, conduct post-mortem reviews to extract valuable insights for future amendments. Such retrospectives help pinpoint both triumphs and failures, offering a clear path to refining forecasting techniques.
- **Fostering Continuous Improvement [44]:** Instill a cuture that values relentless refinement of forecasting accuracy. Encouraging open communication and continual skill enhancement among team members leads to perpetual advancements in forecasting methods, sustaining industry leadership.

By adhering to these comprehensive strategies, power system operators can navigate the complexities of energy forecasting with greater precision and make well-informed decisions. This forward-thinking mindset ensures not only immediate operational benefits but also a sustained competitive edge in the energy sector. Through ongoing dedication to improving forecast evaluation practices, energy providers can adeptly handle the ebbs and flows of market demands, customer needs, and technological advancements, all of which contribute to a more stable and efficient power landscape.

7.5 Performance Metrics for Forecasting Accuracy in Generation, Demand, and Price of Electricity

Performance metrics are indispensable benchmarks that serve to appraise the accuracy and reliability of forecasting models within the energy sector [41, 45]. These metrics are employed to evaluate forecasts related to power generation, demand, and electricity prices—three pivotal components that dictate the operational efficacy and economic sustainability of energy systems. The quality of these forecasts has a substantial impact on decision-making processes for energy providers, grid operators, financial stakeholders, and policymakers [2, 46].

Performance metrics are the benchmarks used to gauge the accuracy and reliability of forecasting models. They can be categorized into metrics for point forecasts, which predict a single expected value, and probabilistic forecasts, which predict a range of possible outcomes.

7.5.1 Criteria for Assessing Accuracy

When evaluating forecasting models, we consider the following criteria as mentioned in Figure 7.4.

- **Bias [47]:** The tendency of a model to consistently overestimate or underestimate the true value. In the context of power generation forecasting, bias can lead to mismatches between forecasted and actual generation levels, impacting grid stability and economic efficiency. Similarly, bias in demand forecasting can result in under or overallocation of resources, affecting energy supply and consumer costs. In electricity price forecasting, bias can influence trading decisions and market efficiency, leading to potential revenue losses for market participants.
- **Scale Sensitivity [5]:** Metrics consider the scale of the data, ensuring that more significant numbers do not inherently produce larger errors. In power generation forecasting, scale sensitivity ensures that forecasting accuracy is maintained across different magnitudes of renewable energy generation,

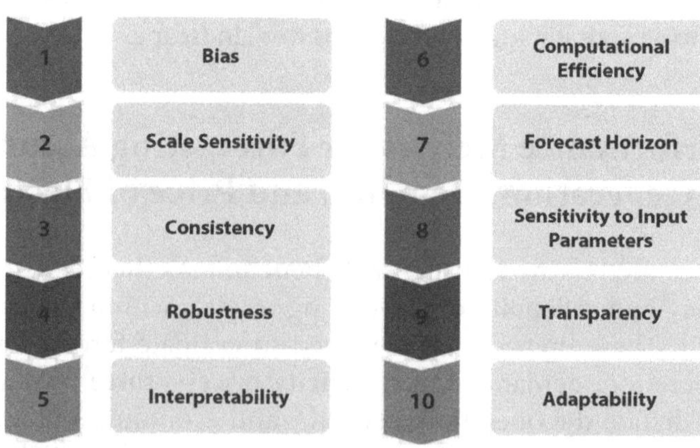

Figure 7.4 Forecasting evaluation criteria.

from small-scale distributed generation to utility-scale power plants. Similarly, in demand forecasting, scale sensitivity accounts for variations in consumption patterns among different user groups and geographic regions. In electricity price forecasting, scale sensitivity ensures that forecasting models are capable of accurately predicting price movements across various market sizes and structures.

- **Consistency** [48]: Models should reliably produce accurate forecasts across various situations and time frames. Consistency is crucial in power generation forecasting to ensure grid stability and optimize renewable energy integration, particularly during periods of high demand or intermittent renewable generation. In demand forecasting, consistency ensures efficient resource allocation and prevents disruptions in supply, enhancing consumer satisfaction and economic productivity. In electricity price forecasting, consistency is essential for market participants to make informed trading decisions and mitigate financial risks in volatile market conditions.
- **Robustness** [49]: The ability of a forecasting model to maintain its accuracy when faced with changes in data distribution, outliers, or unexpected events. Robustness is vital in power generation forecasting to account for fluctuations in renewable resource availability, such as sudden changes in weather patterns or unforeseen equipment failures. In demand forecasting, robustness ensures reliable predictions even in the presence of anomalous consumption patterns or disruptive events like natural disasters or economic downturns. In electricity price forecasting, robustness enables market participants to navigate dynamic market conditions and adapt their trading strategies to mitigate potential losses.
- **Interpretability** [39]: The ease with which the results of a forecasting model can be understood and explained to stakeholders, facilitating informed decision-making. Interpretability is essential in power generation forecasting to communicate forecasted generation levels and grid requirements effectively, enabling grid operators and policymakers to implement appropriate measures for grid stability and renewable energy integration. In demand forecasting, interpretability allows energy suppliers to communicate

anticipated consumption patterns to consumers and adjust supply strategies accordingly. In electricity price forecasting, interpretability assists market participants in understanding the factors driving price movements and making informed trading decisions to optimize profitability.

- **Computational Efficiency [50]:** The computational resources required to train and deploy the forecasting model, ensuring scalability and practicality in real-world applications. Computational efficiency is crucial in power generation forecasting to enable real-time monitoring and control of renewable energy assets, minimizing response times to grid fluctuations and ensuring grid stability. In demand forecasting, computational efficiency allows for timely analysis of large datasets to identify consumption trends and optimize resource allocation. In electricity price forecasting, computational efficiency enables market participants to process vast amounts of market data efficiently and execute trading strategies in rapidly changing market conditions.
- **Forecast Horizon [51]:** The time span over which the forecasted values remain accurate and relevant, considering both short-term and long-term forecasting requirements. Forecast horizon is critical in power generation forecasting to align renewable energy generation forecasts with grid scheduling and maintenance activities, ensuring reliable and efficient operation of the power grid. In demand forecasting, forecast horizon determines the planning horizon for resource allocation and infrastructure development to meet future energy demand growth. In electricity price forecasting, forecast horizon influences trading strategies and risk management practices, with short-term forecasts guiding day-ahead trading decisions and long-term forecasts informing investment planning and hedging strategies.
- **Sensitivity to Input Parameters [52]:** The degree to which the forecasting model's performance is affected by variations in input parameters or assumptions, indicating its stability and reliability. Sensitivity to input parameters is important in power generation forecasting to assess the impact of uncertainties in renewable resource availability, such as changes in weather forecasts or equipment performance. In demand forecasting, sensitivity to

input parameters allows for sensitivity analysis to evaluate the robustness of forecasting models to variations in economic indicators, demographic trends, and policy interventions. In electricity price forecasting, sensitivity to input parameters helps identify key drivers of price movements and assess the potential impact of market factors on forecasting accuracy.

- **Transparency [43]:** The degree to which the inner workings of the forecasting model are transparent and accessible, allowing for scrutiny and validation by experts and stakeholders. Transparency is essential in power generation forecasting to build trust and confidence among grid operators, energy suppliers, and policymakers in the accuracy and reliability of forecasting models. In demand forecasting, transparency fosters collaboration between energy suppliers and consumers, enabling stakeholders to understand the basis for forecasted consumption patterns and collaborate on demand-side management initiatives. In electricity price forecasting, transparency promotes market integrity and fairness by providing market participants with visibility into the factors influencing price forecasts and facilitating regulatory oversight of market operations.
- **Adaptability [50]:** The ability of the forecasting model to adapt and evolve over time in response to changes in data patterns, external factors, and user feedback, ensuring continued relevance and effectiveness. Adaptability is crucial in power generation forecasting to incorporate new data sources, improve model performance, and accommodate changes in renewable energy technology and market dynamics. In demand forecasting, adaptability enables energy suppliers to adjust forecasting methodologies to account for shifting consumer behavior, emerging technologies, and evolving regulatory frameworks. In electricity price forecasting, adaptability allows market participants to refine forecasting models in response to changes in market structure, trading rules, and policy interventions, ensuring accurate and reliable price predictions in dynamic market environments.

7.5.2 Category of Forecasting in in Forecasting Electric Power, Demand, and Price

To comprehensively assess the veracity of forecasts in these areas, the metrics are typically categorized into several types:

- **Statistical Metrics [8, 53–57]:** These include traditional measures of forecast error and accuracy such as mean absolute error, root mean square error, mean absolute percentage error, and the coefficient of determination (R^2). Statistical metrics provide a direct indication of the magnitude and frequency of forecasting errors.
- **Variability Estimation Metrics [42, 56, 58, 59]:** Given that renewable energy sources such as wind and solar have high variability, it is crucial to measure the capability of a forecast model to estimate changes in energy availability. The accuracy of predicting upswings and downswings in power generation significantly influences the stability and balance of the electric grid.
- **Ramping Characterization Metrics [47, 60–63]:** These metrics specifically address the ability of forecast models to predict rapid changes in power generation or demand, known as ramps. Accurate ramping forecasts are imperative for adjusting reserve margins and maintaining grid reliability, especially with the growing penetration of intermittent renewable energy resources.

Each category of metrics illuminates a different aspect of forecast performance, collectively providing a multi-faceted view of model effectiveness. In the dynamic environment of energy markets, where precision in power generation, load anticipation, and price speculation is crucial, the deployment of these performance metrics becomes even more vital. They allow stakeholders to evaluate forecasting tools, refine methodologies, and ultimately lead to more robust and resilient energy systems.

7.5.2.1 Statistical Metrics

Statistical metrics are quantitative measures used to assess the accuracy, precision, and performance of forecasting models in predicting power generation, electricity demand, and electricity prices. These metrics provide valuable insights into the relationship between forecasted values and actual observations, aiding in the evaluation and improvement of forecasting techniques [64]. In power generation forecasting, statistical metrics help quantify the degree of agreement between predicted and observed generation outputs, considering factors such as weather conditions, renewable resource availability, and technological constraints. In demand forecasting, these metrics assess the accuracy of projected electricity consumption patterns across various user groups, considering factors such as seasonal variations, economic activities, and societal trends. For electricity price forecasting, statistical metrics evaluate the reliability of price predictions in electricity markets, accounting for factors such as supply-demand dynamics, fuel costs, regulatory policies, and market structures. By analyzing statistical metrics, stakeholders in the energy sector can assess the performance of forecasting models, identify areas for improvement, and make informed decisions to enhance the reliability and efficiency of energy systems [58, 65]. Table 7.1 presents a summary of statistical metrics found in literature in the field of power generation, electricity demand, and electricity prices.

7.5.2.2 Variability Estimation Metrics

Variability estimation metrics are quantitative measures used to assess the degree of variability or dispersion in power generation, electricity demand, and electricity prices over time. These metrics provide valuable insights into the consistency, stability, and predictability of energy system dynamics. In power generation forecasting, variability estimation metrics help evaluate the reliability of renewable energy sources by quantifying fluctuations in generation output due to factors such as weather variability [81]. In demand forecasting, these metrics aid in understanding the volatility of electricity consumption patterns, which can be influenced by factors such as seasonal variations, economic trends, and technological advancements [25]. For electricity price forecasting, variability estimation metrics provide insights into the volatility of market prices, which can be affected by supply-demand dynamics, fuel costs, policy interventions, and the integration of renewable energy sources [55]. By analyzing variability, stakeholders in the energy sector can better assess risk, optimize resource allocation, and improve decision-making processes. Table 7.2 presents the summary of variability estimation metrics with their potential strengths and weaknesses.

Table 7.1 Summary of statistical metrics in power generation, electricity demand, and electricity price forecasting.

Metrics	Equation	Definition	Strength in forecasting
Pearson's correlation coefficient [66]	$r = \frac{\Sigma_{i=1}^{n}(X_i - \overline{X})(Y_i - \overline{Y})}{\sqrt{\Sigma_{i=1}^{n}(X_i - \overline{X})^2 \, \Sigma_{i=1}^{n}(Y_i - \overline{Y})^2}}$	A statistical measure quantifying the linear relationship between two variables	Indicates the strength and direction of the linear relationship between forecasted and actual values
Root mean square error (RMSE) [43]	$RMSE = \sqrt{\frac{1}{n}\sum_{i=1}^{n}(Forecast_i - Actual_i)^2}$	A measure of the average magnitude of the errors between forecasted and actual values, normalized by the range of the data	Provides a measure of the spread of errors relative to the scale of the data
Maximum absolute error (MaxAE) [41]	$MaxAE = max(\lvert Forecast_i - Actual_i \rvert)$	The maximum absolute difference between forecasted and actual values	Highlights the largest individual error in the forecasting process
Mean absolute error (MAE) [67]	$MAE = \frac{1}{n}\sum_{i=1}^{n}\lvert Forecast_i - Actual_i \rvert$	The average of the absolute errors between forecasted and actual values	Provides a measure of the average magnitude of errors, irrespective of their direction

(Continued)

Table 7.1 Summary of statistical metrics in power generation, electricity demand, and electricity price forecasting. (*Continued*)

Metrics	Equation	Definition	Strength in forecasting
Mean absolute percentage error (MAPE) [39]	$MAPE = \frac{1}{n}\sum_{i=1}^{n}\left\lvert\frac{Forecast_i - Actual_i}{Actual_i}\right\rvert X\,100\%$	The average of the absolute percentage errors between forecasted and actual values	Provides a percentage-based measure of forecast accuracy, allowing for comparison across different datasets or units
Mean bias error (MBE) [68]	$MBE = \frac{1}{n}\sum_{i=1}^{n}(Forecast_i - Actual_i)$	The average difference between forecasted and actual values	Indicates the overall tendency of the forecasts to overestimate or underestimate the actual values
Mean squared error (MSE) [69]	$MBE = \frac{1}{n}\sum_{i=1}^{n}(Forecast_i - Actual_i)^2$	The average of the squared errors between forecasted and actual values	Provides a measure of the average squared deviation from the actual values

(*Continued*)

Table 7.1 Summary of statistical metrics in power generation, electricity demand, and electricity price forecasting. (*Continued*)

Metrics	Equation	Definition	Strength in forecasting
Coefficient of determination (R-squared) [2]	$R^2 = 1 - \frac{\Sigma_{i=1}^{n}(Forecast_i - Actual_i)^2}{\Sigma_{i=1}^{n}(Actual_i - Actual)^2}$	A measure of the proportion of variance explained by the forecasted values	Indicates the goodness of fit of the forecasting model. Feel free to ask if you need more information on any specific metric!
Mean squared percentage error (MSPE) [70]	$MSPE = \frac{1}{n}\sum_{i=1}^{n}\left(\frac{Forecast_i - Actual_i}{Actual_i}\right)^2 X\,100\%$	The average of the squared percentage errors between forecasted and actual values	Provides a measure of the average squared deviation as a percentage of the actual values
Weighted mean absolute percentage error (WMAPE) [71]	$WMAPE = \frac{\Sigma_{i=1}^{n} w_i \left\lvert\frac{Forecast_i - Actual_i}{Actual_i}\right\rvert}{\Sigma_{i=1}^{n} w_i} X\,100\%$	The weighted average of the absolute percentage errors between forecasted and actual values	Accounts for variations in the importance of individual observations, useful when certain observations carry more weight or significance

(*Continued*)

Table 7.1 Summary of statistical metrics in power generation, electricity demand, and electricity price forecasting. (*Continued*)

Metrics	Equation	Definition	Strength in forecasting
Mean absolute scaled error (MASE) [72]	$MASE = \frac{\frac{1}{n}\Sigma_{i=1}^{n}\lvert Forecast_i - Actual_i\rvert}{\frac{1}{m}\Sigma_{i=1}^{m}\lvert Forecast_{i-h} - Actual_{i-h}\rvert} X\, 100\%$	A measure of forecast accuracy relative to a benchmark forecast, such as a simple average or persistence model	Allows for comparison of forecast performance across different time series or datasets
Symmetric mean absolute percentage error (SMAPE) [20]	$SMAPE = \frac{1}{n}\sum_{i=1}^{n}\frac{\lvert Forecast_i - Actual_i\rvert}{\frac{\lvert Forecast_i\rvert + \lvert Actual_i\rvert}{2}} X\, 100\%$	A symmetric measure of forecast accuracy, considering both forecast and actual values	Suitable for comparing forecast performance across different datasets or units
Geometric mean bias deviation (GMBD) [73]	$GMBD = exp\left(\frac{1}{n}\sum_{i=1}^{n} ln\left(\frac{Forecast_i}{Actual_i}\right)\right)$	A multiplicative measure of bias, accounting for the proportional difference between forecasted and actual values	Suitable for datasets with exponential or multiplicative relationships

(*Continued*)

Table 7.1 Summary of statistical metrics in power generation, electricity demand, and electricity price forecasting. (*Continued*)

Metrics	Equation	Definition	Strength in forecasting
Mean percentage error (MPE) [74]	$MPE = \frac{1}{n}\sum_{i=1}^{n} \frac{Forecast_i - Actual_i}{Actual_i} X\,100\%$	General definition: The average of the percentage errors between forecasted and actual values	Provides a measure of the average direction and magnitude of the errors, indicating whether the forecasts tend to overestimate or underestimate
Mean relative error (MRE) [65, 75]	$MRE = \frac{1}{n}\sum_{i=1}^{n}\left(\frac{Forecast_i - Actual_i}{Actual_i}\right)$	The average of the relative errors between forecasted and actual values	Measures the average direction and magnitude of the errors relative to the actual values
Geometric mean squared error (GMSE) [76]	$GMSE = \exp\left(\frac{1}{n}\sum_{i=1}^{n} ln((Forecast_i - Actual_i)^2)\right)$	The geometric mean of the squared errors between forecasted and actual values	Suitable for datasets with skewed distributions or extreme values
Geometric mean absolute error (GMAE) [77]	$GMAE = \exp\left(\frac{1}{n}\sum_{i=1}^{n} ln(\lvert Forecast_i - Actual_i\rvert)\right)$	The geometric mean of the absolute errors between forecasted and actual values	Suitable for datasets with skewed distributions or extreme values

(*Continued*)

Table 7.1 Summary of statistical metrics in power generation, electricity demand, and electricity price forecasting. (*Continued*)

Metrics	Equation	Definition	Strength in forecasting
Normalized mean bias error (NMBE) [78]	$NMBE = \frac{\Sigma_{i=1}^{n}\lvert Forecast_i - Actual_i\rvert}{\Sigma_{i=1}^{n} Actual_i} X100\%$	The normalized average difference between forecasted and actual values	Provides a measure of the average direction and magnitude of the errors relative to the actual values
Weighted root mean square error (WRMSE) [79]	$WRMSE = \sqrt{\frac{\Sigma_{i=1}^{n} w_i(Forecast_i - Actual_i)^2}{\Sigma_{i=1}^{n} w_i}}$	The root mean square error weighted by the importance of individual observations	Accounts for variations in the importance of individual observations, useful when certain observations carry more weight or significance
Weighted mean squared error (WMSE) [79]	$WMSE = \frac{\Sigma_{i=1}^{n} w_i(Forecast_i - Actual_i)^2}{\Sigma_{i=1}^{n} w_i}$	The mean squared error weighted by the importance of individual observations	Accounts for variations in the importance of individual observations, useful when certain observations carry more weight or significance

(*Continued*)

Table 7.1 Summary of statistical metrics in power generation, electricity demand, and electricity price forecasting. (*Continued*)

Metrics	Equation	Definition	Strength in forecasting
Weighted mean bias error (WMBE) [30]	$WMBE = \frac{\Sigma_{i=1}^{n} w_i(Forecast_i - Actual_i)}{\Sigma_{i=1}^{n} w_i}$	The mean bias error weighted by the importance of individual observations	Accounts for variations in the importance of individual observations, useful when certain observations carry more weight or significance
Mean absolute scaled percentage error (MASPE) [31]	$MASPE = \frac{\frac{1}{n}\Sigma_{i=1}^{n}\lvert Forecast_i - Actual_i\rvert}{\frac{1}{n}\Sigma_{i=1}^{n}\lvert Actual_i\rvert} X100\%$	A scaled version of the mean absolute error, considering the relative size of the errors with respect to the actual values	Provides a measure of the average magnitude of errors relative to the scale of the actual values
Geometric mean absolute percentage error (GMAPE) [80]	$GMAPE = exp\left(\frac{1}{n}\sum_{i=1}^{n} ln\left(\left\lvert\frac{Forecast_i - Actual_i}{Actual_i}\right\rvert\right)\right)$	The geometric mean of the absolute percentage errors between forecasted and actual values	Suitable for datasets with skewed distributions or extreme values

(*Continued*)

Table 7.1 Summary of statistical metrics in power generation, electricity demand, and electricity price forecasting. (*Continued*)

Metrics	Equation	Definition	Strength in forecasting
Normalized mean squared error (NMSE) [42]	$NMSE = \frac{\Sigma_{i=1}^{n}(Forecast_i - Actual_i)^2}{\Sigma_{i=1}^{n}(Actual_i - Actual)^2}$	A normalized measure of the mean squared error, indicating the proportion of variance explained by the forecasted values relative to the total variance of the actual values	Provides insight into the proportion of variability in the data captured by the forecasting model, facilitating comparisons across different datasets or units

7.5.2.3 Ramping Characterization Metrics

Ramping characterization metrics are quantitative measures used to assess the dynamics and variability of changes in power generation, electricity demand, and electricity prices over time. These metrics provide valuable insights into the rate, magnitude, duration, frequency, intensity, and other

Table 7.2 Summary of variability estimation metrics in power generation, electricity demand, and electricity price forecasting.

Metric	Definition	Strength	Limitation
Standard deviation [3, 51]	Measures the dispersion or spread of a dataset around its mean	Provides a measure of the variability or volatility in the forecasted values	Does not provide information about the direction or pattern of variability, limiting its interpretability
Mean absolute deviation [49, 50]	Measures the average deviation of individual data points from the mean of the dataset	Offers a straightforward measure of forecast accuracy and variability	Gives equal weight to positive and negative deviations, potentially masking asymmetrical errors in forecasting
Variance [2, 40]	Quantifies the dispersion of data points around the mean and is calculated as the average of the squared differences from the mean	Provides a measure of the overall spread or dispersion of forecasted values	Sensitive to outliers and extreme values, which can distort its interpretation in forecasting applications

(*Continued*)

Table 7.2 Summary of variability estimation metrics in power generation, electricity demand, and electricity price forecasting. (*Continued*)

Metric	Definition	Strength	Limitation
Interquartile range [44, 65]	Measures the spread of the middle 50% of data values in a dataset, representing the variability of the central portion of the distribution	Robust to outliers and extreme values, providing a more robust measure of variability in forecasting applications	Only captures variability within the middle portion of the distribution, potentially overlooking extreme fluctuations
Coefficient of variation [84]	Measures the relative variability of a dataset by expressing the standard deviation as a percentage of the mean	Allows for the comparison of variability across datasets with different scales or units	May be sensitive to changes in the mean value, particularly in datasets with small means, leading to potential misinterpretation of variability
Skewness [50]	Measures the asymmetry of the distribution of data points around the mean	Provides insight into the direction and shape of variability in forecasted values	Sensitive to outliers and extreme values, which can affect its accuracy in capturing the underlying distribution
Kurtosis [3]	Measures the peakedness or flatness of the distribution of data points relative to a normal distribution	Quantifies the degree of concentration or dispersion of forecasted values around the mean	Influenced by outliers and extreme values, potentially leading to misinterpretation of the shape of the distribution

(*Continued*)

Table 7.2 Summary of variability estimation metrics in power generation, electricity demand, and electricity price forecasting. (*Continued*)

Metric	Definition	Strength	Limitation
Autocorrelation [72]	Measures the degree of correlation between a time series and a lagged version of itself	Captures the temporal dependence or persistence in forecasted values	Assumes linear relationship between forecasted values and their lagged counterparts, which may not always hold true

characteristics of ramping events, which are sudden or significant deviations from normal operating conditions [63]. By analyzing ramping behavior, stakeholders in the energy sector can better understand system flexibility, resilience, and reliability, aiding in resource planning, operational decision-making, and risk management [47, 63]. Table 7.3 presents the summary of ramping characterization metrics.

Table 7.3 Summary of ramping characterization metrics in power generation, electricity demand, and electricity price forecasting.

Metric	Definition	Strength	Limitation
Ramping rate [61]	Measures the rate of change in power generation, demand, or prices over a specified time interval	Provides insight into the speed at which changes occur, aiding in assessment of system flexibility	May oversimplify the dynamic nature of ramping events

(*Continued*)

Table 7.3 Summary of ramping characterization metrics in power generation, electricity demand, and electricity price forecasting. (*Continued*)

Metric	Definition	Strength	Limitation
Ramping magnitude [83]	Quantifies the magnitude of changes in power generation, demand, or prices within a given period	Offers a measure of the size or extent of ramping events, helping to identify significant deviations from normal behavior	Does not differentiate between positive and negative ramping events
Ramping duration [84]	Represents the length of time over which ramping events occur in the system	Helps in assessing the persistence or duration of ramping events, aiding in resource planning and system management	May not capture rapid, short-duration ramping events
Ramping frequency [85]	Measures the frequency of occurrence of ramping events within a given timeframe	Provides insights into the frequency distribution of ramping events, aiding in the characterization of system behavior	May not capture rare or extreme ramping events
Ramping intensity [60]	Quantifies the severity of ramping events, often by considering the magnitude and duration of changes	Offers a holistic measure of the impact of ramping events on system operations, facilitating risk assessment and mitigation	May be influenced by the choice of intensity metrics and thresholds

(*Continued*)

Table 7.3 Summary of ramping characterization metrics in power generation, electricity demand, and electricity price forecasting. (*Continued*)

Metric	Definition	Strength	Limitation
Ramping variability [47]	Measures the variability of ramping events over time, considering factors such as amplitude, frequency, and duration	Provides a measure of the consistency or stability of ramping behavior, aiding in the identification of trends or patterns	May be sensitive to changes in data resolution or sampling frequency
Ramping index [86]	Represents an aggregated measure of ramping behavior, often combining multiple ramping characteristics into a single index	Offers a simplified representation of overall ramping activity, facilitating comparisons across different time periods	May obscure underlying variations in individual ramping metrics
Ramping ratio [83]	Compares the magnitude of upward and downward ramping events, assessing the balance or symmetry of ramping behavior in the system	Provides insights into the directional bias or imbalance in ramping events, aiding in the identification of system vulnerabilities	May overlook temporal or contextual factors influencing ramping behavior
Ramping profile [86]	Describes the temporal distribution or shape of ramping events over a specified period, capturing patterns or trends in ramping behavior	Offers a visual representation of ramping dynamics, facilitating qualitative assessment and trend identification	May require subjective interpretation, leading to variability in characterizing ramping profiles

(*Continued*)

Table 7.3 Summary of ramping characterization metrics in power generation, electricity demand, and electricity price forecasting. (*Continued*)

Metric	Definition	Strength	Limitation
Ramping capacity [83]	Represents the system's ability to accommodate ramping events without compromising operational stability or reliability	Provides a measure of resilience or robustness against ramping disturbances, guiding infrastructure investment decisions	May be influenced by system configuration, operational constraints, and resource availability

7.6 Comparative Analysis of Forecasting Methods in Energy Sector

In the pursuit of optimizing operations within the energy sector, a range of forecasting methods is crucial to predict future scenarios in renewable power generation, electricity demand, and market prices. The efficacy of these forecasts is paramount for ensuring the security, reliability, and profitability of energy systems. A comparative analysis of these forecasting methods elucidates the varied approaches and offers insights into their applicability under different conditions and requirements.

Forecasting models span a broad spectrum from time-honored statistical methods to leading-edge artificial intelligence–based algorithms, each with distinct functionalities and suited to particular facets of energy forecasting:

- **Statistical Models:** Foundational approaches like linear and logistic regression are noted for their transparency and are often employed where elucidation of variable interdependencies is critical. They are, however, likely to be inadequate for intricate, nonlinear interrelations that are typical in energy systems.
- **Time-Series Models:** Methods such as ARIMA and exponential smoothing are adept at capturing seasonal and serial trends prevalent in power usage patterns, albeit they might underperform with non-linear and volatile data reflective of today's dynamic energy landscape.

- **Machine Learning Models:** These models, including decision trees and gradient boosting machines, are lauded for handling multifaceted, nonlinear associations and can accommodate a multitude of input variables but entail careful hyperparameter optimizations and are computationally intensive.
- **Dimensionality Reduction and Clustering:** Techniques like PCA and K-means aid in distilling the essence of data, allowing easier pattern recognition and analysis. Their downside is potentially glossing over critical nuances when faced with complex, multivariate datasets.
- **Deep Learning Models:** Sophisticated neural networks, such as CNNs and LSTMs, have the proficiency to decode complex and long-term temporal sequences, providing powerful tools for forecasting in an energy context. Nonetheless, they necessitate significant computation power and large volumes of training data.
- **Hybrid Models:** These emergent methodologies combine multiple forecasting strategies to offset individual weaknesses, promoting a holistic approach that can potentially yield superior predictive performance.

This comparative analysis aims to identify the strengths and limitations inherent in different forecasting approaches, illuminating their potential fit for specific challenges in the energy sector. By balancing the trade-offs between accuracy, computational demand, and interpretability, stakeholders can select and tailor forecasting techniques to streamline energy management and adapt to the increasingly intricate energy market dynamics. Table 7.4 presents a comparative modeling analysis of electric power generation, demand, and price forecasting.

Table 7.4 Comparative analysis of forecasting models in energy sector.

Model	Strength in forecasting	Potential limitation	Application in forecasting
Linear regression [87]	Simple, interpretable; suitable for linear relationships; robust to noise	Limited flexibility in capturing nonlinear patterns; assumes linearity in relationships	Used for linear modeling of relationships between variables

(Continued)

Table 7.4 Comparative analysis of forecasting models in energy sector. (*Continued*)

Model	Strength in forecasting	Potential limitation	Application in forecasting
Logistic regression [87]	Effective for binary classification tasks; handles high-dimensional data well	Assumes linearity in the log odds of the dependent variable; may not capture complex interactions	Applied in classification tasks such as demand response forecasting
Naive Bayes [88]	Fast, simple; handles high-dimensional data well; effective for text data	Assumes independence between predictor variables; may not capture complex relationships	Commonly used in classification tasks, such as customer segmentation
Support vector machines (SVMs) [89]	Effective in high-dimensional spaces; robust to overfitting; suitable for nonlinear relationships	Sensitive to the choice of kernel function; computationally expensive; may not perform well with noisy data	Applied in regression and classification tasks, such as load forecasting
K-Nearest neighbors (KNNs) [90]	Simple, intuitive; non-parametric; suitable for nonlinear relationships	Sensitive to irrelevant features; computationally expensive; performance deteriorates with high-dimensional data	Used in regression and classification tasks, such as load forecasting
Gaussian mixture models (GMMs) [88]	Effective for modeling complex data distributions; suitable for clustering tasks	Sensitive to the number of components; may not capture non-Gaussian distributions; computationally expensive	Applied in clustering tasks, such as demand segmentation

(*Continued*)

Table 7.4 Comparative analysis of forecasting models in energy sector. (*Continued*)

Model	Strength in forecasting	Potential limitation	Application in forecasting
ARIMA (autoregressive integrated moving average) [11]	Effective in modeling time-dependent data; captures temporal dependencies	Assumes linear relationships and stationary data; may not capture long-term dependencies well	Commonly used in time-series forecasting tasks, such as demand forecasting
SARIMA (seasonal ARIMA) [5]	Captures seasonal patterns in time-series data; handles time series with seasonality	Assumes linear relationships and stationary data; may not handle complex seasonality patterns well	Applied in time-series forecasting tasks, especially for seasonal data patterns
WaveNet [5]	Capable of capturing long-range dependencies in time-series data	Computationally expensive; requires large datasets; may not handle noisy data well	Applied in time-series forecasting tasks, particularly for long-range dependencies
Echo state networks (ESNs) [11]	Handles temporal dynamics in time-series data; effective for chaotic and nonlinear time series	Limited interpretability; sensitive to hyperparameters; may not capture long-term dependencies well	Used in time-series forecasting tasks, particularly for nonlinear dynamics

(*Continued*)

Table 7.4 Comparative analysis of forecasting models in energy sector. (*Continued*)

Model	Strength in forecasting	Potential limitation	Application in forecasting
Random forest [22]	Handles nonlinearity; robust to overfitting; effective for high-dimensional data	Prone to overfitting with noisy data; may not capture linear relationships well; computationally expensive	Applied in regression and classification tasks, such as renewable energy generation forecasting
Decision trees [91]	Simple to understand; handles nonlinear relationships; interpretable	Prone to overfitting; sensitive to noisy data; may not capture complex interactions well	Used in classification and regression tasks, such as electricity price forecasting
Gradient boosting [91]	Effective for modeling complex, nonlinear relationships	Sensitive to hyperparameter tuning; computationally expensive; may not handle noisy data well	Applied in regression and classification tasks, such as load forecasting
XGBoost [92]	Handles missing data; effective for large datasets	Sensitive to hyperparameter tuning; computationally expensive; may not handle categorical data well	Used in regression and classification tasks, such as electricity price forecasting

(*Continued*)

Table 7.4 Comparative analysis of forecasting models in energy sector. (*Continued*)

Model	Strength in forecasting	Potential limitation	Application in forecasting
Convolutional neural networks (CNNs) [57]	Effective in capturing spatial patterns in data	Requires large datasets; computationally expensive; may not handle temporal dependencies well	Used in image-based forecasting tasks, such as solar panel output prediction
Recurrent neural networks (RNNs) [30]	Effective for modeling sequential data	Prone to vanishing or exploding gradient problems; may not capture long-range dependencies well	Applied in time-series forecasting tasks, such as electricity demand forecasting
Long short-term memory (LSTM) [37]	Handles long-range dependencies in sequential data	Sensitive to hyperparameter tuning; computationally expensive; may suffer from vanishing gradients	Used in time-series forecasting tasks, particularly for long-range dependencies
Transformer models [33]	Effective for capturing long-range dependencies in sequential data	Computationally efficient; handles variable-length inputs; may require large datasets	Applied in time-series forecasting tasks, particularly for long-range dependencies

(*Continued*)

Table 7.4 Comparative analysis of forecasting models in energy sector. (*Continued*)

Model	Strength in forecasting	Potential limitation	Application in forecasting
Dimensionality reduction and clustering			
Principal component analysis (PCA) [2]	Reduces dimensionality while preserving data variance, interpretable	Assumes linear relationships; may lose interpretability with non-linear data; sensitive to scaling	Applied in feature selection and compression tasks, such as feature extraction for load forecasting
K-means clustering [51]	Simple, scalable; effective for large datasets	Sensitive to initial cluster centroids; may not perform well with non-linear data distributions	Used in customer segmentation, load profiling, and demand response forecasting
Gaussian mixture models (GMMs) [93]	Effective for modeling complex data distributions	Sensitive to the number of components; may not capture non-Gaussian distributions; computationally expensive	Applied in clustering tasks, such as demand segmentation

7.7 Future Directions

As the field of electric power generation, demand, and price forecasting continues to evolve, there are several potential future directions that researchers and practitioners may explore to enhance the evaluation metrics and improve forecasting accuracy.

One area of focus is the development of more sophisticated evaluation metrics tailored to the specific characteristics of renewable energy sources and electricity markets. This includes metrics that account for the intermittent nature of renewable generation, the dynamic behavior of demand patterns, and the complex interactions influencing electricity prices. By refining evaluation metrics to better capture these nuances, forecasters can gain deeper insights into the performance of forecasting models and identify areas for improvement.

Furthermore, advancements in machine learning and artificial intelligence offer opportunities to enhance forecasting accuracy through the development of more adaptive and responsive models. Techniques such as deep learning, reinforcement learning, and ensemble methods hold promise for capturing complex relationships and patterns in the data, thereby improving the predictive capabilities of forecasting models.

However, along with these opportunities come challenges, one significant challenge is the availability and quality of data, particularly in the context of renewable energy forecasting. Incorporating data from diverse sources, such as weather forecasts, grid data, and market information, while ensuring its accuracy and reliability, remains a key hurdle for forecasters.

Moreover, the dynamic nature of electricity markets and the increasing penetration of renewable energy sources introduce additional complexities into forecasting tasks. Variability in weather patterns, changes in regulatory policies, and market dynamics pose challenges for traditional forecasting models and require innovative approaches to address.

Overall, future developments in evaluation metrics and forecasting techniques hold the potential to significantly enhance the accuracy and reliability of electric power generation, demand, and price forecasting. By addressing key challenges and leveraging emerging technologies, forecasters can better anticipate future trends and support the transition to a more sustainable and efficient energy system.

7.8 Conclusion

In the realm of renewable power generation forecasting, the development of robust evaluation criteria and metrics is paramount. Through a comprehensive comparative analysis of diverse forecasting methods, this study has illuminated the varied approaches used to predict future scenarios in renewable power generation, electricity demand, and market prices. Evaluation metrics, ranging from traditional statistical techniques to advanced machine learning algorithms, play a pivotal role in assessing forecast accuracy. However, the evaluation process faces challenges due to the intermittent nature of renewable energy sources and the dynamic behavior of electricity markets. As such, there is a pressing need for more sophisticated evaluation criteria tailored specifically to the unique characteristics of renewable energy forecasting.

Future advancements in evaluation metrics must aim to capture the intricacies of renewable generation, demand patterns, and market dynamics comprehensively. Metrics that account for variability in weather patterns, changes in regulatory policies, and market dynamics will be crucial for accurately assessing forecasting model performance. Despite challenges such as data availability and quality, there is optimism for the future of electric power forecasting. By refining evaluation criteria and leveraging emerging technologies, forecasters can enhance the accuracy and reliability of renewable power generation forecasting. This, in turn, will support the transition to a more sustainable and efficient energy system, paving the way for a greener future.

References

1. Azeem, A., Ismail, I., Jameel, S.M., Harindran, V.R., Electrical Load Forecasting Models for Different Generation Modalities: A Review. *IEEE Access*, 9, 142239–142263, 2021, doi: 10.1109/ACCESS.2021.3120731.
2. Prema, V., Bhaskar, M.S., Almakhles, D., Gowtham, N., Rao, K.U., Critical Review of Data, Models and Performance Metrics for Wind and Solar Power Forecast. *IEEE Access*, 10, 667–688, 2022, doi: 10.1109/ACCESS.2021.3137419.
3. Lu, H., Ma, X., Ma, M., Zhu, S., Energy price prediction using data-driven models: A decade review. *Comput. Sci. Rev.*, 39, 100356, Feb. 2021, doi: 10.1016/J.COSREV.2020.100356.
4. Mahmud, K., Azam, S., Karim, A., Zobaed, S., Shanmugam, B., Mathur, D., Machine Learning Based PV Power Generation Forecasting in Alice Springs. *IEEE Access*, 9, 46117–46128, 2021, doi: 10.1109/ACCESS.2021.3066494.
5. Sharadga, H., Hajimirza, S., Balog, R.S., Time series forecasting of solar power generation for large-scale photovoltaic plants. *Renew. Energy*, 150, 797–807, May 2020, doi: 10.1016/J.RENENE.2019.12.131.
6. Khan, Z.A., Hussain, T., Ullah, W., Baik, S.W., A Trapezoid Attention Mechanism for Power Generation and Consumption Forecasting. *IEEE Trans. Industr. Inform.*, 20, 4, 5750–5762, Apr. 2024, doi: 10.1109/TII.2023.3335453.
7. Aslam, S., Herodotou, H., Mohsin, S.M., Javaid, N., Ashraf, N., Aslam, S., A survey on deep learning methods for power load and renewable energy forecasting in smart microgrids. *Renew. Sustain. Energy Rev.*, 144, 110992, Jul. 2021, doi: 10.1016/J.RSER.2021.110992.
8. Mellit, A., Pavan, A.M., Ogliari, E., Leva, S., Lughi, V., Advanced Methods for Photovoltaic Output Power Forecasting: A Review. *Appl. Sci.*, 10, 2, 487, Jan. 2020, doi: 10.3390/APP10020487.

9. Li, Y., Zhou, L., Gao, P., Yang, B., Han, Y., Lian, C., Short-Term Power Generation Forecasting of a Photovoltaic Plant Based on PSO-BP and GA-BP Neural Networks. *Front. Energy Res.*, 9, 824691, Jan. 2022, doi: 10.3389/FENRG.2021.824691/BIBTEX.
10. Lisi, F. and Shah, I., Forecasting next-day electricity demand and prices based on functional models. *Energy Syst.*, 11, 4, 947–979, Nov. 2020, doi: 10.1007/S12667-019-00356-W/TABLES/9.
11. Amjady, N., Short-term hourly load forecasting using time-series modeling with peak load estimation capability. *IEEE Trans. Power Syst.*, 16, 3, 498–505, Aug. 2001, doi: 10.1109/59.932287.
12. Eskandari, H., Imani, M., Moghaddam, M.P., Convolutional and recurrent neural network based model for short-term load forecasting. *Electr. Power Syst. Res.*, 195, 107173, 2021. doi: https://doi.org/10.1016/j.epsr.2021.107173.
13. Zhang, J., Tan, Z., Wei, Y., An adaptive hybrid model for short term electricity price forecasting. *Appl. Energy*, 258, 114087, Jan. 2020, doi: 10.1016/J.APENERGY.2019.114087.
14. Memarzadeh, G. and Keynia, F., Short-term electricity load and price forecasting by a new optimal LSTM-NN based prediction algorithm. *Electr. Power Syst. Res.*, 192, 106995, Mar. 2021, doi: 10.1016/J.EPSR.2020.106995.
15. Narajewski, M. and Ziel, F., Econometric modelling and forecasting of intraday electricity prices. *J. Commod. Mark.*, 19, 100107, Sep. 2020, doi: 10.1016/J.JCOMM.2019.100107.
16. Fraunholz, C., Kraft, E., Keles, D., Fichtner, W., Advanced price forecasting in agent-based electricity market simulation. *Appl. Energy*, 290, 116688, May 2021, doi: 10.1016/J.APENERGY.2021.116688.
17. Nazir, M.S., *et al.*, Wind Generation Forecasting Methods and Proliferation of Artificial Neural Network: A Review of Five Years Research Trend. *Sustainability*, 12, 9, 3778, May 2020, doi: 10.3390/SU12093778.
18. Yang, W., Sun, S., Hao, Y., Wang, S., A novel machine learning-based electricity price forecasting model based on optimal model selection strategy. *Energy*, 238, 121989, Jan. 2022, doi: 10.1016/J.ENERGY.2021.121989.
19. Malik, H., Fatema, N., Iqbal, A., Intelligent Data-Analytics for Condition Monitoring: Smart Grid Applications, in: *Intelligent Data-Analytics for Condition Monitoring: Smart Grid Applications*, pp. 1–251, Jan. 2021, doi: 10.1016/B978-0-323-85510-5.00011-9.
20. Gebremeskel, D.H., Ahlgren, E.O., Beyene, G.B., Long-term evolution of energy and electricity demand forecasting: The case of Ethiopia. *Energy Strateg. Rev.*, 36, 100671, Jul. 2021, doi: 10.1016/j.esr.2021.100671.
21. Lu, Y., Murzakhanov, I., Chatzivasileiadis, S., Neural network interpretability for forecasting of aggregated renewable generation. *2021 IEEE International Conference on Communications, Control, and Computing Technologies for Smart Grids, SmartGridComm 2021*, pp. 282–288, 2021, doi: 10.1109/SMARTGRIDCOMM51999.2021.9631993.

22. Niu, D., Wang, K., Sun, L., Wu, J., Xu, X., Short-term photovoltaic power generation forecasting based on random forest feature selection and CEEMD: A case study. *Appl. Soft Comput.*, 93, 106389, Aug. 2020, doi: 10.1016/J.ASOC.2020.106389.
23. Kumar, D.S., Yagli, G.M., Kashyap, M., Srinivasan, D., Solar irradiance resource and forecasting: a comprehensive review. *IET Renew. Power Gener.*, 14, 10, 1641–1656, Jul. 2020, doi: 10.1049/IET-RPG.2019.1227.
24. Muzumdar, A.A., Modi, C.N., Vyjayanthi, C., Designing a robust and accurate model for consumer-centric short-term load forecasting in microgrid environment. *IEEE Syst. J.*, 16, 2448–2459, 2021, Accessed: May 11, 2024. doi: 10.1109/JSYST.2021.3073493.
25. Ahmad, N., Ghadi, Y., Adnan, M., Ali, M., Load Forecasting Techniques for Power System: Research Challenges and Survey. *IEEE Access*, 10, 71054–71090, 2022, doi: 10.1109/ACCESS.2022.3187839.
26. Tumiran, T., Sarjiya, S., Putranto, L.M., Putra, E.N., Budi, R.F.S., Nugraha, C.F., Long-Term Electricity Demand Forecast Using Multivariate Regression and End-Use Method: A Study Case of Maluku-Papua Electricity System. *ICT-PEP 2021 - International Conference on Technology and Policy in Energy and Electric Power: Emerging Energy Sustainability, Smart Grid, and Microgrid Technologies for Future Power System, Proceedings*, pp. 258–263, 2021, doi: 10.1109/ICT-PEP53949.2021.9601144.
27. Bhatia, K., Mittal, R., Varanasi, J., Tripathi, M.M., An ensemble approach for electricity price forecasting in markets with renewable energy resources. *Util. Policy*, 70, 101185, Jun. 2021, doi: 10.1016/j.jup.2021.101185.
28. Yang, W., Wang, J., Niu, T., Du, P., A novel system for multi-step electricity price forecasting for electricity market management. *Appl. Soft Comput.*, 88, 106029, Mar. 2020, doi: 10.1016/J.ASOC.2019.106029.
29. Yang, G., Du, S., Duan, Q., Su, J., Short-Term Demand Forecasting Method in Power Markets Based on the KSVM-TCN-GBRT. *Comput. Intell. Neurosci.*, 2022, 6909558, 2022, doi: 10.1155/2022/6909558.
30. Waheed, W., A Data-driven Long-Short Term Load Forecasting with LSTM-RNN Networks A Data-driven Long-Short Term Load Forecasting, pp. 0–22, 2023.
31. Yazdani, M., Torkayesh, A.E., Santibanez-Gonzalez, E.D., Otaghsara, S.K., Evaluation of renewable energy resources using integrated Shannon Entropy—EDAS model. *Sustain. Oper. Comput.*, 1, 35–42, Jan. 2020, doi: 10.1016/J.SUSOC.2020.12.002.
32. Guimarães, T., Costa, L., H.L., A hybrid approach to load forecast at a micro grid level through machine learning algorithms, 2020, ieeexplore.ieee.org, Accessed: May 11, 2024. [Online]. Available: https://ieeexplore.ieee.org/abstract/document/9203308/.
33. Demir, S., Mincev, K., Kok, K., Paterakis, N.G., Data augmentation for time series regression: Applying transformations, autoencoders and adversarial

networks to electricity price forecasting. *Appl. Energy*, 304, 117695, Dec. 2021, doi: 10.1016/J.APENERGY.2021.117695.

34. Hernandez, L., *et al.*, A survey on electric power demand forecasting: Future trends in smart grids, microgrids and smart buildings. *IEEE Commun. Surv. Tutor.*, 16, 3, 1460–1495, 2014, doi: 10.1109/SURV.2014.032014.00094.
35. Mokilane, P., Galpin, J., Sarma Yadavalli, V.S., Debba, P., Koen, R., Sibiya, S., Density forecasting for long-term electricity demand in South Africa using quantile regression. *S. Afr. J. Econ. Manag. Sci.*, 21, 1, 1–14, 2018, doi: 10.4102/sajems.v21i1.1757.
36. Gabrielli, P., Wüthrich, M., Blume, S., Sansavini, G., Data-driven modeling for long-term electricity price forecasting. *Energy*, 244, 123107, Apr. 2022, doi: 10.1016/J.ENERGY.2022.123107.
37. Li, W. and Becker, D.M., Day-ahead electricity price prediction applying hybrid models of LSTM-based deep learning methods and feature selection algorithms under consideration of market coupling. *Energy*, 237, 121543, Dec. 2021, doi: 10.1016/J.ENERGY.2021.121543.
38. Zhang, X., Wang, J., Gao, Y., A hybrid short-term electricity price forecasting framework: Cuckoo search-based feature selection with singular spectrum analysis and SVM. *Energy Econ.*, 81, 899–913, Jun. 2019, doi: 10.1016/j.eneco.2019.05.026.
39. Nahid, F.A., Ongsakul, W., Manjiparambil, N.M., Singh, J.G., Roy, J., Mode decomposition-based short-term multi-step hybrid solar forecasting model for microgrid applications. *Electr. Eng.*, 106, 1–32, 3349–3380, Dec. 2023, https://doi.org/10.1007/s00202-023-02138-1.
40. Al Mamun, A., Sohel, M., Mohammad, N., Haque Sunny, M.S., Dipta, D.R., Hossain, E., A Comprehensive Review of the Load Forecasting Techniques Using Single and Hybrid Predictive Models. *IEEE Access*, 8, 134911–134939, 2020, doi: 10.1109/ACCESS.2020.3010702.
41. Zhang, J., Hodge, B.-M., Florita, A., Lu, S., Hamann, H.F., Banunarayanan, V., Metrics for Evaluating the Accuracy of Solar Power Forecasting. *3rd International Workshop on Integration of Solar Power into Power Systems*, vol. 17436, pp. 1–10, October, 2013, [Online]. Available: www.nrel.gov/publications.%5Cnhttp://www.osti.gov/bridge:%5Cnhttp://www.ntis.gov/help/ordermethods.aspx%5Cnhttp://www.nrel.gov/docs/fy14osti/60142.pdf.
42. Bistline, J., Blanford, G., Mai, T., Merrick, J., Modeling variable renewable energy and storage in the power sector. *Energy Policy*, 156, 112424, Sep. 2021, doi: 10.1016/j.enpol.2021.112424.
43. Nahid, F.A., Ongsakul, W., Manjiparambil, N.M., Short term multi-steps wind speed forecasting for carbon neutral microgrid by decomposition based hybrid model. *Energy Sustain. Dev.*, 73, 87–100, Apr. 2023, doi: 10.1016/j.esd.2023.01.016.
44. Ahmad, T. and Zhang, D., A critical review of comparative global historical energy consumption and future demand: The story told so far. *Energy Rep.*, 6, Elsevier Ltd, 1973–1991, Nov. 01, 2020, doi: 10.1016/j.egyr.2020.07.020.

45. Ghiani, E., Trevisan, R., Pilo, F., Performance Metrics For Renewable Energy Communities. *2022 AEIT International Annual Conference, AEIT 2022*, 2022, doi: 10.23919/AEIT56783.2022.9951739.
46. Zhang, J., *et al.*, A suite of metrics for assessing the performance of solar power forecasting. *Sol. Energy*, 111, 157–175, Jan. 2015, doi: 10.1016/J.SOLENER.2014.10.016.
47. Vallance, L., Charbonnier, B., Paul, N., Dubost, S., Blanc, P., Towards a standardized procedure to assess solar forecast accuracy: A new ramp and time alignment metric. *Sol. Energy*, 150, 408–422, Jul. 2017, doi: 10.1016/J.SOLENER.2017.04.064.
48. Chu, Y., *et al.*, Short-term metropolitan-scale electric load forecasting based on load decomposition and ensemble algorithms. *Energy Build.*, 225, 110343, Oct. 2020, doi: 10.1016/J.ENBUILD.2020.110343.
49. Liu, C., Li, M., Yu, Y., Wu, Z., Gong, H., Cheng, F., A Review of Multitemporal and Multispatial Scales Photovoltaic Forecasting Methods. *IEEE Access*, 10, 35073–35093, 2022, doi: 10.1109/ACCESS.2022.3162206.
50. Román-Portabales, A., López-Nores, M., Pazos-Arias, J.J., Systematic Review of Electricity Demand Forecast Using ANN-Based Machine Learning Algorithms. *Sensors*, 21, 13, 4544, Jul. 2021, doi: 10.3390/S21134544.
51. Verwiebe, P.A., Seim, S., Burges, S., Schulz, L., Müller-Kirchenbauer, J., Modeling Energy Demand—A Systematic Literature Review. *Energies*, 14, 23, 7859, Nov. 2021, doi: 10.3390/EN14237859.
52. Machado, E., Pinto, T., Guedes, V., Morais, H., Electrical Load Demand Forecasting Using Feed-Forward Neural Networks. *Energies*, 14, 22, 7644, Nov. 2021, doi: 10.3390/EN14227644.
53. Ahamed Nahid, F., Ongsakul, W., Madhu Manjiparambil, N., Very Short Term Wind Speed Forecasting Using Convolutional Long Short Term Memory Recurrent Neural Network. *Proceedings of the 2020 International Conference and Utility Exhibition on Energy, Environment and Climate Change, ICUE 2020*, Oct. 2020, doi: 10.1109/ICUE49301.2020.9307061.
54. Bisoi, R., Dash, P.K., Das, P.P., Short-term electricity price forecasting and classification in smart grids using optimized multikernel extreme learning machine. *Neural Comput. Appl.*, 32, 5, 1457–1480, Mar. 2020, doi: 10.1007/S00521-018-3652-5/FIGURES/11.
55. Ribeiro, M.H.D.M., Stefenon, S.F., de Lima, J.D., Nied, A., Mariani, V.C., dos S. Coelho, L., Electricity Price Forecasting Based on Self-Adaptive Decomposition and Heterogeneous Ensemble Learning. *Energies*, 13, 19, 5190, Oct. 2020, doi: 10.3390/EN13195190.
56. Abedinia, O., Lotfi, M., Bagheri, M., Sobhani, B., Shafie-Khah, M., Catalao, J.P.S., Improved EMD-Based Complex Prediction Model for Wind Power Forecasting. *IEEE Trans. Sustain. Energy*, 11, 4, 2790–2802, Oct. 2020, doi: 10.1109/TSTE.2020.2976038.

57. Alhussein, M., Aurangzeb, K., Haider, S.I., Hybrid CNN-LSTM Model for Short-Term Individual Household Load Forecasting. *IEEE Access*, 8, 180544–180557, 2020, doi: 10.1109/ACCESS.2020.3028281.
58. Moroff, N.U., Kurt, E., Kamphues, J., Machine Learning and Statistics: A Study for assessing innovative Demand Forecasting Models. *Procedia Comput. Sci.*, 180, 40–49, Jan. 2021, doi: 10.1016/J.PROCS.2021.01.127.
59. AlKandari, M. and Ahmad, I., Solar power generation forecasting using ensemble approach based on deep learning and statistical methods. *Appl. Comput. Inf.*, 20, 231–250, 2019, https://doi.org/10.1016/j.aci.2019.11.002.
60. Cui, M., Zhang, J., Feng, C., Florita, A.R., Sun, Y., Hodge, B.M., Characterizing and analyzing ramping events in wind power, solar power, load, and netload. *Renew. Energy*, 111, 227–244, Oct. 2017, doi: 10.1016/J.RENENE.2017.04.005.
61. Kuang, H., Jianhui, W., Zhang, H., Zhong, H., Zhu, G., A New Definition Method of Wind Power Ramp Sections. *IEEE Access*, 8, 202058–202066, 2020, doi: 10.1109/ACCESS.2020.3032547.
62. Han, L., Qiao, Y., Li, M., Shi, L., Wind Power Ramp Event Forecasting Based on Feature Extraction and Deep Learning. *Energies*, 13, 23, 6449, Dec. 2020, doi: 10.3390/EN13236449.
63. Cui, Y., *et al.*, Algorithm for identifying wind power ramp events via novel improved dynamic swinging door. *Renew. Energy*, 171, 542–556, Jun. 2021, doi: 10.1016/J.RENENE.2021.02.123.
64. Grandon, T.G., Schwenzer, J., Steens, T., Breuing, J., Electricity Demand Forecasting with Hybrid Statistical and Machine Learning Algorithms: Case Study of Ukraine, Apr. 2023, Accessed: Nov. 19, 2023. [Online]. Available: https://arxiv.org/abs/2304.05174v1.
65. Wazirali, R., Yaghoubi, E., Abujazar, M.S.S., Ahmad, R., Vakili, A.H., State-of-the-art review on energy and load forecasting in microgrids using artificial neural networks, machine learning, and deep learning techniques. *Electr. Power Syst. Res.*, 225, September, 109792, 2023, doi: 10.1016/j.epsr.2023.109792.
66. Al-Jaafreh, T.M. and Al-Odienat, A., The Solar Energy Forecasting by Pearson Correlation using Deep Learning Techniques. *Earth Sci. Hum. Constr.*, 2, 158–163, Aug. 2022, doi: 10.37394/232024.2022.2.19.
67. Wang, W., An, X., Wang, S., Short-term heat load prediction of heat exchange stations based on CEEMDAN-SE-LSTM. *Proceedings of the 34th Chinese Control and Decision Conference, CCDC 2022*, pp. 1668–1673, Aug. 2022, doi: 10.1109/CCDC55256.2022.10033494.
68. Nahid, F.A., Ongsakul, W., Nimal, M.M., Laopaiboon, T., Hybrid Neural Networks for Renewable Energy Forecasting: Solar and Wind Energy Forecasting Using LSTM and RNN, https://services.igi-global.com/resolvedoi/resolve.aspx?doi=10.4018/978-1-7998-9152-9.ch033, pp. 796–818, Jan. 1AD, doi: 10.4018/978-1-7998-9152-9.CH033.
69. Ahamed Nahid, F., Mahmud Chowdhury, H., Nayeem Jahangir, M., Lecturer, S., Solar Radiation Forecasting using Hybrid Convolutional Long Short

Term Memory Neural Network. *J. Res. Phys. Appl. Sci.*, 2, 2, 1–13, 2019, doi: 10.5281/zenodo.3768721.

70. Zhang, K., Cao, H., Thé, J., Yu, H., A hybrid model for multi-step coal price forecasting using decomposition technique and deep learning algorithms. *Appl. Energy*, 306, 118011, Jan. 2022, doi: 10.1016/J.APENERGY.2021.118011.
71. Sun, M., Feng, C., Zhang, J., Probabilistic solar power forecasting based on weather scenario generation. *Appl. Energy*, 266, 114823, May 2020, doi: 10.1016/J.APENERGY.2020.114823.
72. Ziel, F. and Steinert, R., Probabilistic mid- and long-term electricity price forecasting. *Renew. Sustain. Energy Rev.*, 94, Elsevier Ltd, 251–266, Oct. 01, 2018, doi: 10.1016/j.rser.2018.05.038.
73. Silva, J., Praça, I., Pinto, T., Vale, Z., Energy consumption forecasting using ensemble learning algorithms, in: *Advances in Intelligent Systems and Computing*, vol. 1004, pp. 5–13, 2020, doi: 10.1007/978-3-030-23946-6_1.
74. Hong, T., Gui, M., Baran, M.E., Willis, H.L., Modeling and forecasting hourly electric load by multiple linear regression with interactions. *IEEE PES General Meeting, PES 2010*, pp. 1–8, 2010, doi: 10.1109/PES.2010.5589959.
75. Saglam, M., Spataru, C., Karaman, O.A., Forecasting Electricity Demand in Turkey Using Optimization and Machine Learning Algorithms. *Energies (Basel)*, 16, 11, 4499, 2023, doi: 10.3390/en16114499.
76. Dai, Y. and Zhao, P., A hybrid load forecasting model based on support vector machine with intelligent methods for feature selection and parameter optimization. *Appl. Energy*, 279, 115332, May, 2020, doi: 10.1016/j.apenergy.2020.115332.
77. Duan, J., *et al.*, Short-term wind power forecasting using the hybrid model of improved variational mode decomposition and Correntropy Long Short-term memory neural network. *Energy*, 214, 118980, 2021, doi: 10.1016/j.energy.2020.118980.
78. Rawal, K. and Ahmad, A., A Comparative Analysis of Supervised Machine Learning Algorithms for Electricity Demand Forecasting. *ICPC2T 2022 - 2nd International Conference on Power, Control and Computing Technologies, Proceedings*, pp. 1–6, 2022, doi: 10.1109/ICPC2T53885.2022.9776960.
79. Vardhan, B.V.S., Khedkar, M., Srivastava, I., Thakre, P., Bokde, N.D., A Comparative Analysis of Hyperparameter Tuned Stochastic Short Term Load Forecasting for Power System Operator. *Energies (Basel)*, 16, 3, 1–21, 2023, doi: 10.3390/en16031243.
80. Inage, S., Development of an advection model for solar forecasting based on ground data. Part II: Verification of the forecasting model over a wide geographical area. *Sol. Energy*, 180, 257–276, Mar. 2019, doi: 10.1016/J.SOLENER.2018.12.068.
81. Yang, B., *et al.*, Classification and Summarization of Solar Irradiance and Power Forecasting Methods: A Thorough Review. *CSEE J. Power Energy Syst*, 9, 3, 978–995, May 2023, doi: 10.17775/CSEEJPES.2020.04930.

82. Zheng, J., Du, J., Wang, B., Klemeš, J.J., Liao, Q., Liang, Y., A hybrid framework for forecasting power generation of multiple renewable energy sources. *Renew. Sustain. Energy Rev.*, 172, 113046, Feb. 2023, doi: 10.1016/j.rser.2022.113046.
83. Cui, M., Zhang, J., Wang, Q., Krishnan, V., Hodge, B.M., A data-driven methodology for probabilistic wind power ramp forecasting. *IEEE Trans. Smart Grid*, 10, 2, 1326–1338, Mar. 2019, doi: 10.1109/TSG.2017.2763827.
84. Cui, M., Zhang, J., Florita, A., Hodge, B.-M., Ke, D., Sun, Y., Solar Power Ramp Events Detection Using an Optimized Swinging Door Algorithm: Preprint, Aug. 2015, doi: 10.1115/DETC2015-46849.
85. Dorado-Moreno, M., *et al.*, Multi-task learning for the prediction of wind power ramp events with deep neural networks. *Neural Netw.*, 123, 401–411, Mar. 2020, doi: 10.1016/J.NEUNET.2019.12.017.
86. Ela, E. and O'Malley, M., Scheduling and Pricing for Expected Ramp Capability in Real-Time Power Markets. *IEEE Trans. Power Syst.*, 31, 3, 1681–1691, May 2016, doi: 10.1109/TPWRS.2015.2461535.
87. Gochhait, S. and Sharma, D.K., Regression Model-Based Short-Term Load Forecasting for Load Despatch Centre. *J. Appl. Eng. Technol. Sci. (JAETS)*, 4, 2, 693–710, Jun. 2023, doi: 10.37385/JAETS.V4I2.1682.
88. Angeles, E.R., Badreldin, M.D., Santos, A.J.C., Ostia, C.F., Day Ahead Load Forecasting Model Using Gaussian Naïve Bayes with Ensemble Empirical Mode Decomposition. *TENSYMP 2021 - 2021 IEEE Region 10 Symposium*, pp. 1–6, 2021, doi: 10.1109/TENSYMP52854.2021.9550820.
89. Luo, J., Hong, T., Gao, Z., Fang, S.C., A robust support vector regression model for electric load forecasting. *Int. J. Forecast.*, 39, 2, 1005–1020, Apr. 2023, doi: 10.1016/J.IJFORECAST.2022.04.001.
90. Dong, Y., Ma, X., Fu, T., Electrical load forecasting: A deep learning approach based on K-nearest neighbors. *Appl. Soft Comput.*, 99, 106900, Feb. 2021, doi: 10.1016/J.ASOC.2020.106900.
91. Bansal, M., Goyal, A., Choudhary, A., A comparative analysis of K-Nearest Neighbor, Genetic, Support Vector Machine, Decision Tree, and Long Short Term Memory algorithms in machine learning. *Decis. Anal. J.*, 3, May, 100071, 2022, doi: 10.1016/j.dajour.2022.100071.
92. Gokce, M.M. and Duman, E., Performance Comparison of Simple Regression, Random Forest and XGBoost Algorithms for Forecasting Electricity Demand. *3rd International Informatics and Software Engineering Conference, IISEC 2022*, 2022, doi: 10.1109/IISEC56263.2022.9998213.
93. Yadav, A., Bareth, R., Kochar, M., Pazoki, M., Sehiemy, R.A.E., Gaussian process regression-based load forecasting model. *IET Gener. Transm. Distrib.*, 18, 5, 899–910, Mar. 2024, doi: 10.1049/GTD2.12926.

8

Forecasting Electricity Prices Using NNAR Approach: An Emerging Nation Experience

Sonal Gupta[1,2]* and Deepankar Chakrabarti[3]

[1]*School of Engineering, University of Edinburgh, Institute of Energy Systems, Edinburgh, United Kingdom*
[2]*Energy Cluster, School of Business, UPES, Dehradun, Uttarakhand, India*
[3]*Jaipuria Institute of Management Indore, Madhya Pradesh, India*

Abstract

This chapter attempts to examine the forecasting performance of the daily electricity prices of the Indian Energy Exchange using the neural network autoregression (NNAR) model. The main forte of applying NNAR lies in its capability to handle the intricacies and non-linearity of the seasonal data. The data is taken of all the five regions (north, south, east, west, and northeast) further divided into 13 bid areas (A1, A2, E1, E2, N1, N2, N3, S1, S2, S3, W1, W2, and W3) based on their different characteristics. For 25 days, medium-term forecasting is done to check the authenticity and robustness of the NNAR model, followed by a comparison with the actuals. The results demonstrated that NNAR is a robust model with fairly good error statistics values. This chapter study will help the policymakers make a calculated decision to introduce risk management strategies, i.e., electricity derivatives in the Indian power market, which have been efficiently implemented in the global electricity markets.

The chapter is divided into five parts. The first part (i.e., Introduction) gives insight into the Indian Electricity market. Part 2 reviews the literature related to forecasting horizons and the models applied in the Global and Indian Electricity markets. Part 3 reflects on the data series used and the research methodology applied. Part 4 analyses the result of the NNAR and the medium-term forecasting conducted, and Part 5 provides the conclusion.

**Corresponding author*: sgupta@ddn.upes.ac.in

Jai Govind Singh, Rupendra Kumar Pachauri and Sasidharan Sreedharan (eds.) Forecasting Methods for Renewable Power Generation, (219–242) © 2025 Scrivener Publishing LLC

Keywords: Electricity market, NNAR, forecasting, India, machine learning

8.1 Introduction

Globally, the electricity market is well-developed [1]. These markets have employed various tools to reduce risk caused by high volatility in the electricity markets, making power trading competitive and transparent.

Among the advanced electricity markets globally, such as the USA, Australia (NEMMCO New Zealand), and Europe (Nordpool), the European electricity market design is considered more transparent and flexible when launching new products with dynamic regulatory policy. It also has a low level of system integration and basic transmission pricing rules, making it easier for other developing markets to adapt. Both the European and the Indian electricity market follow a voluntary decentralized market model.

In India, a significant quantity of electricity is obtained by long-term power purchase agreements (PPAs) between generating electric utilities and distribution utilities. Long-term contracts are generally bilateral contracts with agreements ranging from 12 years to 25 years, with medium-term agreements ranging from 3 months to 3 years and the span of short-term contracts usually ranging between 15 min to 3 months [2]. Over a while, GenTime and Discoms have shifted from long-term to short-term or medium-term contracts (7–8 years) [3].

Because the market participants have shown a trend to move from long-term PPAs to medium-term contracts, a comparative analysis of both contracts was conducted, stating the advantages and disadvantages of entering into both contracts. The results are shown in Table 8.1 [4–7].

Table 8.1 clearly states that trading via short-term power purchase contracts is better than trading via long-term contracts for lesser prices, meeting immediate electricity requirements, and performing data analysis.

The Discoms would have saved around ₹3200 crores in FY17 [4] if it had bought electricity from the exchange instead of PPA if the exchange prices were lesser than the variable costs for several PPAs signed by the state Discoms.

India has three electricity exchanges: Indian Energy Exchange (IEX), Hindustan Power Exchange (HPX), and Power Exchange India Limited (PXIL). These exchanges span all five regions (northern, southern, western, eastern, and northeastern) of the Indian electricity market. They conduct anonymous and automatic bidding and have an efficient price-discovery mechanism with advanced risk-minimizing tools. These functions help in

Table 8.1 Advantages and disadvantages of short-term and long-term power contracts [8].

		Long-term power purchase agreements		Short-term power purchase agreements	
S. no.	**Particulars**	**Advantages**	**Disadvantages**	**Advantages**	**Disadvantages**
1	Price		Higher	Lower	
2	Availability of transmission corridor	First preference			Last preference
3	Price fluctuations	No fluctuation in tariff for a long period			Determined by market fundamentals [12]
4	Contingency management		Not met	Immediately done due to various products introduced	
5	Data Analysis		Forecasting of prices is difficult for long periods.	Easier employing statistical models	

better electricity management, thereby reducing the demand-supply gap. We have considered IEX prices in our research because it is the first and dominant player in the market.

IEX constitutes a meager of 3.92% of the total electricity traded [9] despite offering lower prices and flexible operations. It has further divided all the five regions mentioned above into 13 bid areas (shown in Figure 8.1 [10] to involve more participation of the power market players such as captive power plants, industry buyers, independent power producers, and state-private utilities in the trading process from each area.

IEX trades in various products such as the day-ahead market (DAM), term-ahead market (TAM), renewable energy certificates (REC), green DAM (G-DAM), and energy-saving certificates (ESCerts) with different years of commencement. The DAM contract is a physical delivery market offering bidding from buyers and sellers for delivery on the next Day.

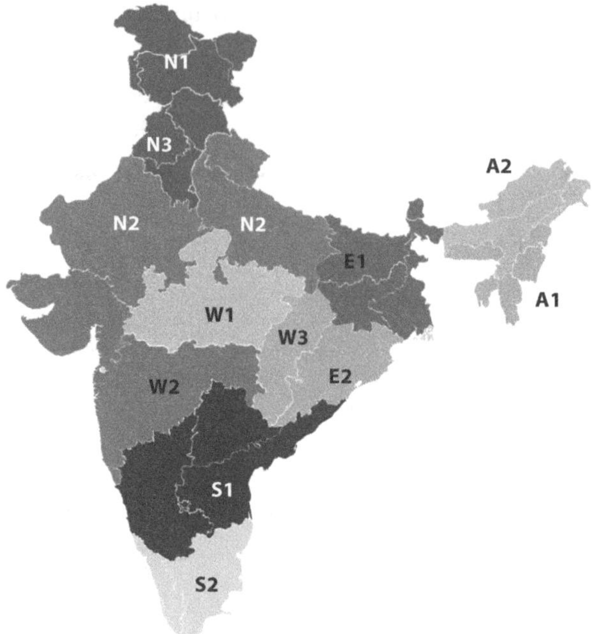

Figure 8.1 IEX bid areas [10].

The contract is entered into 1 day before the actual implementation of the agreement, i.e., if the price and volume for a contract are set on Friday, then it will be delivered on Saturday. DAM constitutes a significant part of IEX trading with 15-min time blocks in 24 h. Hence, forecasting the available data is more manageable than forecasting other products [2, 11].

Electricity prices are usually seasonal at the annual, monthly, weekly, daily, and intra-day levels across the globe [12]. Especially, if we observe the IEX prices of DAM, then each region has displayed uniqueness in its price fluctuation patterns [13]. Prices in the IEX's DAM reflect a high level of instability and extreme spurts ranging from as low as around 7% in the year 2013–2014 to as high as approximately 20% in 2008–2009 (Central Electricity Regulatory Commission, 2008–2017).

The business of each power market participant is affected tremendously due to extreme price movements. For example, the power generator's revenue, suppliers' costs, and the trader's option values are impacted when applying and valuing derivatives, making the application of risk management tools vital for managing energy commodity portfolios. Moreover, industry professionals, regulators, and other power market participants are concerned about the vast price spurts, which could also lead to regulatory failure [14]. All this makes it necessary to apply robust tools to forecast the prices. This would further reduce power losses, thereby leading to a profitable business.

Many researchers and academicians have developed various tools and algorithms for load and price forecasting. Whereas load forecasting is at a progressive stage with a mean absolute percentage error (MAPE) below 3%, price-forecasting techniques are still in their early stages of maturity [15, 16]. Based on the gaps in the literature in this chapter, there is a need to focus on the study of power forecasting techniques. Hence, the application of neural network (NN) autoregression (NNAR) model on the daily DAM prices of the IEX from August 2008 to August 2017 and 25 days of medium forecasting is done to check the robustness of the model.

8.2 Literature Review

As mentioned in [15], there are three types of forecasting horizons as follows:

(a.) Short-term electricity price forecasting (EPF) predicts the prices for a time span ranging from a few minutes to a few days ahead and are helpful in daily operational tasks.
(b.) Medium-term time horizons vary from days to months ahead and are used for balance sheet calculations, risk management, and derivatives pricing and
(c.) Long-term EPF is done for months, quarters, or years and is helpful for investment profitability analysis and planning, such as defining the sites for further generation and the fuel sources.

Studying the above forecasting time horizons plays an important role, and intense studies have been conducted on various forecasting methods, each with its strengths and weaknesses.

Various tools and econometric techniques have been implemented on power prices globally and in India. An intensive literature overview of all the methods has been done by [17–19].

According to them, all the methods were classified into five categories: multi-agent models, fundamental techniques such as parameter-rich fundamental models, and parsimonious structural models, each explained below in detail.

Multi-agent (multi-agent simulation, equilibrium, game theoretic) models are the models that firstly create a simulation between the various units of the system (generating units and companies) networking with each other, and, then, a price process is built by matching the demand and supply in the market. Nash-Cournot, supply function, strategic production cost, and agent-based models are highly used multi-agent models to forecast electricity [20]. Energy policy discussed the different methods to model the participant's bidding behavior in the electricity markets, including the Nash-Cournot framework and the supply function equilibrium approach. Bunn [21] found these methods to need to be more accurate. Also, Koritarov [22] and Weidlich and Veit [23] still need to find relevant agent-based wholesale electricity market models. Although they are flexible tools to study the behavior of the power market participants, some components are required, such as the players, their probable strategies, how they interrelate, and the set of payoffs. Hence, the model mainly considers qualitative issues rather than quantitative results, which is a complicated process.

Fundamental (structural) methods explain the effect of vital physical and economic factors on the power market prices. Parameter-rich fundamental models and parsimonious structural models are the most popular types. Borak and Weron [24] and Fleten *et al.* [25] also concluded that these tedious methods are suitable for medium-term price forecasts, not short-term ones. Also, these methods require various hypotheses about physical and economic relationships in the trading platform, which again requires a lot of adjustment in the data if there is a change in the market forces.

Reduced-form (quantitative and stochastic) models are the models that study the statistical features of the power prices of electricity prices over a period with the aim of derivatives evaluation and risk mitigation. Jump-diffusion and Markov regime switching (MRS) models are the most popular. Methods such as mean-reverting jump-diffusions [26] for short-term forecasting, i.e., the next day's hourly prices, are considered inferior results. Also, Bessec and Bouabdallah [27] have questioned the competence of MRS models in forecasting electricity prices. Kosater and Mosler [28] have suggested that the quantitative methods are helpful for medium-term forecasting of mean daily prices while studying the German EEX (European Energy Exchange) market. They have compared parameter-switching MRS specifications with the mean-reverting diffusion and found that the regime-switching models are more accurate for a forecasting period ranging from 30 to 80 days but unsuitable for a period lesser than that.

Statistical (econometric and technical analysis) approaches involve directly implementing econometric techniques for load and price forecasting. Similar day exponential smoothening, GARCH (generalized autoregressive conditional heteroskedasticity) family, and regression are some of the common types. Few authors have also classified the econometric models as technical analysis tools. Technical analysts check the future performance of an asset by looking at the various charts to identify patterns and give little attention to assessing an asset's intrinsic or fundamental value. While the efficacy and utility of technical analysis in financial markets are often probed, the methods have been tried and tested in the various power markets, where the results have been entirely accurate due to the seasonality in electricity price movements during regular, non-volatile periods [15].

Computational intelligence (artificial intelligence–based, non-parametric, and nonlinear statistical) techniques combine learning, evolution, and fuzziness elements to form methods that help measure volatility and forecasting.

Its types are feedforward NNs, recurrent NNs, fuzzy NNs, and support vector machines. The main forte of computational intelligence tools is their capability to handle the intricacies and non-linearity of the data series. Hence, they are considered one of the best methods of volatility and forecasting [27].

Out of the methods mentioned above, computational intelligence, i.e., artificial intelligence–based techniques, combines elements of learning, evolution, and fuzziness to form strategies that help measure volatility and forecast.

Feedforward NNs [29], recurrent NNs [30], fuzzy NNs [31], and support vector machines [32] are its types. The main forte of computational intelligence tools is their capability to handle the intricacies and non-linearity of the data series. Hence, they are considered one of the best methods of volatility and forecasting [27].

ANN models have been adopted by various researchers worldwide and in India. Zhang *et al.* [33] applied a cascaded NN structure on the New England electricity market clearing price m from 1 May 1999 to 30 November 2000. Guo *et al.* [34] used the cascaded architecture of multiple ANNs to forecast the New England market clearing price from 1 May 2001 to 31 October 2002. A three-layer ANN model has been used to predict short-term and long-term prices in Zhejiang Electricity Market [35]. Zhang and Luh [36] used a decoupled extended Kalman filter combined with ANN to predict the market clearing price of ISO New England's electricity market from 1 July 2000 to 31 May 2001. Amjady and Keynia [37] have applied probabilistic NN (PNN) and a hybrid neuro evolutionary system for forecasting prices of the PJM (Pennsylvania–New

Jersey–Maryland) electricity market for the year 2008. Tan *et al.* [38] used a hybrid method based on wavelet transform (WT), combined with the ARIMA and GARCH model, to forecast prices of the Spanish electricity market for 2002 and the PJM electricity market for 2006. Unsihuay-Vila *et al.* [39] used a hybrid approach of nonlinear chaotic dynamics and evolutionary strategy to predict electricity prices by taking the case of New England, Alberta, and Spain. Also, if we talk about the application of ANN (artificial NN) in the Indian sector, Nargale *et al.* [40] applied Feedforward for a period of five days in the IEX and PXIL from 13 April 2010 to 18 April 2010, which proved to be better than any other model and their MAPE comes out to be 17%.

All these methods have been widely applied in the global and Indian electricity markets. However, NNs for forecasting have yet to be explored in the IEX's bid areas, and that too for a more extended period.

8.3 Data and Methodology

The current chapter forecasts the India Energy Exchange DAM's daily spot prices of all the 13 bid areas (A1, A2, E1, E2, N1, N2, N3, S1, S2, S3, W1, W2, and W3) from August 2008 to August 2017. We have taken 3,312 daily spot electricity price data for A1, E1, N1, S1, W1, and W3. Also, 3,294 and 3,143 daily spot prices of N3 and S2 are taken, respectively. It has been observed that the data of E1 and E2, A1 and A2, N2 and N1, S2 and S3, and W1 and W2 are the same; hence, the data of E1, A1, N1, N3, S1, S2, W1, and W3 are taken into consideration. The data was taken from the India Energy Exchange (IEX site). The descriptive statistics are shown in Table 8.2. Volatility in the S2 is maximum for the investigation period, stating high standard deviations, and lowest in the A1 and W1 with low standard deviations). The cause of this unusual movement in the prices of the southern region is due to transmission grid congestion. The positively skewed data shows the high probability of prices going upward for all regions. Leptokurtic data proposes high price movements in all the regions [8, 14].

The current chapter uses the ANN method, which allows the modeling of complex nonlinear relationships between the input and output variables. The researcher has applied the NAAR model among the various models used under the ANN methods. Here, the lagged values of a time series are used as input to the model, and the output is the predicted values of the time series. The benefit of using NNAR is that it does not restrict its parameters to ensure stationarity. The model $NNAR(p,P,k)_{[m]}$ proposed by Hyndman and Athanasopoulos [41] is used due to seasonality in the

Table 8.2 Descriptive statistics for bid area's electricity prices [8, 13, 14].

Bid area/ statistics	Mean (μ)	Median (μ1/2)	Maximum	Minimum	Std. dev. (σ)	Skewness	Kurtosis
A1	3,411	2,817	13,843	467	1,789	2.27	8.83
E1	3,368	2,774	13,843	467	1,801	2.29	8.9
N1	3,558	2,940	15,448	1016	1,868	2.34	9.23
N3	3,640	2,981	15,448	1016	1,906	2.26	8.61
S1	4,801	4,251	17,533	467	2,312	1.30	5.17
S2	5,111	4,571	17,533	467	2,507	1.06	4.14
W1	3,378	2,791	13,843	467	1,790	2.33	9.05
W3	3,313	2,700	13,843	467	1,818	2.30	8.87

electricity prices. Here, the value of p states the value of ARIMA (autoregressive integrated moving average), and P states the presence or absence of seasonality; k states the number of neurons in the hidden layers.

In the absence of any hidden layer, the $NNAR(p,P,0)_{[m]}$ is equivalent to the SARIMA model denoted as an $ARIMA(p,0,0)\ (P,0,0)_{[m]}$.

Faraway and Chatfield [42] have suggested that an excellent NN model should combine conventional modeling skills and time series. Here, the researcher has taken m = 365 as the research involves the study of daily prices.

The NNAR model is a feedforward NN model with a linear combination function and an activation function (AF). The equation used to explain the linear combination function with neuron j is shown as follows:

$$net_j = \sum_i w_{ij} y_{ij} \tag{8.1}$$

The AF is a sigmoid function defined as

$$f(y) = \frac{1}{1+e^{-y}} \tag{8.2}$$

The inputs of the NNs are combined using a weighted linear function, and the result of the combination is then modified into the nonlinear sigmoid AF, which becomes the input for the next layer. This process reduces

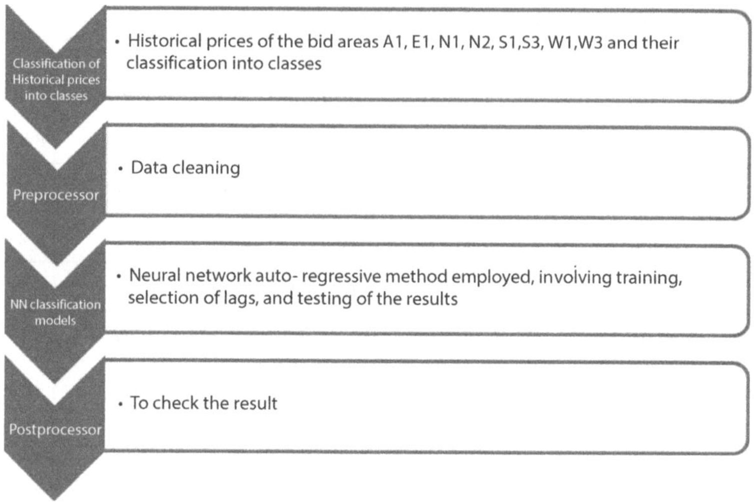

Figure 8.2 Detailed process of neural network autoregressive method.

the effect of extreme input values, making the NN a robust method for outliers. The weights taken at the beginning are random values, further updated using the observed data. The network is trained several times using different random starting points, and the results are averaged [41] to reduce the element of randomness in the predictions done by the method.

In the current data, the researcher trained the model with data from 1 August 2008 to 31 December 2016 (90% of the data) and tested it from 1 January 2017 to 31 August 2017. $NNAR(30,1,16)_{[365]}$ is applied using ARIMA (30,0,0), with one seasonality and 16 hidden layers; 365 states the number of days. Hence, the researcher used a 31-16-1 model wherein 31 daily inputs with 16 hidden layers with 529 weight options were applied to get a single output.

A stepwise approach to applying the NN autoregressive method of ANN is explained in Figure 8.2 [43].

8.4 Data Analysis

While applying the model, the researcher trained the model with data from 1 August 2008 to 31 December 2016 (90% of the data) and tested from 1 January 2017 to 31 August 2017. The NNAR(p,P,K) model is used wherein the value of p states the value of ARIMA (autoregression integrated moving

Table 8.3 Results of NNAR(30,1,16)$_{[365]}$ model.

Bid area	ME	RMSE	MAE	MPE	MAPE	MASE	ACF1	Dataset
A1	–0.99	261.86	188.08	–1.09	6.79	0.55	–0.01	Training
A1	205.65	255.49	225.79	8.29	**9.25**	0.67		Test
E1	–1.49	245.50	178.15	–1.19	6.83	0.55	–0.01	Training
E1	198.99	246.07	212.58	8.01	**8.65**	0.65		Test
N1	0.02	263.54	188.91	–0.97	6.59	0.57	–0.01	Training
N1	469.69	632.70	488.24	15.22	**16.09**	1.46		Test
N3	–0.51	290.97	199.92	–1.14	6.83	0.56	0.01	Training
N3	372.75	531.59	393.79	11.88	**12.86**	1.10		Test
S1	–2.46	468.59	302.61	–1.31	7.18	0.67	–0.02	Training
S1	167.28	348.93	297.22	4.90	**10.32**	0.66		Test
S2	–0.77	709.37	440.43	–2.05	9.27	0.76	0.01	Training
S2	–134.82	260.15	226.15	–5.39	**8.18**	0.39		Test
W1	0.04	227.21	167.93	–0.82	6.18	0.54	–0.01	Training
W1	212.89	259.93	226.46	8.61	**9.25**	0.72		Test
W3	–1.50	247.50	180.37	–1.16	6.94	0.55	–0.01	Training
W3	209.57	262.13	230.71	8.43	**9.44**	0.71		Test

average), and P states the presence of seasonality or not; K states the number of hidden layers. Because NNs usually have no underlying statistical model, AIC/BIC (Akaike information criterion/Bayesian information criterion) does not make sense.

NNAR(30,1,16)$_{[365]}$ using ARIMA (30,0,0) with one seasonality, 16 hidden layers, and 365 number of days is applied wherein 31 daily inputs with 16 hidden layers, 529 weight options, and 68,340 as $\sigma 2$ are used to get a single output. The results of the NNAR model are as follows (Table 8.3):

The NNAR model was run on R, which reflects that there has been a rise in the MAPE between the trained and the tested data for A1, E1, N1, N3, S1, W1, and W3 by approximately 36%, 27%, 144%, 88%, 44%, 50%, and 36%, respectively, and a decrease in S2 by 11%. The extreme rise in the

Table 8.4 Twenty-five days of forecasted values of NNAR(30,1,16)$_{[365]}$ model for A1.

Forecasted values	Lo95	Hi95	Actual	Deviation from the actuals	Period
2,880.96	2,409.12	3,371.19	2,973.4	3.11	01/09/2017
2,918.54	2,301.43	3,488.53	2,616.24	−11.55	02/09/2017
2,646.33	1,992.63	3,351.17	2,658.73	0.47	03/09/2017
3,195.89	2,439.1	3,942.19	3,247.07	1.58	04/09/2017
3,075.73	2,364.21	3,911.34	3,504.84	12.24	05/09/2017
3,041.93	2,348.64	3,900.77	3,731	18.47	06/09/2017
2,884.33	2,133.03	3,797.62	3,949.98	26.98	07/09/2017
2,737.1	2,156.91	3,573.02	4,058.37	32.56	08/09/2017
2,983.55	2,349.42	3,689.11	4,663.61	36.02	09/09/2017
2,641.01	2,004.41	3,422.07	4,397.5	39.94	10/09/2017
2,936.56	2,269.07	3,788.97	5,084.37	42.24	11/09/2017
2,893.09	2,199.76	3,796.25	5,229.85	44.68	12/09/2017
2,963.16	2,317.76	3,803.82	5,138.04	42.33	13/09/2017
3,098.16	2,332.99	4,078.13	5,049.07	38.64	14/09/2017
3,255.89	2,293.45	4,428.77	5,421.67	39.95	15/09/2017
3,199.88	2,341.1	4,369.88	5,532.64	42.16	16/09/2017
2,743.71	1,918.85	3,993.04	3,938.42	30.33	17/09/2017
2,998.44	2,183.78	4,032.68	5,032.88	40.42	18/09/2017
2,811.31	2,012.86	3,958.27	3,856.84	27.11	19/09/2017
2,870.71	2,043.48	4,037.81	4,333.42	33.75	20/09/2017
2,834.73	1,998.14	4,126.11	4,293.64	33.98	21/09/2017
2,758.77	1,944.25	4,274.15	3,557.38	22.45	22/09/2017
2,797.86	2,000.62	4,209.77	3,216.63	13.02	23/09/2017
2,578.66	1,836.12	3,959.41	3,014.56	14.46	24/09/2017
2,881.35	2,105.74	4,129.67	3,699.93	22.12	25/09/2017

Table 8.5 Twenty-five days of forecasted values of NNAR(30,1,16)$_{[365]}$ model for E1.

Forecasted values	Lo95	Hi95	Actual	Deviation from the actuals	Period
2,880.96	2,409.12	3,371.19	2,973.4	3.11	01/09/2017
2,918.54	2,301.43	3,488.53	2,616.24	−11.55	02/09/2017
2,646.33	1,992.63	3,351.17	2,658.73	0.47	03/09/2017
3,195.89	2,439.1	3,942.19	3247.07	1.58	04/09/2017
3,075.73	2,364.21	3,911.34	3504.84	12.24	05/09/2017
3,041.93	2,348.64	3,900.77	3,731	18.47	06/09/2017
2,884.33	2,133.03	3,797.62	3,949.98	26.98	07/09/2017
2,737.1	2,156.91	3,573.02	4,058.37	32.56	08/09/2017
2,983.55	2,349.42	3,689.11	4,663.61	36.02	09/09/2017
2,641.01	2,004.41	3,422.07	4,397.5	39.94	10/09/2017
2,936.56	2,269.07	3,788.97	5,084.37	42.24	11/09/2017
2,893.09	2,199.76	3,796.25	5,229.85	44.68	12/09/2017
2,963.16	2,317.76	3,803.82	5,138.04	42.33	13/09/2017
3,098.16	2,332.99	4,078.13	5,049.07	38.64	14/09/2017
3,255.89	2,293.45	4,428.77	5,421.67	39.95	15/09/2017
3,199.88	2,341.1	4,369.88	5,532.64	42.16	16/09/2017
2,743.71	1,918.85	3,993.04	3,938.42	30.33	17/09/2017
2,998.44	2,183.78	4,032.68	5,032.88	40.42	18/09/2017
2,811.31	2,012.86	3,958.27	3,856.84	27.11	19/09/2017
2,870.71	2,043.48	4,037.81	4,333.42	33.75	20/09/2017
2,834.73	1,998.14	4,126.11	4,293.64	33.98	21/09/2017
2,758.77	1,944.25	4274.15	3,557.38	22.45	22/09/2017
2,797.86	2,000.62	4,209.77	3,216.63	13.02	23/09/2017
2,578.66	1,836.12	3,959.41	3,014.56	14.46	24/09/2017
2,881.35	2,105.74	4,129.67	3,699.93	22.12	25/09/2017

Table 8.6 Twenty-five days of forecasted values of NNAR(30,1,16)$_{[365]}$ model for N1.

Forecasted values	Lo95	Hi95	Actual	Deviation from the actuals	Period
2,926.3	2,499.78	3,441.03	2,973.4	1.58	01/09/2017
2,988.74	2,401.71	3,646.44	2,616.24	−14.24	02/09/2017
2,567.04	1,952.38	3,320.7	2,658.73	3.45	03/09/2017
3,078.21	2,404.64	3,890.18	3,247.07	5.2	04/09/2017
2,875.01	2,144.59	3,719.33	3,504.84	17.97	05/09/2017
2,891.32	2,164.92	3,859.72	3,731	22.51	06/09/2017
2,984.51	2,254.95	4,043.59	3,949.98	24.44	07/09/2017
2,826.22	2,146.41	3,788.32	4,058.37	30.36	08/09/2017
2,925.34	2,272.03	3,811.28	4,663.61	37.27	09/09/2017
2,557.41	1,913.38	3,451.26	4,397.5	41.84	10/09/2017
2,928.42	2,220.37	3,937.92	5,084.37	42.4	11/09/2017
2,836.85	2,138.9	3,899.83	5,229.85	45.76	12/09/2017
2,894.03	2,191.99	3,997.73	5,138.04	43.67	13/09/2017
2,844.15	2,106.55	4,171.22	5,049.07	43.67	14/09/2017
2,908.53	2,179.07	4,387.22	5,421.67	46.35	15/09/2017
2,949.97	2,253.43	4,451.4	5,532.64	46.68	16/09/2017
2,470.17	1,766.43	4,056.39	3,938.42	37.28	17/09/2017
2,695.19	2,038.71	4,092.91	5,032.88	46.45	18/09/2017
2,534.21	1,785.69	3,921.73	3,856.84	34.29	19/09/2017
2,636.12	1,889.19	4,127.53	4,333.42	39.17	20/09/2017
2,669.59	1,919.74	4,237.88	4,293.64	37.82	21/09/2017
2656.86	1,833.05	4,358.31	3,557.38	25.31	22/09/2017
2,638.05	1,828.03	4,325.26	3,216.63	17.99	23/09/2017
2,331.69	1,691.97	4,002.23	3,014.56	22.65	24/09/2017
2,607.45	1,891.76	4,246.61	3,699.93	29.53	25/09/2017

Table 8.7 Twenty-five days of forecasted values of NNAR(30,1,16)$_{[365]}$ model for N2.

2,822.71	2,331.1	3,321.48	2,973.4	5.07	01/09/2017
2,877.18	2,267.54	3,495.13	2,616.24	−9.97	02/09/2017
2,552.51	1,810.8	3,283.61	2,658.73	4	03/09/2017
3,007.88	2,325.12	3,795.72	3,247.07	7.37	04/09/2017
2,951.82	2,225.34	3,779.99	3,504.84	15.78	05/09/2017
3,006.89	2,263.69	3,912.91	3,731	19.41	06/09/2017
2996.73	2,190.32	3,951.44	3,949.98	24.13	07/09/2017
2,916.93	2,104.8	3,844.22	4,058.37	28.13	08/09/2017
2,875.94	2,086.82	3,734.79	4,663.61	38.33	09/09/2017
2,572.66	1,838.26	3,450.83	4,397.5	41.5	10/09/2017
3,004.09	2,233.47	3,933.69	5,084.37	40.92	11/09/2017
2,977.27	2,203.71	3,929.26	5,229.85	43.07	12/09/2017
3,083.96	2,295.49	3,992.88	5,138.04	39.98	13/09/2017
3,050.95	2,242.01	4,019.6	5,049.07	39.57	14/09/2017
3119.98	2,186.36	4,252	5,460.05	42.86	15/09/2017
3150.71	2,180.44	4,422.13	5,532.64	43.05	16/09/2017
2,626.95	1,717.73	3,962.22	3,938.42	33.3	17/09/2017
2,836.15	1,967.02	4,175.39	5,032.88	43.65	18/09/2017
2,752.97	1,914.03	4,046.07	3,856.84	28.62	19/09/2017
2,809.71	1,933.29	4,293.56	4,333.42	35.16	20/09/2017
2,807.92	1,862.17	4,369.36	4,293.64	34.6	21/09/2017
2,810.31	1,817.85	4,387.91	3,557.38	21	22/09/2017
2,853.45	1,858.13	4,335.61	3,216.63	11.29	23/09/2017
2,589.5	1,634.85	4,040.76	3,014.56	14.1	24/09/2017
2,791.26	1,835.77	4,375.31	3,699.93	24.56	25/09/2017

Table 8.8 Twenty-five days of forecasted values of NNAR(30,1,16)$_{[365]}$ model for S1.

Forecasted values	Lo95	Hi95	Actual	Deviation from the actuals	Period
2,866.58	2,331.78	3,398.46	2,973.4	3.59	01/09/2017
3,004.88	2,352.55	3,690.29	2,616.24	−14.85	02/09/2017
2,722.63	1,888.63	3,529.82	2,658.73	−2.4	03/09/2017
3,063.65	2,179.14	3,895.51	3,247.07	5.65	04/09/2017
3,048	2,234.16	3,942.92	3,504.84	13.03	05/09/2017
3,094.26	2,189.48	4,147.7	3,731	17.07	06/09/2017
3,072.55	2,204.67	4,148.29	3,949.98	22.21	07/09/2017
2,942.73	1,995.94	4,099.27	4,058.37	27.49	08/09/2017
2,844.18	1,990.36	3,906.41	4,663.61	39.01	09/09/2017
2,544.59	1,733.58	3,535.82	4,397.5	42.14	10/09/2017
2,921.13	2,147.86	4,041.13	5,084.37	42.55	11/09/2017
2,832.76	1,997.06	3,934.16	5,229.85	45.83	12/09/2017
2,993.73	2,065.15	4,171.02	5,138.04	41.73	13/09/2017
2,934.01	1,998.78	4,254.87	5,049.07	41.89	14/09/2017
3,079.03	2,177.05	4,471.84	5,460.05	43.61	15/09/2017
3,151.68	2,243.15	4,687.37	5,532.64	43.03	16/09/2017
2,716.73	1,856.37	4,402.22	3,938.42	31.02	17/09/2017
2,866.11	1,989.55	4,488.31	5,032.88	43.05	18/09/2017
2,788.73	1,905.83	4,313.6	3,856.84	27.69	19/09/2017
2,832.11	1,887	4,525.51	4,333.42	34.64	20/09/2017
2,806.85	1,844.19	4,587.7	4,293.64	34.63	21/09/2017
2,779.18	1,718.8	4,610.64	3,557.38	21.88	22/09/2017
2,776.61	1,792.16	4,630.43	3,216.63	13.68	23/09/2017
2,565.32	1,653.5	4,534.23	3,014.56	14.9	24/09/2017
2,714.12	1,811.08	4,685.81	3,699.93	26.64	25/09/2017

Table 8.9 Twenty-five days of forecasted values of $NNAR(30,1,16)_{[365]}$ model for S2.

Forecasted values	Lo95	Hi95	Actual	Deviation from the actuals	Period
2,787.36	1,889.09	3,707.74	2,973.4	6.26	01/09/2017
3,012.65	1,804.38	4,200.42	2,616.24	−15.15	02/09/2017
3,064.55	1,742.32	4,379.28	2,658.73	−15.26	03/09/2017
3,123.05	1,837.83	4,460.27	3,247.07	3.82	04/09/2017
3,086.11	1,820.42	4,526.76	3,504.84	11.95	05/09/2017
3,001.02	1,726.07	4,402.8	3,731	19.57	06/09/2017
2,991.01	1,670.95	4,544.86	3,949.98	24.28	07/09/2017
3,018.12	1,594.89	4,721.17	4,058.37	25.63	08/09/2017
3,213.35	1,823.58	4,948.62	4,663.61	31.1	09/09/2017
3,044.06	1,664.25	4,807.26	4,397.5	30.78	10/09/2017
2,906.41	1,497.19	4,763.96	5,084.37	42.84	11/09/2017
2,770.26	1,461.79	4,738.63	5,229.85	47.03	12/09/2017
2,854.32	1,436.89	4,714.13	5,138.04	44.45	13/09/2017
3,007.04	1,656.18	5,010.18	5,049.07	40.44	14/09/2017
3,104.99	1,673.96	5,138.41	5,421.67	42.73	15/09/2017
3,118.05	1,608.2	5,211.78	5,532.64	43.64	16/09/2017
2,902.22	1,443.24	5,002.91	3,938.42	26.31	17/09/2017
2,922.35	1,549.06	5,228.6	5,032.88	41.93	18/09/2017
2,792.38	1,383.52	5,095.12	3,856.84	27.6	19/09/2017
2,902.7	1,572.68	5,240.77	4,333.42	33.02	20/09/2017
2,902.93	1,540.75	5,161.48	4,293.64	32.39	21/09/2017
2,850.21	1,456.88	5,411.66	3,557.38	19.88	22/09/2017
2,850.35	1,501.56	5,507.36	3,216.63	11.39	23/09/2017
2,743.46	1,301.46	5,435.76	3,014.56	8.99	24/09/2017
2,795.54	1,557.8	5,512.51	3,699.93	24.44	25/09/2017

Table 8.10 Twenty-five days of forecasted values of NNAR(30,1,16)$_{[365]}$ model for W1.

Forecasted values	Lo95	Hi95	Actual	Deviation from the actuals	Period
2,769.19	1,461.32	4,024.4	2,973.4	6.87	01/09/2017
2,902.39	1,311.33	4,459.58	2,616.24	−10.94	02/09/2017
2,934.55	1,175.67	4,898.01	2,658.73	−10.37	03/09/2017
2,840.57	973.1	5,133.03	3,247.07	12.52	04/09/2017
2,852.22	986.88	5,249.56	3,504.84	18.62	05/09/2017
2,910.78	900.98	5,587.05	3,731	21.98	06/09/2017
2,858.73	825.46	5,664.53	3,949.98	27.63	07/09/2017
2,780.27	966.44	5,767.22	4,058.37	31.49	08/09/2017
2,761.34	983.99	6,117.28	4,663.61	40.79	09/09/2017
2,581.51	952.89	5,999.9	4,397.5	41.3	10/09/2017
2,657.43	975.09	6,180.33	5,084.37	47.73	11/09/2017
2,628.15	1,137.12	6,243.68	5,229.85	49.75	12/09/2017
2,583.4	1,258.95	6,356.97	5,138.04	49.72	13/09/2017
2,598.43	1,240.48	6,357.17	5,049.07	48.54	14/09/2017
2,594.5	1,349.72	6,407.08	5,421.67	52.15	15/09/2017
2,573.84	1,310.3	6,582.64	5,532.64	53.48	16/09/2017
2,500.27	1,182.61	6,925.14	3,938.42	36.52	17/09/2017
2,552.18	1,410.62	6,843.96	5,032.88	49.29	18/09/2017
2,481.37	1,335.64	6,969.11	3,856.84	35.66	19/09/2017
2,485.49	1,304.41	6,998.95	4,333.42	42.64	20/09/2017
2,479.92	1,388.94	7,239.74	4,293.64	42.24	21/09/2017
2,455.67	1,444.46	7,297.8	3,557.38	30.97	22/09/2017
2,476.73	1,401.41	7,227.97	3,216.63	23	23/09/2017
2,467.62	1,533.55	7,190.54	3,014.56	18.14	24/09/2017
2,479.11	1,638.27	7,219.24	3,699.93	33	25/09/2017

Table 8.11 Twenty-five days of forecasted values of NNAR(30,1,16)$_{[365]}$ model for W3.

Forecasted values	Lo95	Hi95	Actual	Deviation from the actuals	Period
2,843.86	2,413.38	3,310.6	2,973.4	4.36	01/09/2017
2,944.8	2,441.05	3,569.34	2,616.24	−12.56	02/09/2017
2,658.59	2,087.33	3,361.67	2,658.73	0.01	03/09/2017
3,329.69	2,711.36	4,071.84	3,247.07	−2.54	04/09/2017
3,059.86	2,426.59	3,954.29	3,504.84	12.7	05/09/2017
2,997	2,332.9	3,920.16	3,731	19.67	06/09/2017
3,044.54	2,407.74	4,059.58	3,949.98	22.92	07/09/2017
2,913.07	2,287	4,021.97	4,058.37	28.22	08/09/2017
2,952.08	2,370.58	3,932.69	4,663.61	36.7	09/09/2017
2,669.69	2,084.25	3,546.69	4,397.5	39.29	10/09/2017
3,122.09	2,531.77	4,080.73	5,084.37	38.59	11/09/2017
3,022.6	2,326.4	4,049.58	5,229.85	42.2	12/09/2017
3,128.84	2,469.94	4,226.37	5,138.04	39.1	13/09/2017
3,265.71	2,569.43	4,570.44	5,049.07	35.32	14/09/2017
3,242.3	2,469.02	4,756.43	5,421.67	40.2	15/09/2017
3,265.68	2,534.83	4,740.12	5,532.64	40.97	16/09/2017
2,729.13	2,081.1	4,153.43	3,938.42	30.7	17/09/2017
2,959.33	2,368.92	4,371.63	5,032.88	41.2	18/09/2017
2,713.86	2,147.01	4,114.67	3,856.84	29.64	19/09/2017
2,790.7	2,227.43	4,128.36	4,333.42	35.6	20/09/2017
2,936.21	2,270.94	4,478.51	4,293.64	31.62	21/09/2017
2,874.76	2,234.93	4,442.8	3,557.38	19.19	22/09/2017
2,824.31	2,177.21	4,403.06	3,216.63	12.2	23/09/2017
2,536.45	1,908.97	4,103.69	3,014.56	15.86	24/09/2017
2,902.7	2,282.53	4,417.2	3,699.93	21.55	25/09/2017

error in N1 and N3 might be due to overfitting or environmental changes. The northern region of India consists of Jammu and Kashmir, Himachal Pradesh, Chandigarh, Haryana, Uttar Pradesh, Uttaranchal, Rajasthan, Delhi, and Punjab. According to the Annual Report 2017–2018 of the Central Electricity Authority, the percentage of energy not supplied in the northern region is higher compared to the other regions, which could also have been a reason for the higher percentage increase in the error in the northern region.

To check the authenticity and robustness of the NNAR model, forecasting electricity prices for the next 25 days was conducted, as illustrated in Tables 8.4 to 8.11.

In Tables 8.4 to 8.11 (authors' own analysis), it has been observed that the point forecast values of NNAR lie between Lo95 and Hi95, i.e., the hypothesis of the prediction value at 95% confidence interval states that NNAR is a robust model.

8.5 Conclusion

The IEX DAM is subjected to high variations, which have led to massive losses. Despite posing a lesser risk and better efficiency and liquidity, the volume traded through it is still low compared to other forms of trading, i.e., long-term PPAs and short-term contracts. Because much money is involved in the power trading business, applying forecasting techniques is inevitable in adopting appropriate risk management techniques. The current chapter aims to employ the NNAR model to forecast the spot prices of IEX's DAM for the next 25 days.

The literature shows that the industry needs a rule or standard to measure forecasting performance. Still, the Indian electricity DAM prices can be forecasted precisely for all the bid areas with MAPE statistic values hovering around 8%–10%, which is relatively good except for N1 and N2, with MAPE statistics of around 16% and 12%.

The current study only includes the application of the models on the data set from 1 August 2008 to 31 August 2017, although the IEX commenced its operations on 27 June 2008. The data from the first two months has been excluded from the analysis, as, in the beginning, the power market participants observed that the market exhibits unusual behavior.

The researcher used daily data from the Indian Electricity DAM series, which studied spike behavior for 1 day each. In contrast, the electricity market in India has 15-min time blocks, which are averaged to get the daily prices. Each day, consisting of 96-time blocks, has substantial variations yet to be explored, primarily when the Indian electricity market works on the time of the day concept.

While applying the NNAR model, the number of iterations and neurons in the hidden layer has been selected based on ACF (autocorrelation function) plots to avoid overfitting. The authors have applied a single method to forecast electricity prices. In future, different techniques can be used alone or in combination to get better results.

References

1. Bandyopadhayay, A., Roy, S., Ghosh, D., Forecasting day-ahead price of electricity – a dynamic regression approach. *Int. J. Bus. Excell.*, 6, 5, 584–604, 2013.
2. Sinha, P. and Mathur, K., *Empirical Analysis of Developments in the Day Ahead Electricity Markets in India*, MPRA Paper No. 72969, University Library of Munich, Germany, 2016, available at: https://mpra.ub.uni-muenchen.de/72969/.
3. Kujur, A., Why both discoms and power generation cos prefer short-term contracts, 2017, June 22, http://www.moneycontrol.com/news/business/economy/why-both-discoms-and-power-generation-cos-prefer-short-term-contracts-2309231.html.
4. Sinha, A. and Jain, P., Try Short-Term Power Trading, Business Standard, 2017, http://www.bain.com/publications/articles/try-short-term-power-trading-business-standard.aspx.
5. Dewan, S., Tying up Long Term PPAs by Discoms:How much is too much?, Energetica, 2016, February, http://www.energeticaindia.net/download.php?seccion=articles&archivo.pdf.
6. Central Electricity Regulatory Commission, Grant of Connectivity, Long-term Access and Medium-term Access to the inter-State Transmission and related matters, 2009, http://www.cercind.gov.in/Regulations/Final-version-of-Long-term-and-Medium-term -Regulations-2009.pdf.
7. Jog, S., Discoms keen on shorter power purchase pacts instead of 25-year contracts, Business Standard, 2016, May 10, http://www.businessstandard.com/article/economy-policy/discoms-keen-on-shorter-powerpurchase-pacts-instead-of-25-year-contracts-116050900133_1.html.
8. Gupta, S., Chakrabarti, D., Roy, H., Indian Energy Exchange Day- ahead market: The issue of stationarity. *Int. J. Emerg. Technol. Innov. Res.*, 5, 8, 632–638, 2018.
9. Central Electricity Regulatory Commission, Monthly Report on Short-term Transactions of Electricity in India, 2008-2017, http://www.cercind.gov. in/report_MM.html.

10. India Energy Exchange, Bid areas, https://www.iexindia.com/bidareas.aspx?id=qZpXzaO5CaM%3d&mid=IT8b%2bZM5cBA%3d.
11. Ahmad, F. and Alam, M.S., Assessment of power exchange based electricity market in India. *Energy Strateg. Rev.*, 23, 163–177, 2019, 23(December 2018), https://doi.org/10.1016/j.esr.2018.12.012.
12. Girish, G.P. and Vijayalakshmi, S., Determinants of Electricity Price in Competitive Power Market. *Int. J. Bus. Manag.*, 8, 21, 70–75, 2013, https://doi.org/10.5539/ijbm.v8n21p70.
13. Gupta, S. and Bhat, M., Electricity Price spirals now or later: An emerging market experience. *Electr. J.*, 36, 5, 107272, 2023, ISSN: 1040-6190.
14. Gupta, S., Chakrabarti, D., Kumar, R., Empirical analysis of electricity exchange-traded market prices: Evidence from India, in: *Oil, Gas & Energy Quarterly*, pp. 591–614, Matthew Bender, New York, 2023, ISBN 9780820515205.
15. Weron, R., Electricity price forecasting: A review of the state-of-the-art with a look into the future. *Int. J. Forecast.*, 30, 4, 1030–1081, 2014, https://doi.org/10.1016/j.ijforecast.2014.08.008.
16. Abdel-Aal, R.E., Modelling and forecasting electric daily peak loads using abductive networks. *Int. J. Electr. Power Energy Syst.*, 28, 133–141, 2006, 10.1016/j.ijepes.2005.11.006.
17. Aggarwal, S.K., Saini, L.M., Kumar, A., Electrical Power and Energy Systems Electricity price forecasting in deregulated markets: A review and evaluation. *Int. J. Electr. Power Energy Syst.*, 31, 1, 13–22, 2009, https://doi.org/10.1016/j.ijepes.2008.09.003.
18. Weron, R., *Modeling and Forecasting Electricity Loads and Prices: a Statistical Approach*, Wiley Finance Publication, John Wiley and Sons headquarters-Hoboken, New Jersey, 2006.
19. Girish, G.P., Modeling and Forecasting Day-Ahead Hourly Electricity Prices: A Review, 2012.
20. Cabero, J., Baíllo, A., Cerisola, S., Ventosa, M., García-Alcalde, A., Perán, F., Relańo, G., A medium-term integrated risk management model for a hydrothermal generation company. *IEEE Trans. Power Syst.*, 20, 3, 1379–1388, 2005.
21. Bunn, Forecasting loads and prices in competitive power markets. *Proc. IEEE*, 88, 2, 163–169, 20002000.
22. Koritarov, V.S., Real-world market representation with agents. *IEEE Power Energy Mag.*, 2, 4, 39–46, 2004.
23. Weidlich, A. and Veit, D., A critical survey of agent-based wholesale electricity market models. *Energy Econ.*, 30, 1728–1759, 2008.
24. Borak, S. and Weron, R., A semiparametric factor model for electricity forward curve dynamics. *J. Energy Mark.*, 1, 3, 3–16, 2008.
25. Fleten, S.-E., Heggedal, A.M., Siddiqui, A., Transmission capacity between Norway and Germany: a real options analysis. *J. Energy Mark.*, 4, 1, 121–147, 2011.

26. Misiorek, A., Trück, S., Weron, R., Point and interval forecasting of spot electricity prices: Linear vs. non-linear time series models. *Stud. Nonlinear Dyn. Econom.*, 10, 3, 1–36, 2006, Article 2.
27. Bessec, M. and Bouabdallah, O., What causes the forecasting failure of Markov-switching models? A Monte Carlo study. *Stud. Nonlinear Dyn. Econom.*, 9, 2, 1–24, 2005, Article 6.
28. Kosater, P. and Mosler, K., Can Markov regime-switching models improve power-price forecasts? Evidence from German daily power prices. *Appl. Energy*, 83, 943–958, 2006.
29. Chaâbane, N., A hybrid ARFIMA and neural network model for electricity price prediction. *Int. J. Electr. Power Energy Syst.*, 55, 187–194, 2014.
30. Sharma, V. and Srinivasan, D., A hybrid intelligent model based on recurrent neural networks and excitable dynamics for price prediction in deregulated electricity market. *Eng. Appl. Artif. Intell.*, 26, 5-6, 1562–1574, 2013.
31. Azadeh, A., Moghaddam, M., Saffar, M., Seyedmahmoudi, S.H., Optimum Long-Term Electricity Price Forecasting in Noisy and Complex Environments. *Energy Sources*, 8, 235–244, 2013, 10.1080/15567249.2012.678559.
32. Yan, X. and Chowdhury, N.A., Mid-term electricity market clearing price forecasting: A hybrid LSSVM and ARMAX approach. *Int. J. Electr. Power Energy Syst.*, 53, 1, 20–26, 2010.
33. Zhang, L., Luh, P.B., Kasiviswanathan, K., Energy Clearing Price Prediction and Confidence Interval Estimation With Cascaded Neural Networks. *IEEE Power Eng. Rev.*, 18, 1, 99–105, 2003.
34. Guo, J., Member, S., Luh, P.B., Improving Market Clearing Price Prediction by Using a Committee Machine of Neural Networks. *IEEE Trans. Power Syst.*, 19, 4, 1867–1876, 2004.
35. Hu, Z., Yu, Y., Wang, Z., Sun, W., Gan, D., Han, Z., Price forecasting using an integrated approach, April 2004, pp. 28–31.
36. Zhang, L. and Luh, P.B., Neural network-based market clearing price prediction and confidence interval estimation with an improved extended Kalman filter method. *IEEE Trans. Power Syst.*, 20, 1, 59–66, 2005, https://doi.org/10.1109/TPWRS.2004.840416.
37. Amjady, N. and Keynia, F., A new prediction strategy for price spike forecasting of day-ahead electricity markets. *Appl. Soft Comput.*, 11, 6, 4246–4256, 2011, https://doi.org/10.1016/j.asoc.2011.03.024.
38. Tan, Z., Zhang, J., Wang, J., Xu, J., Day-ahead electricity price forecasting using wavelet transform combined with ARIMA and GARCH models. *Appl. Energy*, 87, 11, 3606–3610, 2010, https://doi.org/10.1016/j. apenergy.2010.05.012.
39. Unsihuay-Vila, C., Zambroni de Souza, A., Lima, J., Balestrassi, P., Electricity demand and spot price forecasting using evolutionary computation combined with chaotic nonlinear dynamic model. *Int. J. Electr. Power Energy Syst.*, 32, 108–116, 2010, 10.1016/j.ijepes.2009.06.018.

40. Nargale, K., Kanchan, B., Patil, S., Day ahead price forecasting in deregulated electricity market using Artificial Neural Network, pp. 527–532, 2016, 10.1109/ICEETS.2016.7583810.
41. Hyndman, R.J. and Athanasopoulos, G., *Forecasting: principles and practice*, 2nd edition, OTexts, Melbourne, Australia, 2018, Otexts.com/fpp2.
42. Faraway, J. and Chatfield, C., Time series forecasting with neural networks: a comparative study using the airline data. *J. R. Stat. Soc. Ser. C (Appl. Stat.)*, 47, 2, 231–250, 1998.
43. Anbazhagan, S. and Kumarappan, N., Day-ahead deregulated electricity market price forecasting using neural network input featured by DCT. *Energy Convers. Manage.*, 78, 711–719, 2012, https://doi.org/10.1016/j.enconman.2013.11.031.

9

Machine Learning–Enabled Solar Photovoltaic Energy Forecasting for Modern-Day Grid Integration: A Virtual Power Plant Perspective

Subhajit Roy[1]*, Smriti Jaiswal[2], Manav Sanghi[2], Mriganka Dhar[2], Arif Mohammed[2], Kothalanka K. Pavan[2], D. C. Das[2] and Nidul Sinha[2]

[1]Researcher NIT Silchar, Assam & Ex-HoD CS&DF, ISOAH, Kolkata, India
[2]National Institute of Technology, Silchar, Assam, India

Abstract

The increasing significance of renewable energy sources arises from the depletion of natural resources and the escalating demand for energy. In recent times, solar photovoltaic (PV)–based electricity has garnered substantial attention due to its environmentally friendly and sustainable attributes. This chapter focuses on the vital aspect of predicting the electrical output of solar power plants at different locations, a task intricately linked to local climatic conditions. Given the diverse factors influencing solar power plant performance, including weather patterns, seasonal variations, and the types of solar panels utilized, predicting their electrical production is inherently challenging. Accurate forecasting of solar PV–based electricity generation is indispensable for solar energy to meet demand and evolve into a dependable source of electricity. Through this study, the authors proposed one of the most promising and unique applications of soft computing techniques based on the machine learning (ML) process, which handle the most impactful parameters used for solar power installation purpose. ML-based techniques opened a number of avenues in the field of renewable energy integration with modern-day grid for a promising virtual power plant–based environment by a near to accurate energy generation prediction data to properly distributing the generated energy. The application and inclusion of solar PV with the grid always come under questions because of its availability to ensure dependability on the renewable-based

**Corresponding author*: subhajitroy111@gmail.com

Jai Govind Singh, Rupendra Kumar Pachauri and Sasidharan Sreedharan (eds.) Forecasting Methods for Renewable Power Generation, (243–278) © 2025 Scrivener Publishing LLC

energy generation. The application of ML-based techniques ensures the resiliency of the renewable energy generation to the modern grid system. From the understanding of the VPP, it became more important to plan to achieve a proper clustering of dispersed generation achieved by the mini- to micro-generation like solar PV-based generation. ML-based techniques help these predicting strategies to ensure coordination of renewable energy sources and thus to enhance overall grid performance. The chapter discusses the applications of ML algorithms and their impact on enhancing power extraction from solar PV–based electricity producers within the framework of solar PV–based energy forecasting.

Keywords: Solar PV forecasting, renewable energy integration, machine learning, grid reliability, virtual power plants, grid integration

9.1 Introduction

The transformation of electricity generation toward sustainable energy systems is undergoing a significant change driven by global concerns about climate change [1]. Photovoltaic (PV) systems are important in both large-scale grid-connected systems and residential applications. As per the 2050 projection, PV capacity may contribute 22% of global electricity generation, marking a deliberate move for more sustainable option of energy sources [2]. The variability of PV output leads to voltage and frequency changes, which pose challenges for integrating solar energy into power networks [3]. It is much needed to make accurate forecasting to be an important tool to manage these problems effectively [4]. Accurate forecasting mollifies operational challenges associated with high PV perforation rates, streamlines integration, and reduces storage requirements. Solar PV (SPV) is essential for optimizing grid performance to align consumption patterns with electricity output patterns affected by variable climate.

This research focuses on SPV power forecasting, specifically incorporating machine learning capabilities, within a virtual power plant (VPP) [5]. The expected upgradation in the basis of PV based (ML) generation by 2050, showcasing the need for more advanced approaches to maximize the use of solar energy. The ML-based approaches help to determine the scattering of sunrays. Accurate foreboding facilitates include the proactive grid management, eliminating the need for supportive generation, thereby reducing financial twig and improving resource utilization. ML-based techniques help to predict accurately the complicated sunray availability pattern on a time-to-time basis [6]. For the purpose to achieve a SPV-based VPP system, ML-based techniques play a crucial rule. Both the modernized techniques used for ML and VPP help the integration of renewable-based generators with better reliability and help to achieve potential of a sustainable and resilient energy future [7]. The significance of different dispersed renewable energy sources to achieve a low-carbon–based future is immensely high, and SPV plays a crucial contribution, predicting almost a 22% share of the global clean energy generation by 2050 [8]. SPV comes with several benefits, which includes reduced greenhouse gas emissions, reduced dependence on fossil fuels, and reduced environmental damage. However, the biggest issue lies for the SPV-based generation with its reliability and requires proper predicting techniques for commercial usage.

For the purpose of better grid management and allowing utilities to optimally match the required power supply and consumption, as well as reducing the need for standby capacity, advanced technologies and control mechanisms are essential. Apart from improving the accuracy and flexibility, ML-based approaches also address the complex and nonlinear linkages in SPV-based generation with the help of past training data to predict similar future events [9]. VPP based on ML-enabled SPV power forecasting continues to be helpful in the broader picture [10, 11]. The idea of VPP-based models depends on the optimizing different integrated renewable energy sources as a single entity with the help of modern computational approaches where ML-based strategies can play a crucial role. Proper integration of ML-based strategies for VPP models can both improve grid stability and create opportunities for further integration of renewable energy sources [12].

The synergy between renewable energy, ML, and VPPs holds the key to unlocking the full potential of sustainable energy. This article aims to contribute meaningfully to the ongoing dialogue by exploring case studies, current developments, and specific ML techniques in subsequent sections, shedding light on the intricacies and leveraging synergies for a resilient and sustainable energy future.

9.2 Literature Review

The listed references provide a diverse overview of studies related to climate change, renewable energy, and ML in the context of energy forecasting. Here is a brief summary of each reference:

[13] "Uncertainty energy planning of net-zero energy communities with peer-to-peer energy trading and green vehicle storage

considering climate changes by 2050 with machine learning methods" - Liu Jia, Zhou Yuekuan, and Hongxing Yang

Summary: The study examines epistemic and aleatory uncertainties of climate change, focusing on meteorological parameters affecting wind and SPV. The goal is to facilitate research on peer-to-peer energy trade and achieve carbon neutrality by 2050, employing ML for wind and PV power generation models.

[14] "Impact of climate change on electricity systems and markets – A review of models and forecasts" - Torben K. Mideksa and Steffen Kallbekken

Summary: This review evaluates advancements in understanding how climate change affects electrical systems, exploring implications on renewable energy sources and temperature-responsive demand forecasting. The review identifies unknowns in areas such as broader climate change effects, economic modeling of extreme weather events, and system-wide impacts on power production.

[15] "A systematic analysis of meteorological variables for PV output power estimation" - Dev Soumyabrata, Visser Lennard, and Murhaf Hossari

Summary: The study systematically analyzes meteorological variables' impact on PV output power estimation, acknowledging advancements in understanding localized climate effects on renewable energy and demand forecasting. However, uncertainties persist in system-wide effects and economic modeling of extreme weather events.

[16] "Linear Regression model in estimating solar radiation in Perlis" - Irwan Y.M., Irwanto M., and Farhana Z. I.

Summary: This study investigates the relationship between temperature and solar radiation in Perlis, Malaysia, using a linear regression model. It evaluates solar radiation levels based on scatter plots, showing a linear trend throughout the year 2006.

[17] "Tackling climate change with machine learning" - David Rolnick and Lynn H. Kaack

Summary: The study emphasizes the significant impact of ML in addressing climate change, showcasing applications in scientific discoveries and remote sensing for automated monitoring, with the potential to advance the discipline and benefit society.

[18] "Predictive analysis using machine learning: trends and methods" - Bokonda Patrick Loola and Ouazzani Touhami Khadija

Summary: The study examines widely used ML techniques, including random forests, artificial neural networks, decision trees, and linear regression, across various sectors such as education, building, botany, social science, and medicine.

[19] "Solar power forecasting using artificial neural networks" - Abuella Mohamed and Chowdhury Badrul

Summary: This research presents an artificial neural network (ANN) model for forecasting solar power generation, comparing its performance with multiple linear regression and persistence models, and conducting sensitivity analysis on input variables to determine the best choice.

[20] "Probabilistic energy forecasting: Global Energy Forecasting Competition 2014 and beyond" - Tao Hong, Pierre Pinson, and Hyndman Rob J.

Summary: The paper focuses on the Global Energy Forecasting Competition 2014 (GEFCom2014), a competition centered on building probabilistic energy forecasting models across load, price, wind, and solar tracks, involving 581 competitors from 61 countries.

[21] "Solar forecasting methods for renewable energy integration" - Rich Inman and Coimbra Carlos F.M.

Summary: The study emphasizes the need for extremely precise solar forecasting systems for scheduling, dispatching, regulating, and unit commitment over various time periods. It investigates the theoretical foundations of solar forecasting techniques and explores effective utility-scale applications for forecasting solar resource and electricity output.

9.3 Application of Machine Learning to Tackle Climatic Constraints

The use of ML-based techniques to address global problems is quite significant, in the context of climate change and the development of sustainable energy sources. Already different fields have actively started collaborating with the ML community by recognizing its potential to enhance social welfare. ML can significantly improve the understanding and management of green energy sources. A significant challenge in the field of renewable energy, exacerbated by climate change, is the integration of solar power into smart grids. The unpredictability of solar power generation due to constantly changing weather patterns makes accurate forecasting essential for efficient energy management [22]. For the purpose to ensure a high standard of living for electricity energy users with the help of sustainable sources like SPV, accurate forecasting is much needed.

This project deals with the unpredictability problem of SPV where incorporation of ML-based techniques shows promise like it has shown in lots of other fields effectively. As the time passes, ML proves to be an even more powerful tool for the researchers. The ML-based techniques have shown its effectiveness over the training models that the designers used to train and later to testify the effectiveness on the actual system.

Here, in this project, the meteorological variables significantly impact solar power generation, taking into consideration to build the training dataset, and it is the first objective of the project. The second goal is the original work by using relevant weather data analysis to predict SPV power at particular locations; this should be considered as an essential step for seamlessly integrating solar power into grid. The third goal focuses on finding the best ML-based methods for weather-driven solar power output forecasting. In this step, the researchers work on a thorough investigation of different types of ML models, from different ANN techniques and decision trees to more complex algorithms such as random forests [23, 24]. The goal is to identify algorithms that strike the right balance between accuracy and efficiency and able to generate the efficient result. Among the global challenges of tackling climate change and transitioning to renewable energy, ML is playing an increasingly important role [25]. It should be mentioned that the collaborative efforts of the ML-based strategies addressing the challenges of climate change are crucial to the advancement of the sustainable growth for the betterment of humanity.

9.4 Application of ML in Solar PV–Based Generation

9.4.1 Importance of Solar PV in Modern Electrification System

Solving complex issues of energy security and environmental concerns is important, making SPV energy a priority. Increasing overconsumption: Climate change impacts increasingly highlight the urgent need to shift to renewable energy sources. Solar PV clean: It stands as a sustainable energy solution; using unlimited solar energy to reduce greenhouse gas emissions reduces heat emissions and reduces dependence on limited fossil fuels [26].

The importance of SPV extends beyond its positive environmental impact. It encourages the transfer of electricity generation, diversifies the energy mix, and thus increases energy security. Distributed SPV systems empower communities, reducing their vulnerability to central power outages. Furthermore, the vast and abundant availability of solar energy provides a scalable solution for growing global energy needs.

SPVs are more than just a technical feat. This transformation and sustainability is emerging as an important way to shape the future of energy. This technology addresses the challenges posed by climate change and provides reliable and accessible energy for communities around the world. The adoption of SPV technologies is becoming a cornerstone in the quest for a cleaner and more resilient energy environment. As we navigate the challenges of a changing climate, the adoption of SPV is an important step toward a sustainable energy model [27].

9.4.2 Working of Solar PV

Because PV cells act as a key component in converting sunlight into usable energy, they play an important role in the performance of solar panels. These cells exhibit an interesting dance of charges within the layer structure, indicating the complexity of their structure. Technically referred to as "silicon sandwich slices," these cells are examples of microscopically integrated materials [28]. Surfaces with more phosphorus give off extra electrons, resulting in a negative charge, whereas lower layers with more boron give off positive charges with fewer electrons. This fixed arrangement gives way to atoms fixed types, similar to well-designed electric circuits.

The transition period occurs when photons from sunlight are disconnected from the electronic contacts in these PV cells. These released electrons generate energy flow as the released electrons go to the negative charge of the solar cell as shown in Figure 9.1. This process is not only beautiful in its simplicity, but it is also powerful, providing the electricity needed for solar panels [29].

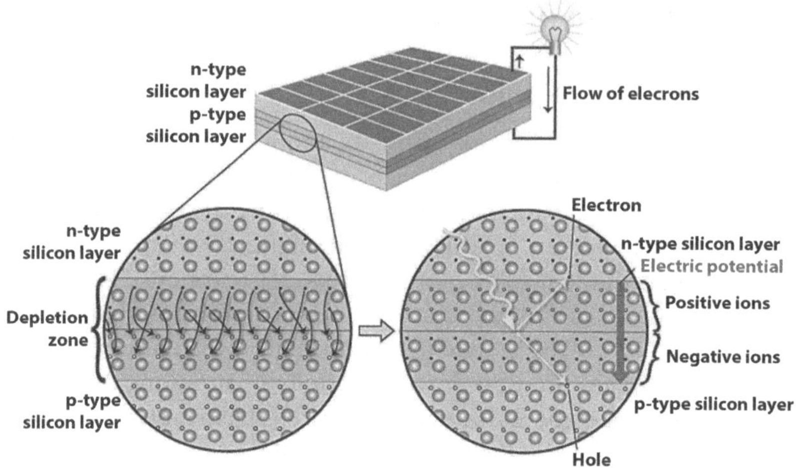

Figure 9.1 Illustration of a solar cell's operation, showcasing the movement of electrons between the layers of silicon [29].

Essential to this solar cell is an overlapping conductive metal plate. These plates acting as drivers gathered electrons and are connected by the hammer, which are gathered for different purposes, initiated for different purposes, and, through integration, it emphasizes that PVs play a key role in the global transition to renewable and sustainable energy [30]. The harmonious combination of creativity and science inherent in these solar cells is proof of that humanity has become wise in the pursuit of a sustainable and clean energy future.

9.4.3 Factors Affecting PV Power Generation

The complex web of factors affecting the efficiency of solar PV generation as listed in Table 9.1 closely monitors the subtle interplay between weather conditions and the conversion of sunlight into electricity (an important component given as 39%). Increased cloud cover reducing the amount solar radiation absorption translates directly into reduced solar irradiance, affecting the optimal performance of PV cells [31]. The relative humidity, expressed as a percentage, silently affects the absorption and transmission of light, causing the solar panels to operate. Because the delicate environmental water balance plays an important role in generating energy, changes in this balance can alter the amount of energy produced per mile of capacity; it emerges as the unsung hero in solar energy efficiency, ensuring uninterrupted sunlight paths for solar cells. Wind speed may not have a

Table 9.1 Different weather-related factors influence SPV-based generation.

Weather features	Unit	Weather features	Unit
Cloud coverage	% range	Relative humidity	%
Visibility	Miles	Wind speed	Mph
Temperature	°C	Station pressure	inchHg
Dew point	°C	Altimeter	inchHg

very distinctive impact over SPV-based energy generation, but strong wind poses critical structural impact of solar panels, which leads to unavailability of SPV-based power during a certain situation and require prediction strategies as well [32]. Another important factor for SPV-based generation is operating temperature measured in degrees Celsius, which affects the thermal conductivity of solar cells, and, in return, the amount of electricity generation by the panels may differ. There are few other parameters affecting the weather-based condition for electricity generation from SPV; these are metrics such as altimeter readings of mercury (inchHg) to understand the effective pressure, as well as dew point and station pressure that are contributing intricately to these multifaceted elements. Collectively, these variables provide the idea of complexity of the test system. A careful examination of these issues implies a commitment to efficient use of solar energy in line with the larger objective of promoting clean and sustainable power generation. To get the maximum output from SPV-based generation and achieve the highest efficiency from the system with enhanced reliability, these meteorological parameters help to predict well in advance if properly monitored with the help of modern computational strategies.

Cloud Coverage: Cloud coverage in the sky plays the most crucial impact on the electricity generation output from SPV-based system. This examination should include various cloud properties, including their individual characteristics and geographic distribution. Low, dense clouds, acting as reflective shields in nature, play an important role in deflecting solar radiation and contribute significantly to cooling the Earth's surface [33]. Conversely, tall, thin clouds engage in a dual interaction with solar radiation, reflecting sunlight and absorbing and redirecting some of Earth's emitted infrared radiation, warming the surface in the process. The composition, size, and height of cloud particles appear to be key factors in determining whether a particular cloud will have a cooling or warming effect. Despite the delicate balance of these opposing effects, a more comprehensive research perspective reveals that, when considering the cumulative effect of clouds on Earth's climate, the cooling effect generally prevails. Figure 9.2 [34] provides a visual representation of the sky cover (oktas) range on a scale of 1–10, nicely illustrating the complex interplay of clouds on the Earth's radiation budget.

Temperature: A pivotal factor intricately woven into the efficiency dynamics of solar cells, particularly those reliant on semiconductor materials, is temperature. In semiconductor materials, the band gap undergoes narrowing with temperature changes, demanding less energy for electron transitions. This intricate connection between temperature and voltage output is visually depicted in Figure 9.3 [35], showcasing the nuanced sensitivity of solar cell open circuit voltage to fluctuations in temperature.

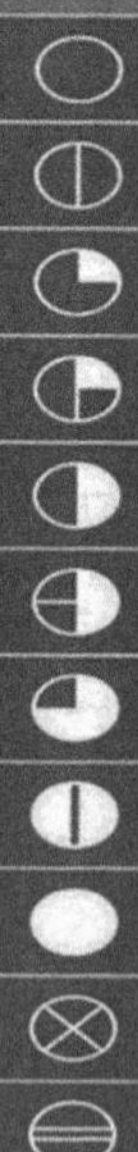

Sky cover (oktas)	Symbol	Name	Abbreviation
0		Sky clear	SKC
1		Few clouds	FEW
2		Few clouds	FEW
3		Scattered clouds	SCT
4		Scattered clouds	SCT
5		Broken clouds	BKN
6		Broken clouds	BKN
7		Broken clouds	BKN
8		Overcast	OVC
(9)		Sky obscured	—
(/)		Not measured	—

Figure 9.2 Variation in sky cover (oktas) graded on a scale of 1–10.

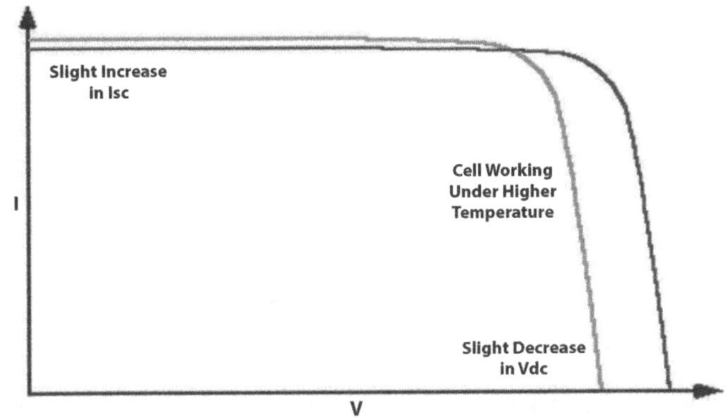

Figure 9.3 Correlation between temperature and voltage output illustrated in a graph.

The complex interplay between temperature and the efficiency of solar panels underscores the importance of cooling systems, with wind speed emerging as a crucial factor. Observations indicate that a mere 1°C decrease in solar panel temperature leads to a 0.05% boost in efficiency, despite the seemingly modest impact on overall productivity [36]. This underscores the critical role that the cooling effect of the wind plays in optimizing the effectiveness of solar panels.

Humidity: Humidity poses multifaceted challenges to the efficiency of solar panels, manifesting in both short- and long-term impacts. Minute water vapor or droplets accumulating on solar panels can act as refractive elements, diverting sunlight away from the cells and diminishing their energy production capacity [37]. Simultaneously, there is a significant concern about the enduring consequences of hot and humid weather on the degradation of solar panels. While both thin-film modules and crystalline silicon solar cells exhibit limitations, it is worth noting that cadmium telluride solar cells show better performance, boasting an average 5% increase, particularly in tropical regions [38]. In Figure 9.4 the graph is showing the relative change in SPV-based power generation with change of humidity level [39].

Air Pressure: A detailed examination of the data reveals a fascinating correlation between air pressure and solar parameters: elevated air pressure correlates with increased output current, voltage, and solar illumination/intensity. This correlation stems from the gravitational field created by the weight of the Earth's atmosphere, enhancing the downward pressure on solar radiation particles. Figure 9.5 vividly illustrates this phenomenon,

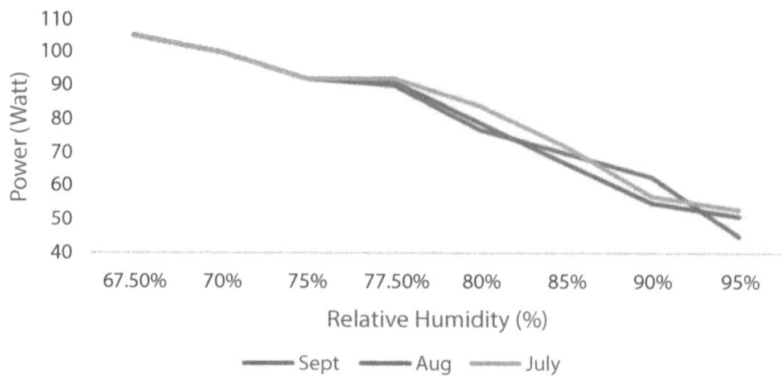

Figure 9.4 Graph displaying the relationship between relative humidity and power (W).

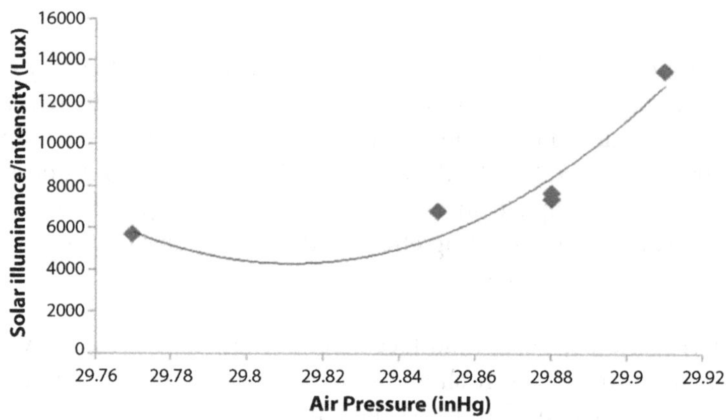

Figure 9.5 Shots in solar luminance (Lux) in relation to changes in air pressure measured in inches of mercury (Hg).

displaying the variation in sun illuminance (Lux) in relation to air pressure measured in inches of mercury (Hg) [40].

9.5 Design of a Predictive ML Model

The subsequent section of the article delves into the utilization of computational intelligence for forecasting climatic changes and assessing their negative impact on SPV-based generation. Various ML-based techniques are explored in this context, aiming to identify the most suitable approach through a comparative study.

9.5.1 Kth-Nearest Neighbor (KNN) Algorithm

An illustration of a nonparametric supervised learning classifier is the K-nearest neighbor (KNN) method, employing the concept of closeness to make predictions or classifications for grouping individual data points [41]. KNN proves to be a highly versatile technique applicable to both regression and classification problems. Its optimal performance in classification tasks is observed when nearby data points share similarities as it depicts in the Figure 9.6 [42]. The effectiveness of KNN lies in its simplicity and intuitive understanding—the idea that closely situated data points are likely to belong to the same category. Working of KNN (selecting the parameter K) [43]: In determining the classification or regression value for a specific query point, the KNN algorithm relies on a crucial parameter, K. Selecting the appropriate value for K is a nuanced decision, as it can make the model either overly flexible or excessively rigid. Smaller K values heighten variance at the expense of introducing bias, whereas larger K values bring about strong bias but lower variance [44]. The optimal K value is largely contingent on the characteristics of the input data; datasets with substantial noise or outliers may benefit from larger K values [45]. Opting for an odd value for K is advisable to resolve ties in classification scenarios, and the use of cross-validation techniques becomes pivotal in determining the most suitable K for a given dataset [46].

Feedforward Neural Network (FNN): As a fundamental component of artificial neural networks, the FNN orchestrates the flow of information in a unidirectional manner, commencing from the input layer, traversing one or more hidden layers, and culminating at the output layer as shown in Figure 9.7 [47]. Widely applied across a variety of ML tasks, it finds prevalence in regression situations. Integral key components of a FNN include the following:

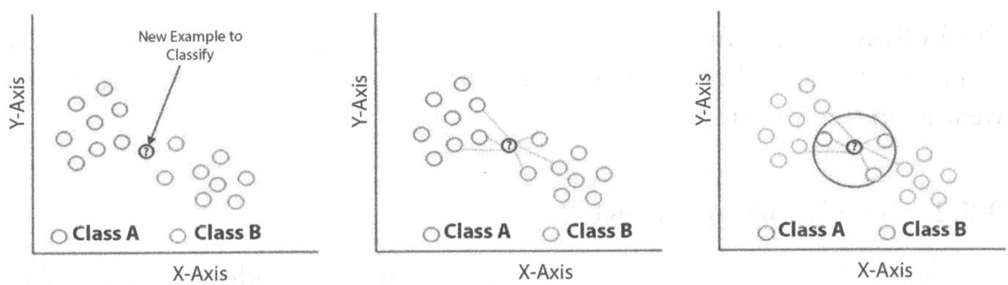

Figure 9.6 Functionality of the KNN algorithm illustrated.

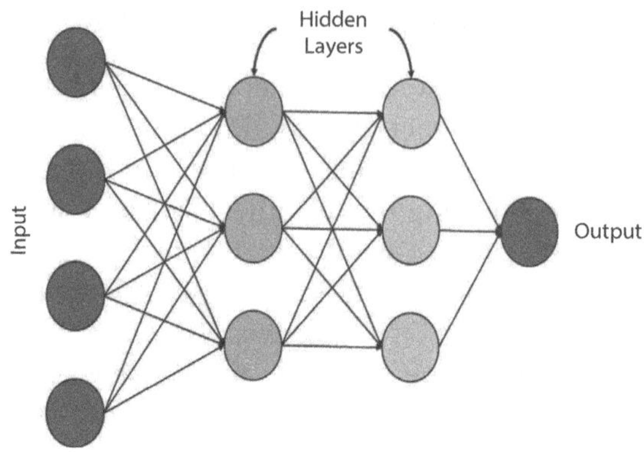

Figure 9.7 Representation of a feedforward neural network.

Architecture: The schematic design dictates the architecture of the network, specifying the size and number of hidden layers, including the number of input and output nodes.

Activation Function: Non-linearity in the system is being introduced because of activation functions, i.e., why a proper selection of activation function for the hidden and output layers, such as ReLU or sigmoid, must be chosen [48].

Training Process: Training process is the most significant step for any ANN model, where the model undergoes training through different optimization techniques such as Adam or stochastic gradient descent (SGD) [49].

Adjusting Hyperparameters: There are few parameters of the model tune to achieve a more robust and efficient system, which includes hyperparameter values—regularization techniques such as learning rate, batch size, number of hidden units, and dropout [50].

Prediction: After the completion of the training, a well-trained FNN achieves the capability to correctly predict future estimation related to the weather-based prediction [51].

9.5.2 The Random Forest Regressor

The Figure 9.8 visualizing a random forest regressor model which an effective tool for ensemble learning, which combines different types of decision

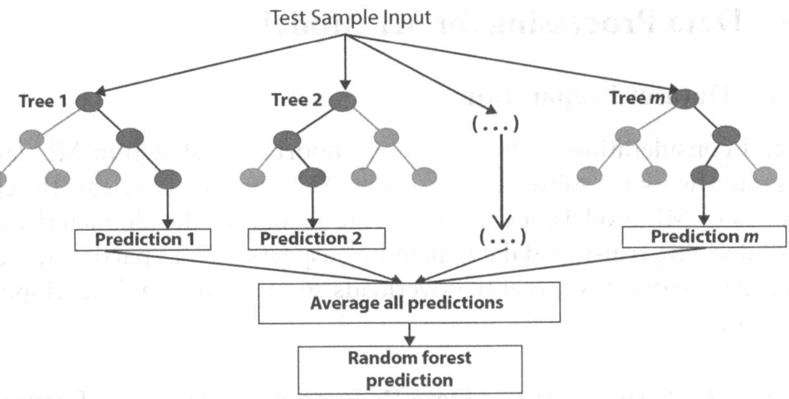

Figure 9.8 Visual depiction of the random forest [58].

trees to provide a predictive result [52]. Its appeal lies in its resilience, flexibility, and adept handling of high-dimensional datasets. Key features of the regressor model are as follows:

Combination of Decision Trees: This consists of a number of trees that are different from each other in the aspect of random subsets including training data and features [53].

Tree Construction: Each decision tree is carefully compiled using various methods such as CART (classification and regression trees), which grow until it satisfies predefined stopping criteria mentioned by the designer such as maximum depth or minimum number of samples per leaf [54].

Randomness and Diversity: Randomness of the model is a key ingredient for the algorithm where careful use of data subsets and features introduces randomness and also prevents overfitting, which enhances diversity among the trees [55].

Aggregate Prediction: In this stage, the model delivers a predicted result generated by the model through the average individual estimation in the previous steps [56].

9.6 Data Processing for ML Model

9.6.1 Dataset Preparation

Careful consideration of datasets is an important first step in ML, laying the groundwork for subsequent research. As researchers navigate the complexities of ML models, it is important to recognize the characteristics of the underlying real-world data, including aspects such as partiality, inconsistencies, errors, and potential overloads so, planning, and development included.

9.6.2 The Importance of Data Processing in Machine Learning

Data preprocessing is an important and fundamental step at the beginning of the journey to build ML models. Real-world data, inherently prone to error, inconsistency, and outlier sensitivity, present complexity, compounded by the absence of trait-specific values or trends. In response to these challenges, data preprocessing emerged as a comprehensive approach designed to clean, more refined, and organized raw data as they ensure adequacy. Through the Figure 9.9 it has been represented the missing data for each characteristics of the proposed work.

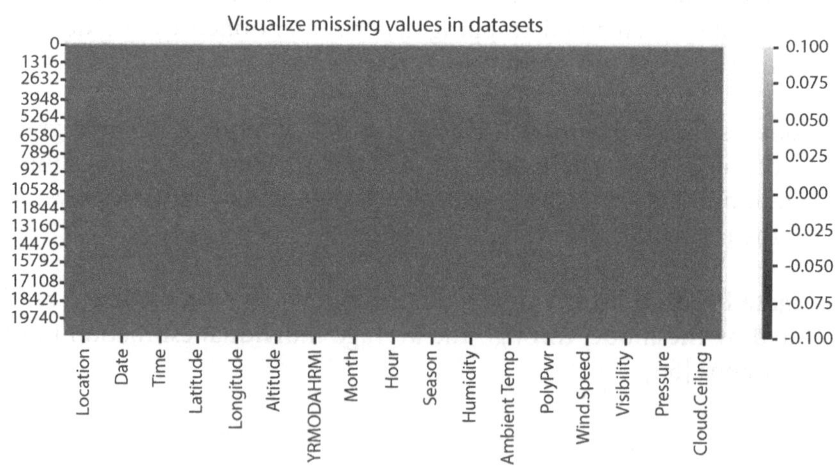

Figure 9.9 A representation of the missing data for each characteristic.

9.6.3 Steps Involved in Data Preprocessing

Acquiring the Proper Dataset: Data preprocessing in ML begins with acquiring datasets carefully crafted to meet the specific requirements of the model. This important initial step necessitates the collection of a relevant and carefully selected dataset necessary for building and training ML models. The process involves gathering information from various sources and combining it efficiently in a manner commensurate with the complexity of the specific use case.

Importing Necessary Python Libraries: As ML plays an important role in preprocessing datasets, Python libraries contribute significantly. For seamless integration of mathematical operations, data visualization and data manipulation into code, core libraries such as NumPy, Matplotlib, and Pandas are essential [59].

a. **NumPy:** Serving as a fundamental tool for scientific computing, NumPy facilitates the manipulation of large multidimensional arrays and matrices, streamlining the necessary mathematical operations [60].
b. **Pandas:** Positioned as a powerful open-source Python toolkit for data processing and analysis, Pandas provides powerful data structures as well as user-friendly features for importing and manipulating records [61].
c. **Matplotlib:** Functioning as a Python toolkit for 2D graph plotting, Matplotlib empowers researchers to generate figures suitable for publication across various platforms [62].

```
import pandas as pd, import matplotlib.pyplot as plt
import numpy as np
import seaborn as sns
from warnings import simplefilter
from sklearn.preprocessing import StandardScaler, OneHotEncoder
from sklearn.model_selection import train_test_split,
RandomizedSearchCV
from sklearn.neighbors import KNeighborsRegressor
```

Identifying and Handling Missing Values: Addressing missing values in an efficient manner is an essential component of preprocessing data [63]. Maintaining the integrity of ML projects requires precise identification and management of missing values in order to avoid drawing erroneous conclusions.

```
df = pd.read_csv('/content/Pasion et al dataset.csv')
```

Encoding Categorical Data: Before being fed into ML models, categorical data—information that falls into discrete categories—must be encoded. ML models can better understand categorical variables, thanks to two popular encoding techniques: label encoding and one-hot encoding [64].

```
df.info()
<class 'pandas.core.frame.DataFrame">
RangeIndex: 19784 entries, 0 to 19783
Data columns (total 17 columns):

#    Column           Non-Null Count     Dtype
---  ------           --------------     -----
0    Location         19784 non-null     object
1    Date             19784 non-null     int64
2    Time             19784 non-null     int64
3    Latitude         19784 non-null     float64
4    Longitude        19784 non-null     float64
5    Altitude         19784 non-null     int64
6    YRMODAHRMI       19784 non-null     float64
7    Month            19783 non-null     float64
8    Hour             19783 non-null     float64
9    Season           19783 non-null     object
10   Humidity         19783 non-null     float64
11   Ambient Temp     19783 non-null     float64
12   PolyPwr          19783 non-null     float64
13   Wind.Speed       19783 non-null     float64
14   Visibility       19783 non-null     float64
15   Pressure         19783 non-null     float64
16   Cloud.Ceiling    19783 non-null     float64
```

9.6.4 Visualizing the Dataset

```
#only for heat map
final_df = pd.get_dummies (df, columns = ['Location'], drop_first = True)
final_df = pd.get_dummies(final_df, columns = ['Season'],drop_first = True)
final_df.shape
final_df.head()
```

Location_MNANG	Location_Malmstrom	Location_March AFB	Location_Offutt	Location_Peterson	Location_Travis	Location_USAFA	Season_Spring	Season_Summer	Season_Winter
0	0	0	0	0	0	0	0	0	1
0	0	0	0	0	0	0	0	0	1
0	0	0	0	0	0	0	0	0	1
0	0	0	0	0	0	0	0	0	1
0	0	0	0	0	0	0	0	0	1

```
# Plot histogram of target variable
plt.figure(figsize=(12,5))
sns.histplot(df.PolyPwr);
plt.xlabel('PolyPwr (W)');
```

Splitting the Dataset: The careful separation of the dataset is essential to the ML models' training and testing stages. An 80-20 split, which is widely used, typically uses 80% of the dataset for training and 20% for testing [65].

```
input_feat = list(set(selected_columns). difference (set([target_label])))
df_X = final_df[input_feat].reset_index(drop=True)
df_y = final_df[target_label]
x_train, x_test, y_train, y_test = train_test_split(df_X, df_y, test_
size=8.2, random_state=42)
```

Feature Scaling: Feature scaling, a crucial step in normalizing independent variables within a given range, is the final stage of data preprocessing in ML [66]. By bringing variables down to a similar scale, this approach makes fair comparisons possible. The two different types of scaling are as follows:

a. Standardization:

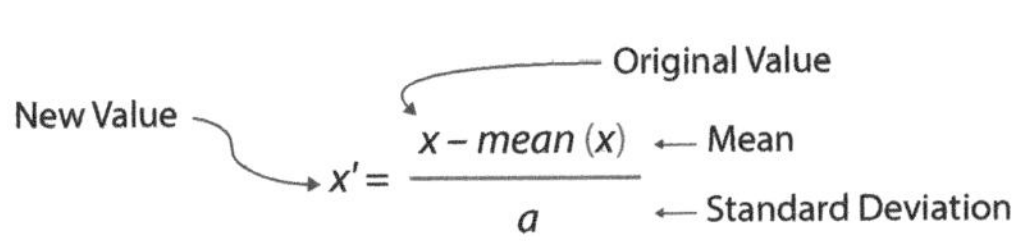

b. Normalization

New Value — Original Value

$$x' = \frac{x - min\,(x)}{\max(x) - \min(x)}$$

```
Performing Standard Scaling
scalar StandardScaler()
scalar.fit(x_train)
x_train_scaled = scalar.transform(x_train)
x_test_scaled = scalar.transform(x_test)
x_test.head()
```

Feature Selection from the Correlation Analysis Heat Map

```
selected_columns = ['Latitude', 'Humidity', 'Ambient Temp', 'PolyPwr',
'Wind. Speed'','Visibility', 'Pressure', 'Cloud.Ceiling', 'Location_Grissom',
'Location_Hill Weber', 'Location JDMT', 'Location_Kahului', Location_
MNANG', 'Location_Malmstrom', 'Location_March AFB', 'Location
Offutt', 'Location_Peterson', 'Location_Travis', 'Location_USAFA', 'Sea-
son_Spring', 'Season_Summer', 'Season_Winter', 'sine mon', 'cos_mon',
'sine_hr', 'cos_hr']
```

An Overview Based on the Feature Selection: Because of its strong link with pressure, which is more dynamic and stays constant at a particular location, altitude is disregarded. Because longitude has no correlation at all with the target variable, it is deleted. Strong correlations with manufactured cyclic aspects but low correlations with the target variable lead to the exclusion of time, hour, date, etc. With the representation showing in Figures 9.10 and 9.11 showing the distribution of solar PV output frequencies across various solar power plants and a proper representation of heat map illustrating correlations, which help to choosing the majorly impacting factors for the ML based model.

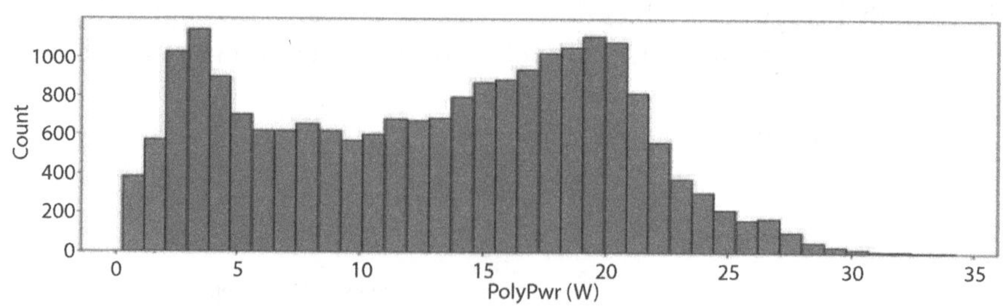

Figure 9.10 Distribution of solar PV output frequencies across various solar power plants.

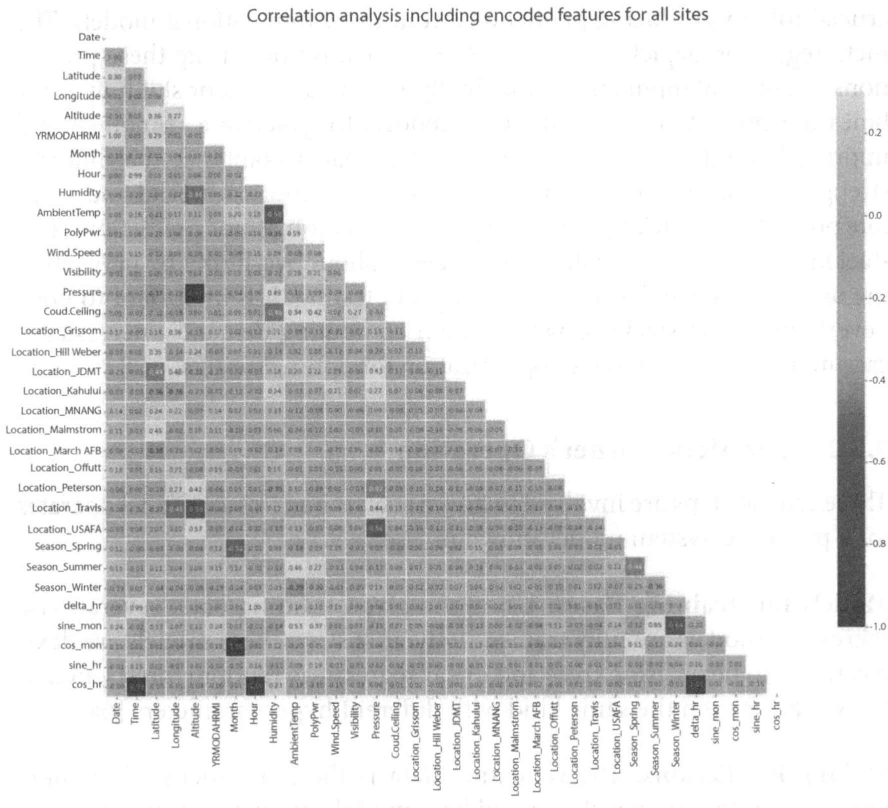

Figure 9.11 Heat map illustrating correlations.

9.7 MetaLearner Model

9.7.1 Dataset Preparation

The stacking regressor is depicted through Figure 9.12, showing a sophisticated ensemble learning method often referred to as stacked generalization, which is a critical tool for combining many regression models to improve prediction accuracy [67]. The basic idea is to train a number of regression models, also known as base models, whose predictions are used as inputs to a final regression model, called a meta-regressor, to produce the final result. The dataset is strategically divided into two or more subsets for use in the stacking regressor model. A separate subset is used for training multiple types of basic regression models or the same type with different hyperparameters. The other subset—referred to as the validation set—then plays a crucial role in producing predictions from these foundational models. The meta-regressor, depicted in Figure 9.13, is then trained using these predictions as essential input attributes. Finally, the meta-regressor skillfully combines the predictions from the base models to generate a combined and improved result. Stacking's main benefit is that it combines the different strengths of various regression models to create a more powerful and accurate predictive model. By using a separate validation set for predictions, this stacking process further alleviates concerns about overfitting by preventing the use of the same dataset for training and testing. It is important to note, nevertheless, that stacking, with its high computing demands, necessitates careful model and parameter optimization for optimal results.

9.7.2 The MetaLearner's Operation

Three crucial steps are involved in how a stacking regressor model operates for a proposed system model shown in Figure 9.12.

Models for Training Bases: Using the given training data, several base regression models are fully trained in the first step. These models are flexible; they can be decision trees and linear regression, random forests, neural networks, or even the same kind with different hyperparameters [68].

Making Predictions: The validation data is the main focus of the next phase, when the previously trained base models use it as a canvas to paint their predictions. These insightful forecasts are carefully chosen to serve as the meta-regressor's input features.

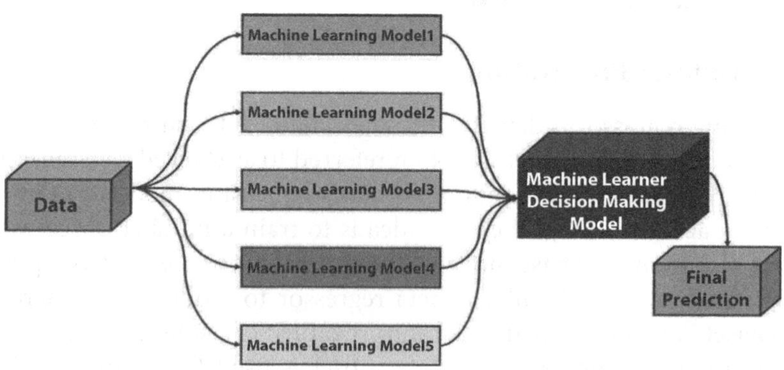

Figure 9.12 Operational process of the MetaLearner model.

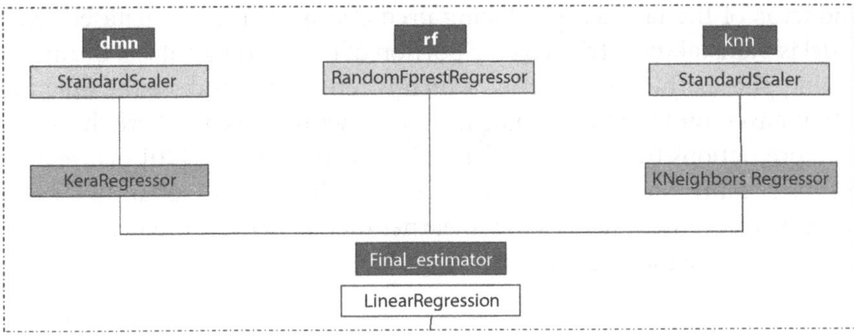

Figure 9.13 Diagram displaying the workflow of the stacking regressor.

Training the Meta-Regressor: The last stage involves using the projected values from the basic models to train the meta-regressor, which is an essential part of the stacking model [69]. The meta-regressor combines these predictions to create the final result in an elegant manner. It can be as basic as a linear regression model or as complex as a decision tree or neural network. As a standard and prudent procedure, cross-validation is used when training the stacking regressor model. This method effectively reduces the danger of overfitting and promotes a more robust model by segmenting the data into folds and training base models on a variety of combinations of training and validation data. The combined forecasts from multiple folds yield the final predictions. As a result, the stacking regressor model strategically outperforms each individual multiple regression model. It is essentially a strong and flexible combination of multiple regression models. Using many kinds of base models guarantees a thorough understanding of various parts of the data, which translates into increased accuracy.

9.7.3 Holding the Model in Place

In this section, we will take a closer look at stacking in practice, using a regression problem related to PV power generation. Three different methods are highlighted: random forest, neural network, and K-nearest neighbor regression. The training dataset assumes a crucial role in the training process of random forest, neural network, and KNN regression models. The validation dataset is then subjected to these models' predictive might. The forecasts derived from every model provide priceless information to the MetaLearner, which is a crucial component of the stacking paradigm. With careful training on the subtleties of the validation dataset, the MetaLearner becomes the conductor of the last act, producing predictions on the test dataset. Every model is painstakingly trained on a portion of the training dataset during the training phase. The MetaLearner, which is refined to produce the final prediction based on the many inputs from distinct models, is where the subsequent predictions from each model find their convergence [70]. As we move on to the testing step, every model examines the test dataset and makes its own distinct predictions. The MetaLearner then uses these predictions as the threads to create the final prediction.

```
## Define the base models
basee = list()
base.append(('dnn', dl_model))
base.append(('rf', rf_model))
base.append(('knn', knn_model))
# Define meta learner model
base1 = Linear Regression()
# Define the stacking ensemble
stacked_model = stackingRegressor (estimators-based, final_estima-
tor-base1, cv-3, passthrough=True)
# Fit the model on the training data
stacked_model.fit(x_train, y_train)
```

From a broad perspective, the MetaLearner functions as a unifying factor, combining the unique advantages of several models—KNN regression, neural network, and random forest—to generate an output that performs better than any of the individual models alone.

9.8 Result and Discussion

The following in Table 9.2, a well established and tested dataset that was used to assess the effectiveness of various modeled systems in predicting weather conditions and achieving an efficient SPV-based generation system through ML.

9.8.1 KNN Model

It is evident from Figure 9.14 with an R2 score of 0.6184, the KNN model shows a modest level of predictive power; roughly 61.84% of the variance in the target variable is explained. With a mean absolute error (MAE) of

Table 9.2 Dataset used for training and testing for ML model in this project (first five rows) [57].

Location	Date	Time	Latitude	Longitude	Altitude	YRMODAHRMI	Month	Hr	Season	Humidity	Ambient temp	PolyPur	Wind speed	Visibility	Pressure	Cloud ceiling
0 Camp Murray	20171203	1145	47.11	-122.57	84	2.017120e + 11	12	11	Winter	81.71997	12.86919	2.42769	5	10	1010.6	722
1 Camp Murray	20171203	1315	47.11	-122.57	84	2.017120 + 11	12	13	Winter	96.64917	9.66415	2.46273	0	10	1011.3	23
2 Camp Murray	20171203	1330	47.11	-122.57	84	2017120 + 11	12	13	Winter	93.61572	15.44983	4.46836	5	10	1011.6	32
3 Camp Murray	20171204	1230	47.11	-122.57	84	2017120 + 11	12	12	Winter	77.21558	10.36659	1.65364	5	2.0	1024.4	6
4 Camp Murray	20171204	1415	47.11	-122.57	84	2017120 + 11	12	14	Winter	54.80347	16.85471	6.57939	3	3.0	1023.7	9

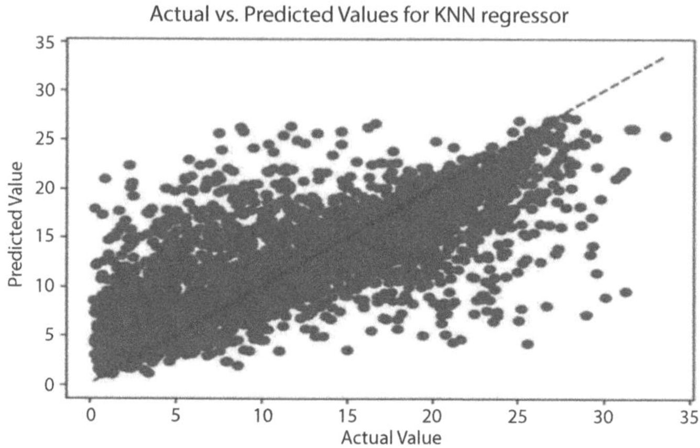

Figure 9.14 Comparison between actual and predicted values using KNN regressor.

2.9709, the model's predictions are, on average, 2.97 units off from the real values. With a little bigger error magnitude and a modest level of precision, the root mean square error (RMSE) is 4.4028.

9.8.2 Feedforward Neural Network (FNN) Model

The FNN model outperforms the KNN model as shown in Figure 9.15, accounting for roughly 66.22% of the variance with an R2 score of

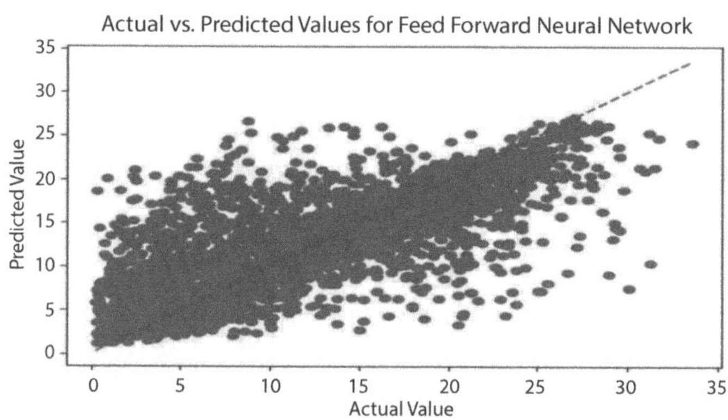

Figure 9.15 Evaluation metrics for the feedforward neural network's performance.

0.6622. In comparison to the KNN model, the lower MAE of 2.7087 suggests a slightly improved accuracy, whereas the RMSE of 4.1420 shows a moderate reduction in error size.

9.8.3 Random Forest Model

The Figure 9.16 shows that with an R2 value of 0.6700, the random forest model consistently demonstrates a high degree of explanatory power. The RMSE of 4.0940 indicates accuracy levels comparable to those in the FNN model, whereas the MAE of 2.7844 indicates a moderate average absolute prediction error.

9.8.4 MetaLearner Model

With the highest R2 score of 0.6783, the MetaLearner model performs better than any other model as shown in Figure 9.17. It can be inferred that the MetaLearner accounts for roughly 67.83% of the variance by aggregating predictions from many models. A decreased MAE of 2.6912 and an RMSE of 4.0422, which show increased accuracy and a little reduced error magnitude, further corroborate the model's performance.

Specific Observations:
Examining the data derived through the results unveils several observations that illuminate the performance metrics of different regression models:

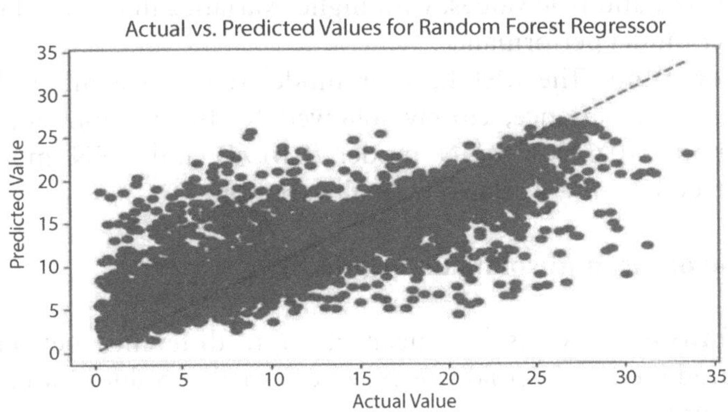

Figure 9.16 Evaluation metrics for the random forest regressor's performance.

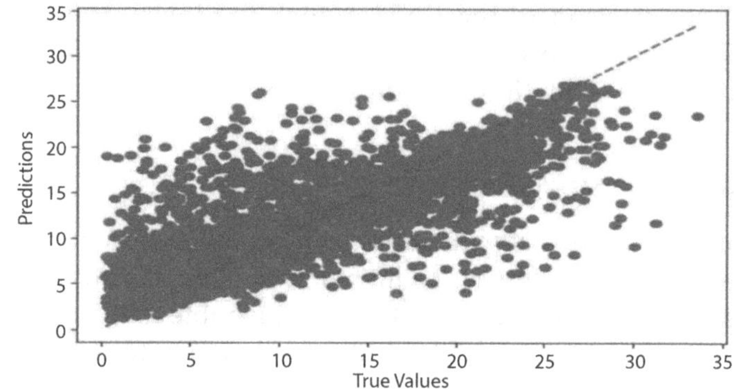

Figure 9.17 Evaluation metrics for the MetaLearner's performance.

Analysis of R2 Score:

Definition: A higher R2 score signifies a better fit, measuring the percentage of the dependent variable's variance that independent variables account for.

Observation: The random forest model (0.6700) and the MetaLearner model (0.6783) boast nearly identical R2 scores, suggesting superior fitting for the former. The R2 values of the KNN model (0.6184) and the FNN model (0.6622) are slightly lower.

Evaluation of Variance:

Definition: Variance represents the difference between predicted and true values, with higher variance indicating better model performance.

Observation: The MetaLearner model (0.6788) exhibits the greatest variance, closely followed by the random forest model (0.6700). KNN model (0.6196) and FNN model (0.6633) show a relatively small variation.

Evaluation of Mean Absolute Error (MAE):

Definition: MAE is the mean absolute difference between predicted and actual values, where smaller values indicate more accuracy.

Observations: The models with the lowest MAE are MetaLearner (2.6912), FNN (2.7087), random forest (2.7844), and KNN (2.9709).

Root Mean Square Error (RMSE): Analyzing:

Definition: RMSE measures model accuracy by taking the square root of the mean squared difference between actual and projected values, with smaller values indicating higher accuracy.

Observations: Random forest model (4.0940), FNN model (4.1420), KNN model, and MetaLearner model (4.0422) all exhibit the lowest RMSEs (4.4028).

These results give a fair idea of the model's performance, where the MetaLearner model consistently showing good results across different research parameters and confirming its effectiveness in predicting target outcomes for the purpose of a robust predictive methodology.

9.9 Conclusion

In conclusion, it can be stated that the authors performed extensive testing with the proposed methods discussed earlier in the chapter and found that MetaLearner model performed better in comparison to the other regression models considered in the study. The evaluated observations from R2 score, variance, MAE, and RMSE for the MetaLearner model stand out with better results than those for the FNNs and KNN-based models. However, it is important to proceed with caution and recognize the simplicity of model selection. Although the MetaLearner model showcased a better result on the test system, practical applicability must be considered to choose the most effective algorithm, and their factors such as problem domain complexity, dataset characteristics, and any specific requirements or constraints associated with the implementation must be carefully considered. Although MetaLearner clearly leads in the model selection scenario, the wider context in which it operates is important. Future developments in ML should prioritize a comprehensive understanding of multiple contextual parameters to ensure the applicability of developed models for a given specific real-world scenario. Lastly, the application of ML-based strategies to include the renewable-based electricity energy generation helps to predict the energy generation capability and also eases the load management possibility to make a resilient and robust electrical network that supports the implementation of VPP model. So, the ML-based techniques are helping the VPP models to be realized in the modernized grid structure.

References

1. Stephens, J.C., Energy Democracy: Redistributing power to the people through renewable transformation. *Environ.: Sci. Policy Sustain. Dev.*, 61, 2, 4–13, 2019, https://doi.org/10.1080/00139157.2019.1564212.
2. Chenic, A.Ş., Creţu, A.I., Burlacu, A., Moroianu, N., Vîrjan, D., Huru, D., Stanef-Puică, M.R., Enăchescu, V.A., Logical Analysis on the Strategy for a Sustainable Transition of the World to Green Energy—2050. Smart Cities and Villages Coupled to Renewable Energy Sources with Low Carbon Footprint. *Sustainability*, 14, 14, 8622, 2022, https://doi.org/10.3390/su14148622.
3. Roy, S., Shukla, R.D., Chakrabarti, R., Nath, S., De, A., Pandev, A., Cost Effective Design & Control of DC-DC Converter for Solar PV System with MPPT, in: *2019 International Conference on Energy Management for Green Environment (UEMGREEN)*, IEEE, pp. 1–6, 2019, https://doi.org/10.1109/uemgreen46813.2019.9221571.
4. Sweeney, C., Bessa, R.J., Browell, J., Pinson, P., The future of forecasting for renewable energy. *Wiley Interdiscip. Rev.: Energy Environ.*, 9, 2, e365, 2020, https://doi.org/10.1002/wene.365.
5. Bzdok, D. and Meyer–Lindenberg, A., Machine Learning for Precision Psychiatry: Opportunities and challenges. *Biol. Psychiatry: Cognit. Neurosci. Neuroimaging*, 3, 3, 223–230, 2018, https://doi.org/10.1016/j.bpsc.2017.11.007.
6. Roy, S., Das, D.C., Sinha, N., Shukla, R.D., A Systematic Review of Islanding Detection Approaches in Microgrids, in: *2023 IEEE Silchar Subsection Conference (SILCON)*, IEEE, pp. 1–6, 2023, https://doi.org/10.1109/SILCON59133.2023.10405116.
7. Roy, S., Dulal Chandra, D., Sinha, N., Optimizing Smart City Virtual Power Plants with V2G Integration for Improved Grid Resilience. *2024 IEEE International Conference on Interdisciplinary Approaches in Technology and Management for Social Innovation (IATMSI)*, IEEE, pp. 1–6, 2024, https://doi.org/10.1109/IATMSI60426.2024.10502468.
8. Lowe, R. and Drummond, P., Solar, wind and logistic substitution in global energy supply to 2050 – Barriers and implications. *Renew. Sustain. Energy Rev.*, 153, 111720, 2022, https://doi.org/10.1016/j.rser.2021.111720.
9. Prado-Rujas, I., García-Dopico, A., Serrano, E., Pérez, M.S., A flexible and robust Deep Learning-Based system for solar irradiance forecasting. *IEEE Access*, 9, 12348–12361, 2021, https://doi.org/10.1109/access.2021.3051839.

10. Frederiksen, C.A.F. and Cai, Z., Novel machine learning approach for solar photovoltaic energy output forecast using extra-terrestrial solar irradiance. *Appl. Energy*, 306, 118152, 2022, https://doi.org/10.1016/j.apenergy.2021.118152.
11. Faisal, L., Rama, V.S.B., Roy, S., Nath, S., Modelling of electric vehicle using modified SEPIC converter configuration to enhance DC–DC converter performance using MATLAB. In: *Lecture notes in electrical engineering*, pp. 643–653, Springer-Singapore, 2022. doi: 10.1007/978-981-19-4975-3_51.
12. Yavuz, L., Önen, A., Muyeen, S.M., Kamwa, I., Transformation of microgrid to virtual power plant – a comprehensive review. *IET Gener. Transm. Distrib.*, 13, 11, 1994–2005, 2019, https://doi.org/10.1049/iet-gtd.2018.5649.
13. Liu, J., Zhou, Y., Yang, H., Wu, H., Uncertainty energy planning of net-zero energy communities with peer-to-peer energy trading and green vehicle storage considering climate changes by 2050 with machine learning methods. *Appl. Energy*, 321, 119394, 2022, https://doi.org/10.1016/j.apenergy.2022.119394.
14. Mideksa, T.K. and Kallbekken, S., The impact of climate change on the electricity market: A review. *Energy Policy*, 38, 7, 3579–3585, 2010, https://doi.org/10.1016/j.enpol.2010.02.035.
15. AlSkaif, T., Dev, S., Visser, L., Hossari, M., Van Sark, W., A systematic analysis of meteorological variables for PV output power estimation. *Renew. Energy*, 153, 12–22, 2020, https://doi.org/10.1016/j.renene.2020.01.150.
16. Ibrahim, S.Z., Ismail, D., Irwan, Y., Irwanto, M., Gomesh, N., Farhana, Z., Linear regression model in estimating solar radiation in Perlis. *Energy Procedia*, 18, 1402–1412, 2012, https://doi.org/10.1016/j.egypro.2012.05.156.
17. Rolnick, D., Donti, P.L., Kaack, L.H., Kochanski, K., Lacoste, A., Sankaran, K., Ross, A.S., Milojevic-Dupont, N., Jaques, N., Waldman-Brown, A., Luccioni, A.S., Maharaj, T., Sherwin, E.D., Mukkavilli, S.K., Kording, K.P., Gomes, C.P., Ng, A.Y., Hassabis, D., Platt, J., Bengio, Y., Tackling Climate Change with Machine Learning, arXiv (Cornell University), 2019, https://doi.org/10.48550/arxiv.1906.05433.
18. Bokonda, P.L., Ouazzani-Touhami, K., Souissi, N., Predictive analysis using machine learning: Review of trends and methods, in: *2020 International Symposium on Advanced Electrical and Communication Technologies (ISAECT)*, November, IEEE, pp. 1–6, 2020, https://doi.org/10.1109/isaect50560.2020.9523703.
19. Abuella, M. and Chowdhury, B., Solar power forecasting using artificial neural networks. *2015 North American Power Symposium (NAPS)*, pp. 1–5, 2015, https://doi.org/10.1109/naps.2015.7335176.
20. Hong, T., Pinson, P., Fan, S., Zareipour, H., Troccoli, A., Hyndman, R.J., Probabilistic energy forecasting: Global Energy Forecasting Competition 2014 and beyond. *Int. J. Forecast.*, 32, 3, 896–913, 2016, https://doi.org/10.1016/j.ijforecast.2016.02.001.

21. Inman, R.H., Pedro, H.T., Coimbra, C.F., Solar forecasting methods for renewable energy integration. *Prog. Energy Combust. Sci.*, 39, 6, 535–576, 2013, https://doi.org/10.1016/j.pecs.2013.06.002.
22. Agüera-Pérez, A., Palomares-Salas, J.C., De La Rosa, J.J.G., Florencias-Oliveros, O., Weather forecasts for microgrid energy management: Review, discussion and recommendations. *Appl. Energy*, 228, 265–278, 2018, https://doi.org/10.1016/j.apenergy.2018.06.087.
23. Roy, S., Shukla, R.D., Bhowmik, P., Nandy, N., Kole, A., Ghosh, K., Predictive Modelling and Simulation of Vehicle-to-Grid Systems Using Hidden Markov Algorithm and Microgrid Integration. *2023 International Conference on IoT, Communication and Automation Technology (ICICAT)*, pp. 1–6, 2023, https://doi.org/10.1109/icicat57735.2023.10263674.
24. Han, T., Jiang, D., Zhao, Q., Wang, L., Yin, K., Comparison of random forest, artificial neural networks and support vector machine for intelligent diagnosis of rotating machinery. *Trans. Inst. Meas. Control*, 40, 8, 2681–2693, 2017, https://doi.org/10.1177/0142331217708242.
25. Roy, S., Implementation of Model to Analyse the Performance of Microturbine as in Microgrid Comparison with Fuel Cell. *Int. J. Energy Optim. Eng.*, 5, 3, 19–42, 2016, https://doi.org/10.4018/ijeoe.2016070102.
26. Kabeyi, M.J.B. and Olanrewaju, O.A., Sustainable energy transition for renewable and low carbon grid electricity generation and supply. *Front. Energy Res.*, 9, 1–45, 2022, https://doi.org/10.3389/fenrg.2021.743114.
27. Shukla, R.D., Singh, N., Roy, S., Power Electronics for Solar Photovoltaic System: configuration, topologies, and control, in: *Handbook of Renewable Energy Technology & Systems*, pp. 235–263, WORLD SCIENTIFIC (EUROPE), Covent Garden, London, 2021, https://doi.org/10.1142/9781786349033_0009.
28. Jhou, J., Gaurav, A., Chang, C., Lin, C., Enhanced efficiency of semitransparent perovskite solar cells via Double-Sided Sandwich evaporation technique for four terminal Perovskite-Silicon tandem application. *Nanomaterials*, 12, 9, 1569, 2022, https://doi.org/10.3390/nano12091569.
29. American Chemical Society, How a solar cell works, American Chemical Society, 2013, https://www.acs.org/education/resources/highschool/chemmatters/past-issues/archive-2013-2014/how-a-solar-cell-works.html.
30. Adnan, N. and Shahrina, N., A comprehensive approach: Diffusion of environment-friendly energy technologies in residential photovoltaic markets. *Sustain. Energy Technol. Assess.*, 46, 101289, 2021, https://doi.org/10.1016/j.seta.2021.101289.
31. Hasan, K., Yousuf, S.B., Tushar, M.S.H.K., Das, B., Das, P., Islam, M.S., Effects of different environmental and operational factors on the PV performance: A comprehensive review. *Energy Sci. Eng.*, 10, 2, 656–675, 2021, https://doi.org/10.1002/ese3.1043.

32. Sathkumara, T.M., Waidyasekara, A., Victar, H., The feasibility of transparent solar panels for high-rise building façade in Sri Lanka. *Constr. Innov. Inf. Process Manage.*, 2023, https://doi.org/10.1108/ci-06-2023-0150.
33. Bromley, B.C., Khan, S.H., Kenyon, S.J., Dust as a solar shield. *PLOS Clim.*, 2, 2, e0000133, 2023, https://doi.org/10.1371/journal.pclm.0000133.
34. Okta, must-know measurement unit of clouds amount, Windy app, https://windy.app/blog/okta-cloud-cover.html.
35. Honsberg, C. and Bowden, S., Effect of Temperature, PVEDucation, (accessed on 08th Jan 2024), https://www.pveducation.org/pvcdrom/solar-cell-operation/effect-of-temperature.
36. Mahdavi, A., Farhadi, M., Gorji-Bandpy, M., Mahmoudi, A.H., A review of passive cooling of photovoltaic devices. *Clean. Eng. Technol.*, 11, 100579, 2022, https://doi.org/10.1016/j.clet.2022.100579.
37. Prasad, R.M. and Krishnamoorthy, A., Barriers restraining the performance in solar photovoltaic and thermal segments. *2017 IEEE International Conference on Power, Control, Signals and Instrumentation Engineering (ICPCSI)*, pp. 861–867, 2017, https://doi.org/10.1109/icpcsi.2017.8391834.
38. Cell Press, Climate conditions affect solar cell performance more than expected. ScienceDaily, 2017, December 13, Retrieved January 1, 2024 from www.sciencedaily.com/releases/2017/12/171213143725.htm.
39. Tripathi, A.K., Ray, S., Aruna, M., Prasad, S.B.S., Evaluation of solar PV panel performance under humid atmosphere. *Mater. Today Proc.*, 45, 5916–5920, 2021.
40. Amajama, J., Effect of Air Pressure on the Output of Photovoltaic Panel and Solar Illuminance. *Int. J. Sci. Eng. Appl. Sci. (IJSEAS)*, 2, 8, 139–144, 2016.
41. Cunningham, P. and Delany, S.J., K-Nearest Neighbour Classifiers - a tutorial. *ACM Comput. Surv.*, 54, 6, 1–25, 2021, https://doi.org/10.1145/3459665.
42. Lopez-Bernal, D., Balderas, D., Ponce, P., Molina, A., Education 4.0: Teaching the basics of KNN, LDA and Simple Perceptron Algorithms for binary classification problems. *Future Internet,* 13, 193, 2021. https://doi.org/10.3390/fi13080193.
43. What is the k-nearest neighbors algorithm?, IBM, Retrieved January 1, 2024 https://www.ibm.com/topics/knn.
44. Mehta, P., Bukov, M., Wang, C.H., Day, A.G.R., Richardson, C., Fisher, C.K., Schwab, D.J., A high-bias, low-variance introduction to Machine Learning for physicists. *Phys. Rep.*, 810, 1–124, 2019, https://doi.org/10.1016/j.physrep.2019.03.001.
45. Alfeilat, H.A.A., Hassanat, A.B.A., Lasassmeh, O., Tarawneh, A.S., Alhasanat, M.B., Salman, H.E., Prasath, V.B.S., Effects of distance measure choice on K-Nearest Neighbor Classifier performance: a review. *Big Data*, 7, 4, 221–248, 2019, https://doi.org/10.1089/big.2018.0175.
46. Carpenter, K.A., Machine learning-based virtual screening and its Applications to Alzheimer's drug discovery: a review. *Curr. Pharm. Des.*, 24, 28, 3347–3358, 2018, https://doi.org/10.2174/1381612824666180607124038.

47. Bengio, Y., Feedforward deep architectures for classification and synthesis, 2018, March 21, https://papyrus.bib.umontreal.ca/xmlui/handle/1866/20501.
48. Banerjee, C., Mukherjee, T., Pasiliao, E.L., The Multi-phase ReLU Activation Function, in: *proc. ACM SE '20*, pp. 239–242, 2020, https://doi.org/10.1145/3374135.3385313.
49. Cui, X., Zhang, W., Tüske, Z., Picheny, M., Evolutionary stochastic gradient descent for optimization of deep neural networks, in: *Proc. NeurIPS 2018*, pp. 1–11, 2018, https://proceedings.neurips.cc/paper_files/paper/2018.
50. Garbin, C., Zhu, X., Marques, O., Dropout vs. batch normalization: an empirical study of their impact to deep learning. *Multimed. Tools Appl.*, 79, 19–20, 12777–12815, 2020, https://doi.org/10.1007/s11042-019-08453-9.
51. Aggarwal, C., *Neural networks and deep learning*, Springer Cham, Springer Nature, Switzerland AG, 2018, https://doi.org/10.1007/978-3-319-94463-0.
52. Abuella, M. and Chowdhury, B., Random forest ensemble of support vector regression models for solar power forecasting, in: *Proc. 2017 IEEE Power & Energy Society Innovative Smart Grid Technologies Conference (ISGT)*, pp. 1–5, 2017, https://doi.org/10.1109/isgt.2017.8086027.
53. Fratello, M. and Tagliaferri, R., Decision trees and random forests, in: *Encyclopedia of Bioinformatic & Computational Biology*, pp. 374–383, 2019, https://doi.org/10.1016/b978-0-12-809633-8.20337-3.
54. Shaik, A.B. and Srinivasan, S., A brief survey on random forest ensembles in classification model, in: *Lecture notes in networks and systems*, pp. 253–260, 2018, https://doi.org/10.1007/978-981-13-2354-6_27.
55. Feng, W., Ma, C., Zhao, G., Zhang, R., FSRF:An Improved Random Forest for Classification. *2020 IEEE International Conference on Advances in Electrical Engineering and Computer Applications (AEECA)*, pp. 173–178, 2020, https://doi.org/10.1109/aeeca49918.2020.9213456.
56. Akbulut, U., Çifçi, M.A., Aslan, Z., Hybrid modeling for stream flow estimation: integrating machine learning and federated learning. *Appl. Sci.*, 13, 18, 10203, 2023, https://doi.org/10.3390/app131810203.
57. Zeini, H.A., Al-Jeznawi, D., Imran, H., Bernardo, L., Al-Khafaji, Z.S., Ostrowski, K., Random Forest Algorithm for the strength Prediction of Geopolymer Stabilized Clayey Soil. *Sustainability*, 15, 2, 1408, 2023, https://doi.org/10.3390/su15021408.
58. McKinney, W., *Python for data analysis*, O'Reilly Media, Inc., Gravenstein, California, USA, 2022.
59. Cakmak, U.M. and Cuhadaroglu, M., *Mastering Numerical Computing with NumPy: Master scientific computing and perform complex operations with ease*, Packt Publishing Ltd., Birmingham, UK, 2018.
60. Pajankar, A., Hands-on Matplotlib, New York, USA, 2021, doi: 10.1007/978-1-4842-7410-1.
61. Yim, A., Chung, C., Yu, A., *MatPlotlib for Python developers: Effective techniques for data visualization with Python, 2nd Edition*, Packt Publishing Ltd, Birmingham, UK, 2018.

62. Kang, M. and Tian, J., Machine Learning: Data Pre-processing, in: *Prognostics and Health Management of Electronics: Fundamentals, Machine Learning, and the Internet of Things*, pp. 111–130, 2018, https://doi.org/10.1002/9781119515326.ch5.
63. Dahouda, M.K. and Joe, I., A Deep-Learned embedding technique for categorical features encoding. *IEEE Access*, 9, 114381–114391, 2021, https://doi.org/10.1109/access.2021.3104357.
64. Rácz, A., Bajusz, D., Héberger, K., Effect of dataset size and Train/Test split ratios in QSAR/QSPR Multiclass Classification. *Molecules*, 26, 4, 1111, 2021, https://doi.org/10.3390/molecules26041111.
65. Singh, D. and Singh, B., Investigating the impact of data normalization on classification performance. *Appl. Soft Comput.*, 97, 105524, 2020, https://doi.org/10.1016/j.asoc.2019.105524.
66. Naimi, A.I. and Balzer, L.B., Stacked generalization: an introduction to super learning. *Eur. J. Epidemiol.*, 33, 5, 459–464, 2018, https://doi.org/10.1007/s10654-018-0390-z.
67. Ahmad, M.W., Reynolds, J., Rezgui, Y., Predictive modelling for solar thermal energy systems: A comparison of support vector regression, random forest, extra trees and regression trees. *J. Clean. Prod.*, 203, 810–821, 2018, https://doi.org/10.1016/j.jclepro.2018.08.207.
68. Sehrawat, N., Vashisht, S., Singh, A., Solar irradiance forecasting models using machine learning techniques and digital twin: A case study with comparison. *Int. J. Intell. Netw.*, 4, 90–102, 2023, https://doi.org/10.1016/j.ijin.2023.04.001.
69. Chu, X., Cai, F., Cui, C., Hu, M., Li, L., Qin, Q., Adaptive recommendation model using meta-learning for population-based algorithms. *Inf. Sci.*, 476, 192–210, 2019, https://doi.org/10.1016/j.ins.2018.10.013.
70. Horizontal Photovoltaic Power Output data, 2021, March 28, https://www.kaggle.com/datasets/saurabhshahane/northern-hemisphere-horizontal-photovoltaic?resource=download.

10

Scenario Analysis and Practical Approach of Deep Learning and Machine Learning Techniques in the Renewable Energy Sector

Supriya[1*], Ashutosh Shukla[2], Priyanka Sharma[3] and Rupendra Kumar Pachauri[2]

[1]COER University, Department of AI & ML, School of Computing, Roorkee, India
[2]Electrical Cluster, School of Advanced Engineering, UPES, Dehradun, India
[3]Department of Mathematics, Shivalik College of Engineering, Dehradun, Uttarakhand, India

Abstract

Renewable energy is one of the most prominent sources of generating clean electricity, a cost-competitive alternative to conventional fossil fuels across the globe for both residential and commercial consumption. Solar energy is the most abundantly available and cleanest energy resource known for a long time. Recently, researchers and practitioners have been efficiently working on power generation by solar photovoltaic (PV) systems, addressing challenges like uneven solar radiation, expensive storage capacity, and varying climatic conditions. Nowadays, lots of machine learning (ML) and deep learning (DL) techniques, including support vector machine, artificial neural network, random forest, convolutional neural network, and recurrent neural network, have been extensively practice to regulate these challenges. These approaches have enormous potential to take solar energy to the next level and satisfy global energy needs. These forecasting techniques and models leverage ML and DL to make accurate predictions for power generation, demand, and electricity pricing involving various renewable energy sources. With this in perspective, a survey of research work discussed the practical applications of these sophisticated techniques in various solar PV system domains like solar power prediction, identification of anomalies or irregularities in the performance

**Corresponding author*: supriya@coer.ac.in

Jai Govind Singh, Rupendra Kumar Pachauri and Sasidharan Sreedharan (eds.) Forecasting Methods for Renewable Power Generation, (279–310) © 2025 Scrivener Publishing LLC

of solar PV systems, predictive maintenance, energy consumption forecasting, and energy trading and pricing. Accordingly, the premise of the research work is to discuss a case study for identifying clean and dusty (surface covered with dust, snow, bird drops, etc.) PV solar panels. Images and comparative analysis of various classification techniques of ML and DL based on evaluation metrics like evaluation metrics, confusion matrix, and F1-score have been performed in this study so that an optimal schedule to monitor and clean these panels can be put into practice to enhance the efficiency and reduce maintenance costs of these modules.

***Keywords*:** Renewable energy, photovoltaic system, soiling effects, convolutional neural network (CNN), no free lunch (NFL) theorem

10.1 Introduction

India is currently among the world's top importers of energy, with approximately 85% of the fuel and energy required coming from outside the country, driving up the cost of energy resources above all other costs. There is a great deal of promise to meet demand while reducing reliance on imports with the potential to harness solar energy.

The Indian Renewable Energy Development Agency Limited claims that the country has an abundance of solar energy capable of producing five trillion kilowatt-hours (kW) of clean energy. Country receives 4–7 kWh of solar insolation per square meter per day and 300 sunny days per year over most of the country. Because India is a developing country with high power consumption, adopting this renewable energy source will become mandatory rather than just necessary [1, 2].

Deep learning (DL) and machine learning (ML) techniques are increasingly being utilized in various aspects of solar photovoltaic (PV) cell technology to improve efficiency, monitoring, and maintenance [5]. Here are several areas where these techniques are applied.

- Solar Power Prediction: These models can be employed to predict the energy output of solar PV systems based on historical weather data, solar radiation patterns, energy management strategies, and other relevant parameters, for optimizing energy production.
- Anomaly Detection: These algorithms can identify anomalies or irregularities in the performance of solar PV systems. This includes the detection of faulty equipment, shading issues, or other factors affecting efficiency.

- Predictive Maintenance: These models can predict when maintenance is needed by analyzing historical performance data. This can help in scheduling maintenance activities and minimizing downtime.
- System Configuration: Machine learning can assist in optimizing the layout and configuration of solar PV systems based on factors such as site conditions, shading analysis, and expected energy production.
- Materials Science: Deep learning can be applied to accelerate the discovery and optimization of materials used in solar cells. This involves predicting the properties of materials and their performance in different conditions.
- Energy Consumption Forecasting: Machine learning models can be used to predict the energy consumption patterns of buildings or facilities, helping to optimize the sizing of solar PV systems for specific applications.
- Energy Trading and Pricing: These algorithms can analyze market data to predict energy prices, enabling solar PV system operators to make informed decisions about energy trading and selling excess energy to the grid.

Given its ability to meet India's increasing energy needs, slow down environmental deterioration, and promote energy security, renewable energy is extremely vital. With ambitious objectives for the expansion of renewable energy capacity, such as the target of 450 gigawatts (GW) by 2030, India hopes to meet its energy needs in an environmentally responsible way, strengthen its energy independence, and create job opportunities and sustainable development [7].

India has an enormous potential for producing renewable energy from a wide range of sources, including biomass, sun, wind, small hydro, and cogeneration bagasse. As of the end of March 2022, the country's overall potential for generating renewable energy is projected to be 1,490,727 MW. This comprises 748,990 MW (50.24%) of solar power potential, 695,509 MW (46.66%) of wind power at a hub height of 120 meters, 21,134 MW (1.42%) of SHP (small-hydro power) potential, 17,538 MW (1.18%) of biomass power, 5,000 MW (0.34%) from bagasse-based cogeneration in sugar mills, and 2,556 MW (0.17%) from waste to energy [8]. Figures 10.1 and 10.2 explore about the renewable energy potential in Indian context as follows.

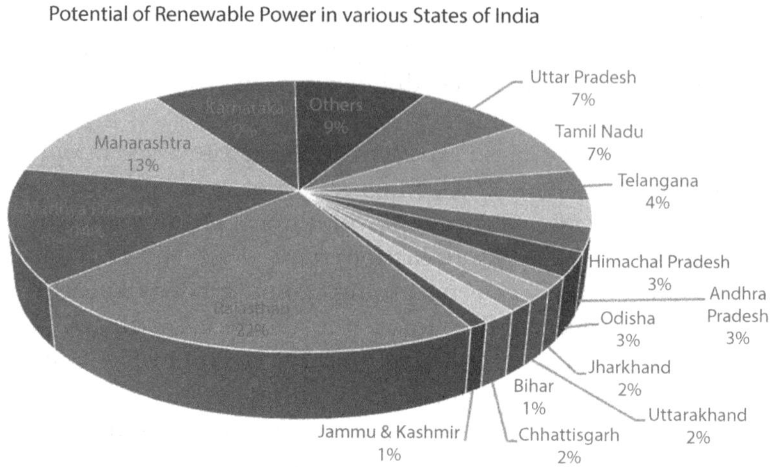

Figure 10.1 Potential of renewable power in various states of India.

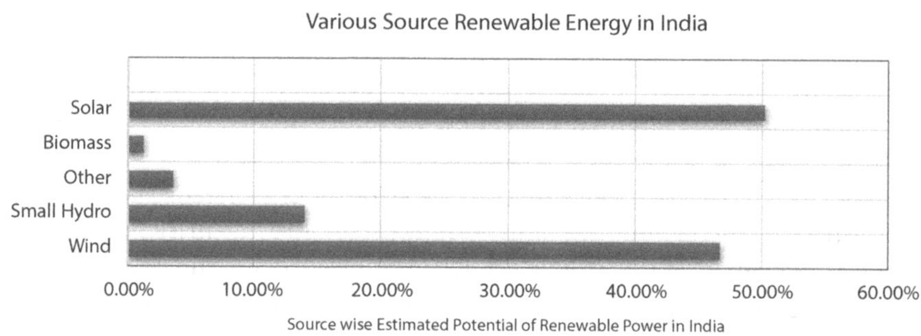

Figure 10.2 Various renewable energy sources in India.

The geographical distribution of India's estimated state-by-state renewable power potential is depicted in the pie chart. With the largest share of over 22% (271,219 MW), Rajasthan is expected to have a total potential of 1,171.86 GW. Karnataka and Maharashtra come next, with 14% and 9% of the total (166,743 MW and 154,162 MW, respectively). In India, these three states account for almost half of the potential for renewable energy [19].

This chart shows the potential of renewable power in India, broken down by source. The x-axis shows the different renewable energy sources, and the y-axis shows the percentage of potential renewable power generated by each source. The chart shows that the majority of India's renewable power potential comes from solar (60%), followed by wind (14%), biomass

(10%), and other sources (16%). The development of renewable energy in India is still in its early stages, but the potential is very large. The chart shows that there is a wide range of renewable energy sources with potential in India, and the government is committed to developing these sources. The Indian government has set ambitious targets for 500 GW by 2030 [20].

Working of the solar PV panels consists of numerous individual solar cells that are made from semiconductor material, often silicon. When sunlight hits the solar cells, the energy from the photons excites electrons in the semiconductor material, creating an electric field. This electric field allows electrons to flow through the solar cell, generating direct current (DC) electricity. The DC electricity then flows through an inverter, which converts it into alternating current (AC) electricity that can be used to power homes and businesses [21]. The amount of electricity generated by a solar PV system depends on several factors, including the size of the system, the amount of sunlight the system receives, and the efficiency of the solar panels, which, in turn, depends on several environmental and system configuration factors shown in the figure that depicts that several factors can affect the performance of a solar PV system, which can be broadly categorized into sunlight irradiation, system design and installation factors, climatic conditions, soiling effects, etc. There are several mythologies related for PV solar forecasting like statistical, image based, ML, and DL for forccasting the generation of power and its consumption. Lots of ML and DL techniques, including support vector machine (SVM), artificial neural network (ANN), random forest (RF), convolutional neural network (CNN), and recurrent neural network (RNN) models, have been extensively practice to address the challenges like uneven solar radiation, expensive storage capacity, and varying climatic conditions [29, 30].

This research work is divided into six sections. The first section, as discussed, presents an introductory segment that analyzes the current scenario of the renewable energy sector in India, the proportion of various renewable energy sources for generation of clean energy, the working of solar PV systems, and their challenges. In the second section, more than 40 research papers were identified and studied. A comparative study of a few recent papers was conducted on the basis of their models and performance on specific datasets. The third section describes the essential steps required for building an intelligent system for solar PV analysis. The fourth section explains some popular ML and DL techniques for solar PV classification, such as SVM, RF, and CNN models used to address the challenges of solar PV panels. A case study is presented in the fifth section, focusing on identifying clean and dusty (surface covered with dust, snow, bird droppings, etc.) PV solar panels. Images and comparative analyses of

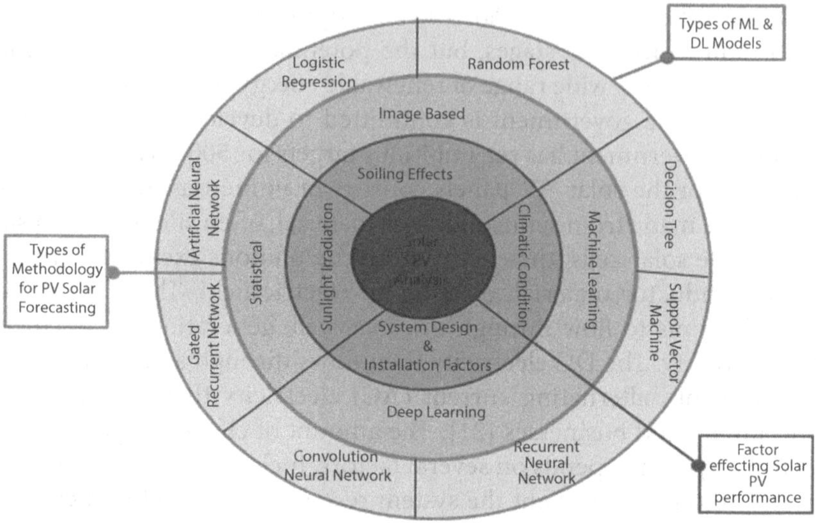

Figure 10.3 Scenario analysis of deep learning and machine learning techniques in solar PV.

various ML and DL classification techniques are provided on the basis of evaluation metrics such as confusion matrix and F1-score. The last section summarizes the conclusions of this research work and identifies its future scope. Figure 10.3 is given for elucidating the DL and ML techniques used in solar PV technology.

10.1.1 Literature Survey

A comparative study of few latest related publications has been carried out and presented in the taxonomy data as shown in Table 10.1. The models and their performances on specific dataset have been summarized as follows.

Through this literature survey, it can be seen that many researchers have used ML and DL for renewable energy in recent years. CNN has been used in a wide range of applications, and they are particularly well suited for solar PV image processing tasks because they can learn to extract spatial features from these images. LSTM networks are the next most popular model, being well suited for tasks that involve sequential data, such as text and time series. Similarly, Generative Adversarial Networks (GANs) are well suited for tasks that involve generating new data, such as images and text features of solar PV systems.

Table 10.1 Taxonomy of the available literature.

S. no.	Aspects investigated	Methodology	Data source	Accuracy	Research gap	Authors
1.	Solar power prediction	CNN-Bi-LSTM hybrid neural network model	Advanced Metering Infrastructure (AMI) system in a solar power plant	78%	Conceals the precise measurements that were employed to assess the proposed model's achievement	Chien-Hsing Lee *et al.* [1]
2.	Solar power prediction	GRU-CNN hybrid neural network model	Private datasets collected from specific power grid operators	--	Best configurations for achieving the highest forecasting accuracy	Lizhen Wu *et al.* [2]
3.	Forecasting of electricity production	Extreme learning machine learning	National Renewable Energy Laboratory (NREL)	--	Need for efficient feature engineering techniques to extract meaningful information from large datasets	Gabriel *et al.* [3]

(Continued)

Table 10.1 Taxonomy of the available literature. (*Continued*)

S. no.	Aspects investigated	Methodology	Data source	Accuracy	Research gap	Authors
4.	Solar power production forecasting	Fuzzy c-means (FCM), fuzzy LSTM neural network, adaptive neuro-fuzzy inference system (ANFIS), and ANFIS with grid partition (GP)	Historical data from a 100-kW PV plant in Turkey	LSTM model was considered as best algo with RMSE, 60.66 kWh.	The model can be used in additional data sources, such as satellite images.	Arif Ozbek *et al.* [4]
5.	Solar power production forecasting	Long short-term memory network	Solar Zenith Angle (SZA) dataset collected from Queensland, Australia	LSTM model demonstrates superior accuracy compared to existing models such as the extreme learning machine (ELM) and the ANN forecast: 0.89.	• Short-term forecasting (up to 1 hour). • Performance for longer forecast horizons would be beneficial.	S. S. Prasad *et al.* [5]

(*Continued*)

Table 10.1 Taxonomy of the available literature. (*Continued*)

S. no.	Aspects investigated	Methodology	Data source	Accuracy	Research gap	Authors
6.	Predicting the radiation effect on solar power generation	SMOreg, linear regression, KNN, and multilayer perceptron	Photovoltaic Geographical Information System (PVGIS)	SMOreg =0.98, linear regression = 0.99	Lack of studies focusing on the integration of big data with physical models for improved forecasting accuracy	B. Kalaiselvi *et al.* [6]
7.	Configurating solar PV module	Support vector machine (SVM) and naïve Bayes (NB)	Publically available dataset	Precision, recall, and F1-score of roof of SVM were 91.2%, 98.6%, and 94.7%.	Lack of insights about the factors influencing forecasting accuracy	Eslam Muhammed *et al.* [7]
8.	Monitoring of solar PV cell	NA	National Oceanic and Atmospheric Administration (NOAA)	-	Lack of Model Transparency and Explain ability	Geetha, S. *et al.* [8]

(*Continued*)

Table 10.1 Taxonomy of the available literature. (*Continued*)

S. no.	Aspects investigated	Methodology	Data source	Accuracy	Research gap	Authors
9.	Photovoltaic system capacity estimation using feeder load	Long short-term memory network	Global Energy Observatory (GEO)	RMSE values occur between 20% and 80% penetration of PV systems	Challenges in integrating data from diverse sources with varying temporal and spatial resolutions	Tang, Lingxi *et al.* [9]
10.	Forecasting regional photovoltaic power generation	Fuzzy logic	Data recorded from 12 dwellings in the central east coast of Italy	-	Ethical concerns regarding data collection	Arif Ozbek *et al.* [10]
11.	Reinforcement agent for electricity trading	Q-learning	• Hourly electricity demand data for a virtual building community in Sweden	-	Implementation of RL in real-world systems raises ethical concerns regarding fairness, bias, and accountability	Lucio Ciabattoni *et al.* [8]

(*Continued*)

Table 10.1 Taxonomy of the available literature. (*Continued*)

S. no.	Aspects investigated	Methodology	Data source	Accuracy	Research gap	Authors
12.	Predicting the household electricity consumption patterns	Based on fuzzy logic inference system	Electricity demand data from 12 dwellings in central-east Italy	-	Need for interpretable models that can provide insights into the factors influencing forecasting accuracy	Lucio Ciabattoni *et al.* [9]
13.	Generate global solar radiation (GSR) forecasts.	Hybrid approach that combines multi-layer perceptrons (MLP) and CNN	Photovoltaic Geographical Information System (PVGIS)	-	Robust data pipelines	Sujan Ghimire *et al.* [11]
14.	Soiling identification	Transfer learning	-	97%	Lack of a study on the effectiveness of transfer learning for soiling detection	Banerjee *et al.* [12]
15.	Soiling classification	Hybrid approach (remote sensing and machine learning)	1000 Solar Panel Soiling Image Dataset	95%	The proposed method is computationally expensive.	Yadav, A *et al.* [13]

(*Continued*)

Table 10.1 Taxonomy of the available literature. (*Continued*)

S. no.	Aspects investigated	Methodology	Data source	Accuracy	Research gap	Authors
16.	Soiling classification	Machine learning and convolutional neural network (CNN)	1,000 Solar PV Panel Images Dataset	90%	The model is not generalizable to different types of solar panels.	A. K. Mishra *et al.* [14]
17.	Soiling classification	Deep learning	200 images of soiled solar panels	96%	The proposed method does not consider the impact of different locations in India on the accuracy of classification.	Singh, S. *et al.* [15]
18.	Estimating solar PV soiling loss	Deep learning solar QRNN is a probabilistic model that combines deep learning and quantile regression.	Solar PV Panel Images Dataset	94.70%	The model does not account for the effects of different types of soiling, such as bird droppings and snow.	Wenjie *et al.* [16]

(*Continued*)

Table 10.1 Taxonomy of the available literature. (*Continued*)

S. no.	Aspects investigated	Methodology	Data source	Accuracy	Research gap	Authors
19.	Soiling classification	Transformer	3,029 images of soiled solar panels	94.00%	The proposed method is computationally efficient, but it is not as accurate as CNN-based methods.	Chen *et al.* [17]
20.	Soiling classification	Convolutional neural network (CNN) with transfer learning	PVGIS-Soiling-2022	96.50%	The proposed method is computationally expensive and requires a large amount of training data.	Zhang, Y. *et al.* [18]
21.	Soiling classification	Convolutional neural network (CNN)	PVGIS-Soiling-2022	98%	Lack of a comprehensive soiling dataset	Kim, J. *et al.* [19]

(*Continued*)

Table 10.1 Taxonomy of the available literature. (*Continued*)

S. no.	Aspects investigated	Methodology	Data source	Accuracy	Research gap	Authors
22.	Soiling prediction	Support vector machine, decision tree, random forest, and gradient boosting	1000 Solar Panel Soiling Image Dataset	90% (SVM)	The model is not yet tested for long-term prediction.	A. K. Mishra *et al.* [20]
23.	Mitigation technique	Experimental study	1000 images of soiled solar panels	85%	The study does not consider the long-term stability of the nanoparticle coating.	N. M. Al-Ghamdi *et al.* [21]
24.	Soiling classification	Convolutional neural network (CNN) with attention mechanism	10 solar panels	98.00%	The proposed method does not consider the effects of different lighting conditions.	Liu, Y., Wang *et al.* [22]
25.	Soiling detection	Generative adversarial network (GAN)	PVGIS-Soiling-2021	97%	The proposed method is not robust to variations in the appearance of soiling.	Yang, Y. *et al.* [23]

10.2 Building an Intelligent System for Solar PV Analyzer

This section demonstrates all the vital steps for building an intelligent system for solar PV analyzer is given in Figure 10.4.

Pre-processing is the premier stage of processing, where the electrical and environmental parameters are monitored and controlled to ensure optimal system performance and safety it involves transforming the data into a format that the model can use to make predictions. Data processing can be performed in a variety of ways, and the most effective method will depend on the particular dataset and ML task at hand. Nonetheless, a few popular preparation methods are as follows: **Cleaning the data** involves removing errors such as missing values, inconsistencies, and outliers. **Scaling the data** means ensuring that all features are on the same standard scale. Encoding categorical data means converting non-numerical data, such as text or images, into a format that the model can understand [24, 25].

Among these parameters, a feature vector is created that select correlated features. The number of elements defines the dimensionality of the vector. These values in the vector can be numerical (like temperature and sunlight irradiation) or categorical (climate like hot and dry, moderate,

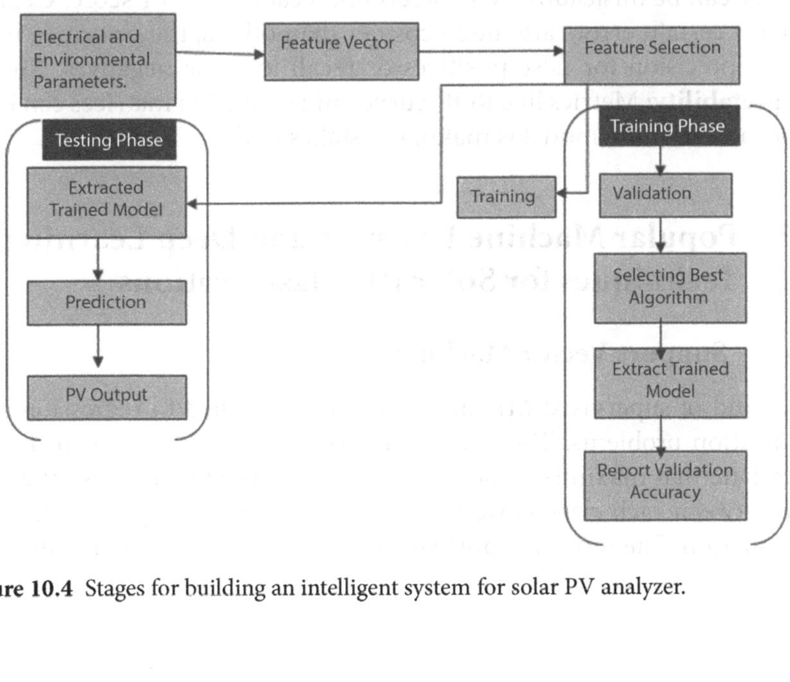

Figure 10.4 Stages for building an intelligent system for solar PV analyzer.

and composite). Feature Selection: In order to select the most relevant and informative features for your model, feature selection is an essential stage in the process of ML. There are two main categories of feature selection techniques: Filter methods: These methods evaluate features based on statistical measures, like correlation or information gain. They are fast and efficient but may not capture complex relationships between features. Wrapper methods: These methods use the actual ML model to assess the performance of different feature subsets. They are more accurate but computationally expensive, especially for complex models.

Selecting Best Algorithm: there is no single algorithm that can outperform all others across all possible learning tasks. It depends on the specific task, the data you have, and your desired outcome according to "no free lunch" (NFL) theorem. Experiment, analysis, and hypertuning of parameters are way to mastering the art of choosing the right algorithm for the real world of ML. Few popular classification algorithms are logistic regression, SVM, RFs, K-nearest neighbors, naive Bayes, etc. For regression, we can use linear regression, support vector regression, RF regression, neural network regression, etc.

Report Validation Accuracy: For calculating the validation accuracy report evaluation metrics, mean squared error (MSE), R-squared, mean absolute error (MAE), root mean squared error (RMSE), etc., are used for regression; use accuracy, precision, recall, ROC curve, AUC, and confusion matrix for classification. Picking the correct metric depends on several factors: **Data imbalance:** If one class is much more frequent than others, then accuracy can be misleading. Use precision, recall, or the F1 score. **Cost of errors:** If certain errors are much costlier than others, the prioritize metrics like precision for false positives or recall for false negatives. **Model interpretability:** Metrics like ROC curves and confusion matrices can help understand how the model is making mistakes [25].

10.3 Popular Machine Learning and Deep Learning Techniques for Solar PV Classifications

10.3.1 Support Vector Machines

It is a kind of supervised ML algorithm that is applied to regression and classification problems. The main objective of this algorithm is to find a hyperplane that maximizes the margin between the two classes. The distance between each class' closest data point and the hyperplane is known as the margin. The term "support vector" refers to the data points that are nearest to the hyperplane.

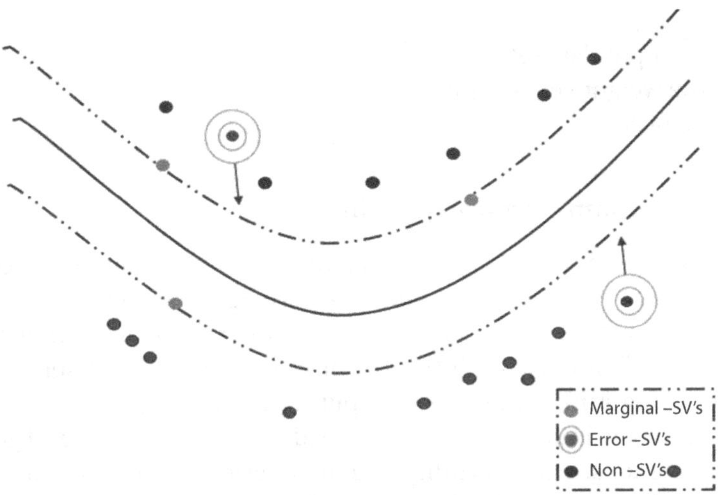

Figure 10.5 Decision boundary through support vectors.

Above figure shows the support decision boundary. The decision boundary is a line that separates the data into two classes. In this case, it is a non-linear boundary. This is because there is a space between the support vectors on each side of the boundary [26].

The support vectors as shown in Figure 10.5 are the data points that are closest to the decision boundary. They are the most important data points, because they are the ones that determine the shape of the boundary. If you remove any of the support vectors, then the decision boundary will change. The margin is the distance between the decision boundary and the support vectors. A larger margin means that the model is more confident in its predictions.

Underfitting can happen if the SVM is trained on too few data points or if the regularization parameter is set too high.

There, the best hyperplane is represented as follows:

$$w_g^T(d_s) + b_i = 0 \tag{10.1}$$

and support vector hyperplanes are represented by

$$w_g^T(d_s) + b_i = -1 \tag{10.2}$$

for lower and

$$w_g^T(d_s) + bi = 1 \tag{10.3}$$

for upper support vector hyperplanes,

where
d_s is the input data set,
w_g is the weight vector, and
b_i is the bias.

10.3.2 Random Forest Algorithm

A powerful ensemble method for classification tasks that combines multiple decision trees is Random Forest. It leverages the collective power of multiple decision trees to improve accuracy and reduce overfitting. By creating a "forest" of trees, each trained on a slightly different subset of the data and using random features to split nodes, the algorithm minimizes the impact of any individual tree's bias, leading to more robust and generalizable predictions. This versatility and high performance make RFs a popular choice for various classification tasks across diverse domains. High accuracy: RFs are known for their high accuracy, robustness, feature ranking, versatility, and scalability. Random forest classifier shown in Figure

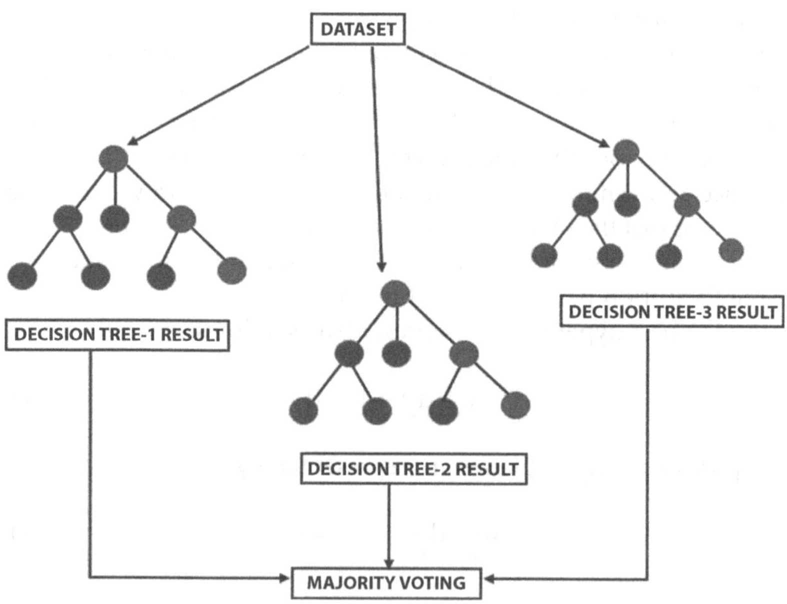

Figure 10.6 Random forest classifier.

10.6 is suitable for situations when we have a large dataset, and interpretability is not a major concern [27].

10.4 Convolutional Neural Network

Convolutional neural network is a type of neural network specifically made to process visual data, like pictures and videos. CNNs, however, are also effective with textual data, particularly in text classification and natural language processing as shown in Figure 10.7. CNN uses an image to extract its spatial features. The arrangement of pixels in an image and their interrelationships are referred to as spatial features. They support precise object identification, as well as the determination of an object's position and relationship to other objects in an image. Semantic segmentation, object detection, image classification, and style transfer are just a few of the many computer vision applications it has. CNNs are capable of effectively learning complex visual patterns because they use specialized filters to extract and analyze spatial features within an image. The filters, referred to as kernels, are convolved throughout the image to capture feature map [28].

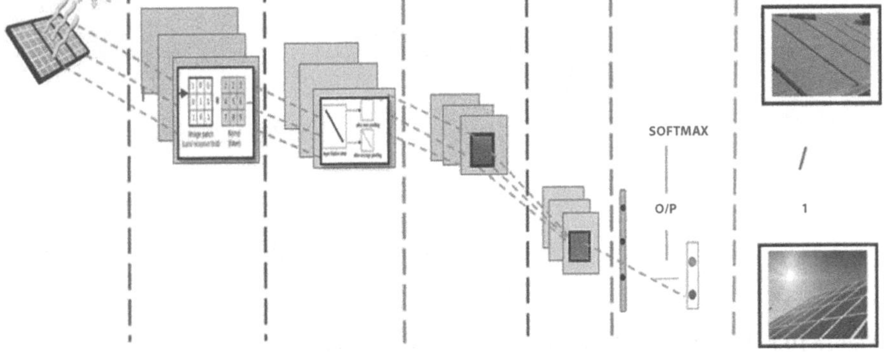

Figure 10.7 Stacks of layer in CNN.

Figure 10.7 illustrates the functionality of a CNN designed to identify clean and dusty images of solar PV panels using stacks of layers. The first layer is a convolutional layer that generates a feature map by performing convolution operations on the pixel values of the image and the filter. The output is then fed into pooling to extract important features. Afterward, a fully connected layer with a sigmoid activation function is utilized to classify binary outputs. Further technical aspects of the model are discussed in the fifth section, where a case study is presented.

Here, the important parameters are as follows:
Input dimension of the image matrix of size

$$(h \times w \times d.) \tag{10.4}$$

with d channels, filter of size

$$(f_h \times f_w \times d) \tag{10.5}$$

and output image obtain is

$$(h-f_h+1) \times (w-f_w+1) \times 1 \tag{10.6}$$

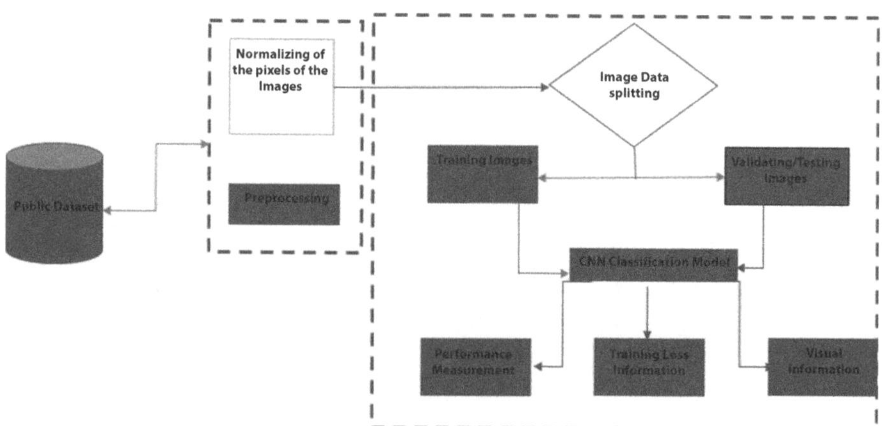

Figure 10.8 Workflow to the solar PV identification system.

where h is the no. of horizontal pixels of the input image,

Figure 10.9 (a) Dusty solar PV module. (b) Clean solar PV module.

w is the no. of vertical pixels of the input image,
d is the no. of channels,
f_h is the no. of horizontal pixels of the filter, and
f_w is the no. of vertical pixels of the filter.

10.5 Case Study

The classification between clean and dusty solar PV modules has been carried out using SVMs, RFs, and DL models as per the methodology given in Figure 10.8. The total number of image datasets is 542, out of which 302 are clean, as shown in Figure 10.9b of the solar PV module, whereas the remaining 240 are dusty, as shown in Figure 10.9a. The whole image dataset is first converted into a textual database through the gray level concurrence matrix (GLCM) method, which is an efficient statistical method for extracting textural information from images. Different combinations of gray levels found within the images have been shown by GLCM. The extractions of textual features from the images by GLCM were useful to identify the different regions in the images. For comparing the values of neighboring pixels, a co-occurrence matrix is generated.

The number of pixels brightness value (gray level) is proportional to the total number of rows and columns. Using mathematical equations, the second-order GLCM characteristics (ASM, entropy, contrast, correlation, variance, IDM, etc.) are computed while taking into consideration the spatial relationship between any two intensity levels of the ROI as shown in Table 10.2.

Table 10.2 Dataset extracted through GLCM.

Feature	Description	Formula
Contrast	Measurement of local variation in the gray-level co-occurrence matrix, where i and j are the row and column indices of the GLCM, and $p(i, j)$ is the probability of the occurrence of the pair (i, j) in the GLCM.	$\sum (i, j)^2 p(i, j)$
Energy	Measurement of the overall uniformity in the GLCM, where $\sum E(i, j)$ represents the pixel location (i, j)	$\sum E(i, j)^2$
Homogeneity	Measurement of how closely the elements in the GLCM are distributed to the diagonal, where $H(i, j)$ represents the pixel location (i, j)	$H = \sum_{i=0}^{A-1} \sum_{j=0}^{A-1} \{H(i, j)\}^2$

(*Continued*)

Table 10.2 Dataset extracted through GLCM. (*Continued*)

Feature	Description	Formula
Correlation	Measurement of the linear dependency between pairs of pixel intensities, where u_p and u_q are the mean values of the ith and jth rows and columns of the GLCM, and σ_p, σ_q are their standard deviations.	$Correlation = \sum_{i=0}^{A-1}\sum_{j=0}^{A-1} \frac{\{i*j\}*P(i,j)-\{u_p*u_q\}}{\sigma_p*\sigma_q}$
Dissimilarity	Measurement of the difference between the pairs of pixel intensities in the GLCM	$\sum \lvert i-j \rvert p(i,j)$
Coarseness	The granularity or roughness of an image is measured.	$C = \frac{1}{1+\sum (i-j)^2 p(i,j)}$
ASM	A measure of the overall uniformity of the GLCM, where i and j are the spatial coordinates of the function $p(i,j)$, and A is the gray tone.	$ASM = \sum_{i=0}^{A-1}\sum_{j=0}^{A-1} P_{i,j}^2$

(*Continued*)

Table 10.2 Dataset extracted through GLCM. (*Continued*)

Feature	Description	Formula
Mean	The average intensity value of the image.	$Mean = \sum_{i=0}^{2A-2} iP_{p+q}(i)$
Variance	Where $P(i, j)$ is the probability of occurrence of the pixel pair (i, j) in the GLCM and μ is the mean value of the GLCM	$Var(X) = \sum\sum(i, j)[P(i, j) - \mu]^2$
Standard Deviation	The statistical measure that characterizes the amount of variation or dispersion in a set of values, where μ represents the mean value of h_{ij}	$S.D = \sqrt{\sum_i \sum_j (i-\mu)^2 h_{ij}}$
Entropy	The measure that characterizes the amount of randomness or uncertainty in a set of values, where $En(i, j)$ represent the pixel at location (i, j)	$Entropy = \sum_{i=0}^{A-1}\sum_{j=0}^{A-1} En(i, j) * \log_2(En(i, j))$

(*Continued*)

Table 10.2 Dataset extracted through GLCM. (*Continued*)

Feature	Description	Formula
Skewness	The measure that characterizes the asymmetry of a distribution of values, where $P(i, j)$ is the probability of occurrence of gray-level pair (i, j) in the GLCM.	$Skewness = \frac{1}{PQ}\sum_{i=1}^{P}\sum_{j=1}^{Q}\left(\frac{p(i,j)-\mu}{\sigma}\right)^3$
Kurtosis	The measure that characterizes the peakedness of a distribution of values.	$K = \left\{\frac{1}{PQ}\sum_{i=1}^{P}\sum_{j=1}^{Q}\left(\frac{p(i,j)-\mu}{\sigma}\right)^4\right\} - 3$

The total attributes are 14, of which 13 are input and 1 is categorical output (clean = 1 and dusty = 0); therefore, the size of the dataset is 542 × 14. After pre-processing the data, it has been split into training (70%) and testing (30%) of the data for implementing the SVM and RF methods. In the case of DL, the CNN method is adopted for the classification. Here, the image can be directly taken as input to develop a convolution layer, followed by pooling, padding, and a flattening layer. The comparative analysis of the above three methods is shown in Table 10.3.

Table 10.3 shows the results of the accuracy, sensitivity, precision, specificity, and F1-score of three models: CNN, RF, and SVM. The CNN model has the highest accuracy (98.29%), followed by the SVM model (98.01%) and the RF model (97.53%). The CNN model also has the highest sensitivity (97.6%) and precision (98.5%), whereas the SVM model has the highest specificity (99.2%). The F1-score is a measure of the balance between precision and recall, and the CNN model has the highest F1-score (0.980), followed by the SVM model (0.977) and the RF model (0.971).

Overall, the CNN model appears to be the best-performing model on this task. However, it is important to note that the performance of a model

Table 10.3 Performance evaluation of ML models.

Model	Accuracy % (Validation)	TP (%)	TN (%)	FP (%)	FN (%)	Sensitivity	Precision	Specificity	F1-score
CNN	98.29877725	97.6	98.8	1.2	2.4	0.976	0.985	0.988	0.980
RF	97.52791069	95.2	99.4	0.6	4.8	0.952	0.991	0.994	0.971
SVM	98.00637959	96.6	99.2	0.8	3.4	0.965	0.989	0.992	0.977

can vary depending on the specific dataset and task. Here are some additional points to consider:

- The sample size of the dataset used to train the models is not given.
- The specific features used in the models are not given.
- The cost of training and deploying each model is not given.

These factors should all be taken into account when choosing a model for a particular task.

10.6 Conclusion and Future Scope

After analyzing the current landscape of the renewable energy sector, one can conclude that DL and ML technologies possess significant potential to revolutionize this clean energy sector. Several practical applications can be designed and implemented in the solar PV sector, which can demonstrably improve cost, efficiency, and reliability. By utilizing these models for optimizing PV cell design and fabrication, predicting PV performance, detecting and diagnosing PV faults, improving solar resource forecasting, and automating PV system operations, the integration of DL and ML has the potential to unlock a new era of solar energy utilization. This paves the way for a more sustainable and efficient future of the PV industry.

Innovative and impactful applications can be seen in coming years, and, with the implementation of these advanced technologies continue to develop, we can expect to see even more. The future scope of these advanced technologies is vast and promising in the renewable energy sector. By harnessing the power of these technologies, we can create sustainable and efficient applications for the clean energy sector. This includes accurately predicting renewable energy generation, demand, and pricing through ensemble learning techniques and new technologies like RNNs, generalized regression neural networks (GRNNs), and LSTMs. These advanced models are ideal candidates for predicting renewable energy due to their ability to process sequential data and learn long-term dependencies.

Appendix: Pseudocode of Algorithms

Appendix- A: Support Vector Machine

Input:

d_s : Training data features
y: Training data labels

Process:

Step 1: Initialize parameters

Initialize weight vector w_g and bias b_i
Set learning rate α
Set regularization parameter C
Set convergence threshold ε

Step 2: Gradient Descent or Optimization

Repeat until convergence:
For each training example ($d_s i_i$, yi) in X, y:
If yi * (w_g * d_s i + b_i) < 1:
Update w = w_g - α * (w_g - C * yi * d_s i)
Update b_i = b_i + α * C * yi
Else:
Update w_g = w_g - α * w_g

Step 3: Store learned parameters (w_g, b_i)

Output:

Updated w_g, b_i

Appendix- B: Random Forest

Input:

Training dataset with features (X) and corresponding labels (Y)
Number of trees to create (num_trees)
Number of features to consider at each split (num_features)

Process: In iterative loop, till all trees are visited:

Step 1: Generate a random sample from the dataset with replacement.
Step 2: Randomly select a subset of features.
Step 3: Build a decision tree using the sampled data and selected features.
Step 4: Add the tree to the forest.

Output:

Random Forest model

Appendix-C: Convolutional Neural Network

Input:
Training dataset Images with Label
Process:
Step: Create a sequential model or define layers:
- Convolutional layers (Conv2D)
- Pooling layers (MaxPooling2D, AveragePooling2D)
- Activation functions (ReLU, Sigmoid, Tanh, etc.)
- Fully connected layers (Dense)
- Output layer with appropriate activation
(softmax for multi-class, sigmoid for binary)

Output:
Accurate CNN model

References

1. Lee, C.-H., Thanh, P.N., Yeh, C.-T., Cho, M.-Y., Three-Phase Load Prediction-Based Hybrid Convolution Neural Network Combined Bidirectional Long Short-Term Memory in Solar Power Plant. *Int. Trans. Electr. Energy Syst.*, 2022, Article ID 2870668, 1, 2022. https://doi.org/10.1155/2022/2870668.
2. Wu, L., Kong, C., Hao, X., Chen, W., A Short-Term Load Forecasting Method Based on GRU-CNN Hybrid Neural Network Model. *Math. Probl. Eng.*, 2020, Article ID 1428104, 10, 2020. https://doi.org/10.1155/2020/1428104.
3. Viscondi, G.F. and Alves-Souza, S.N., A Systematic Literature Review on big data for solar photovoltaic electricity generation forecasting. *Sustain. Energy Technol. Assess.*, 31, 54–63, 2019. ISSN 2213-1388, https://doi.org/10.1016/j.seta.2018.11.008.
4. Ozbek, A., Yildirim, A., Bilgili, M., Deep learning approach for one-hour ahead forecasting of energy production in a solar-PV plant. *Energy Sources Part A*, 44, 4, 10465–10480, 2022, doi: DOI: 10.1080/15567036.2021.1924316.
5. Prasad, S. S., Deo, R.C., Downs, N., Igoe, D., Parisi, A.V., Soar, J., Cloud Affected Solar UV Prediction With Three-Phase Wavelet Hybrid Convolutional Long Short-Term Memory Network Multi-Step Forecast System. *IEEE Access*, 10, 24704–24720, 2022, doi: 10.1109/ACCESS.2022.3153475.
6. Kalaiselvi, B., Karthik, B., Kumaravel, A., Variant Mode Data Analytics in Predicting the Radiation Effect on Solar Power Generation using Machine Learning Algorithms. *2022 IEEE International Conference on Data Science and Information System (ICDSIS)*, Hassan, India, pp. 1–6, 2022, doi: 10.1109/ICDSIS55133.2022.9915814.

7. Muhammed, E., El-Shazly, A., Morsy, S., Building rooftop extraction using machine learning algorithms for solar photovoltaic potential estimation. *Sustainability*, 15, 14, 11004, 2023.
8. Vedulla, G. and Geetha, A., Dust accumulation on solar photovoltaic panels: An investigation study on power loss and efficiency reduction. *Therm. Sci.*, 00, 112, 2023.
9. Tang, L., Hua, W., Ashtine, M., Wallom, D., Distributed Photovoltaic System Capacity Estimation Using Feeder Load Data Based on Deep Learning. Available at SSRN: https://ssrn.com/abstract=4442265, http://dx.doi.org/10.2139%20/ssrn.%204442265, http://dx.doi.org/10.2139/ssrn. 4442265.
10. Ozbek, A., Yildirim, A., Bilgili, M., Deep learning approach for one-hour ahead forecasting of energy production in a solar-PV plant. *Energy Sources, Part A: Recovery, Utilization, and Environmental Effects*, 44, 4, 10465–10480, 2022.
11. Das, S., Mandal, A., Saha, S., Soiling detection of solar panels using a transfer learning approach. *J. Renew. Energy*, 183, 1388–1397, 2023, doi: 10.1016/j.renene.2023.01.117.
12. Ciabattoni, L., Grisostomi, M., Ippoliti, G., Longhi, S., Fuzzy logic home energy consumption modeling for residential photovoltaic plant sizing in the new Italianscenario. *Energy*, 360–442, 74, 359–367, 2014, https://doi.org/10.1016/ j.energy.2014.06.100.
13. Ghimire, S., Deo, R.C., Raj, N., Mi, J., Deep solar radiation forecasting with convolutional neural network and long short-term memory network algorithms. *Appl. Energy*, 253, 113541, 2019.
14. Mudike, R., Barbate, I., Tripathi, A.K., Reddy, Y.B.K., Banerjee, C., Soiling effect on the performance of solar radiometers in composite climatic zone of India, in: *2019 International Conference on Electrical, Electronics and Computer Engineering (UPCON)*, pp. 1–6, IEEE, 2019, November.
15. Yadav, A., Singh, S., Kumar, P., A hybrid approach for soiling classification on solar PV panels in India. *J. Clean. Prod.*, 348, 129827, 2023, doi:10.1016/j.jclepro.2023.129827.
16. Mishra, A.K., Mishra, R.K., Mishra, S.K., A machine learning approach for soiling detection in solar photovoltaic systems. *Sol. Energy*, 211, 122251, 2023.
17. Singh, S., Srivastava, S., Kumar, P., A deep learning approach for soiling classification on solar PV panels in India. *Renew. Energy*, 186, 1425–1434, 2023, doi:10.1016/j.renene.2023.02.059.
18. Zhang, W., Archana, V., Gandhi, O., Rodríguez-Gallegos, C.D., Quan, H., Yang, D., Srinivasan, D., Soiling Edge: PV Soiling Power Loss Estimation at the Edge Using Surveillance Cameras. *IEEE Trans. Sustain. Energy*, 2023.
19. Chen, Y., Zhang, C., Zhang, Y., A transformer-based approach for solar photovoltaic panel soiling classification. *Appl. Energy*, 267, 115287, 2022.

20. Mishra, A.K., Mishra, R.K., Mishra, S.K., A machine learning approach for soiling prediction in solar photovoltaic systems. *Sol. Energy*, 210, 122241, 2022.
21. Al-Ghamdi, N.M., Abdelfattah, A.M.A., Abdelkareem, M.A.M., A novel soiling mitigation technique using nanoparticles for solar photovoltaic systems in desert regions. *Sol. Energy Mater. Sol. Cells*, 230, 111297, 2022.
22. Liu, Y., Wang, W., Zhang, Y., Zhang, C., A convolutional neural network with attention mechanism for solar photovoltaic panel soiling classification. *IEEE Access*, 10, 154441–154450, 2022.
23. Yang, Y., Zhang, C., Zhang, Y., A generative adversarial network approach for solar photovoltaic panel soiling detection. *Appl. Energy*, 278, 115484, 2022.
24. Sharma, A., Singh, R.K., Kumar, A., A review of solar PV analyser tools and techniques. *Int. J. Renew. Energy Res.*, 12, 3, 2345–2354, 2022.
25. Li, H., Zhang, Y., Wang, J., Chen, Y., A hybrid forecasting model for renewable energy based on support vector machine and recurrent neural network. *Energy*, 249, 123732, 2022.
26. Sharma, V., Chaurasia, P., Sharma, G., Performance evaluation of support vector machine for photovoltaic power generation forecasting. *Mater. Today Proc.*, 55, 41–46, 2022.
27. Li, Y., Chen, X., Wang, J., Application of random forest regression for short-term wind speed forecasting based on empirical mode decomposition and chaotic dynamic reconstruction. *Renew. Energy*, 189, 914–926, 2022, doi: 10.1016/j.renene.2022.05.187.
28. Zhao, Y., Zhang, J., Ren, Y., Wang, W., Guo, Z., A hybrid model based on random forest and convolutional neural network for short-term solar power forecasting. *Energy Convers. Manage.*, 267, 115944, 2022, doi: 10.1016/j.enconman.2022.115944.
29. Pachauri, R.K., Imperative role of optimization techniques to reconfigure solar photovoltaic array systems to diminish shading effects. *Int. J. Circuit Theory Appl.*, 50, 1, 317–353, 2022.
30. Pachauri, R.K., Pandey, J.K., Sharmu, A., Nautiyal, O.P., Ram, M. (Eds.), *Applied Soft Computing and Embedded System Applications in Solar Energy*, CRC Press, 2021.

11

Application of Artificial Intelligence and Machine Learning in Assessing Solar Energy Potential

Ajay Mittal

Energy Cluster, School of Advanced Engineering, UPES, Dehradun, India

Abstract
The utilization of renewable energy is growing through the deployment of photovoltaic (PV) solar system on residences and industrial establishments. The recent technological advancement and cost-effectiveness of PV systems has facilitated the expansion of the solar sector. The utilization and expansion of solar energy not only mitigates the effects of climate change by decreasing carbon emissions but also reduces too much reliance on fossil fuels. This chapter presents the notion of artificial intelligence (AI) and highlights its significance in the efficient design, analysis, and estimation of solar energy potential and conversion. The chapter also presents the application of AI in different domains within the field of solar energy and the prevailing AI techniques employed in solar energy research.

Keywords: Artificial intelligence, solar energy, optimization, machine learning

11.1 Introduction

It is remarkable that our planet has a wide array of energy sources, categorized as conventional and non-conventional, that can fulfil our primary and secondary energy needs, particularly for generating electricity. Conventional energy sources such as coal, crude oil, and natural gas are getting depleted, whereas the non-conventional energy sources such as wind energy, solar energy, and hydro energy have not been fully harnessed

Email: ajay.mittal@ddn.upes.ac.in

Jai Govind Singh, Rupendra Kumar Pachauri and Sasidharan Sreedharan (eds.) Forecasting Methods for Renewable Power Generation, (311–320) © 2025 Scrivener Publishing LLC

and utilized [1, 2]. Non-conventional energy sources possess the capacity to alleviate the ecological repercussions associated with conventional energy sources while also diversifying the energy portfolio, thereby potentially fortifying energy security and guaranteeing the enduring accessibility of sustainable energy resources [3, 4]. The renewable energy industry is characterized by substantial unpredictability, which arises from factors such as variations in meteorological conditions, shifts in energy consumption, and limitations in technology [5, 6]. Integrating renewable energy sources into the current energy infrastructure presents numerous barriers due to these challenges [5, 6]. The implementation of MI algorithms in solar energy forecasting provides numerous benefits compared to manual techniques, such as improved precision, scalability, and efficiency [7, 8]. Machine learning (ML) techniques utilize automated data analysis and pattern identification to fully exploit the capabilities of solar energy forecasting. This allows for more efficient resource management, optimal solar system, and augmentation of solar energy infrastructure integration [7]. The fusion of machine learning (ML) and artificial intelligence (AI) with renewable energy sources has the potential to completely transform the energy industry [5].

11.2 Interconnections Between Deep Learning (DL), Machine Learning (ML), and Artificial Intelligence (AI)

AI is the discipline that involves the science and engineering ideas used to create intelligent systems, particularly computer programs. Deep learning (DL)

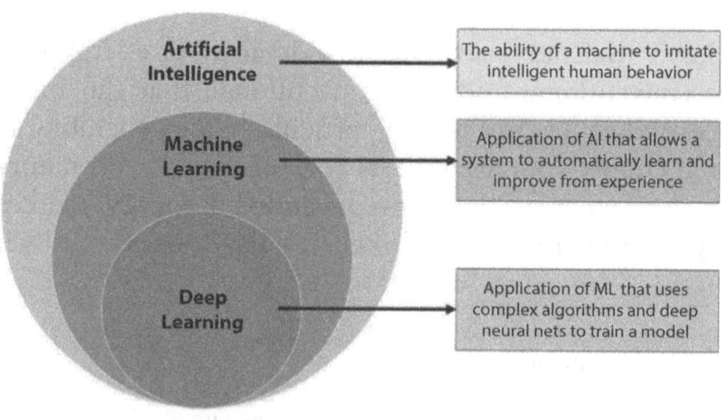

Figure 11.1 Relationships between AI, machine learning, and deep learning [9].

is a specific field within the larger discipline of ML, which is a subset of AI [9]. ML and DL are both regarded as sub-disciplines encompassed under the greater field of AI, as depicted in Figure 11.1.

11.3 Applications of Artificial Intelligence in Assessing Solar Energy Potential

Artificial intelligence possesses an extensive variety of potential applications in the realm of solar power management. The subsequent overview of most of the applications provides a clear understanding of the topic matter.

11.3.1 Predictive Modeling and Evaluation of Solar Systems

Predictive modeling can enhance the efficiency of solar PV systems by assessing historical and real-time data on solar radiation, meteorological conditions, energy consumption, and system performance. This analysis facilitates the optimization of solar system design, operation, and maintenance [7, 9–12].

11.3.2 Selection of Optimal Locations for Solar Installations

The assessment and choice of prospective sites for solar farms are of the utmost importance due to the adjacent conditions have a substantial impact on the efficiency, energy generating capacity, and determine the optimal placement for futuristic development. The utilization of AI aids in the analysis of vast geographical and environmental data, enabling the discovery of sites and regions that offer ideal solar resources and circumstances [13–16].

11.3.3 Design and Fabrication of Solar Cells

Machine learning methodologies have exhibited significant efficiency in facilitating the development and manufacturing of solar cells. Utilizing ML methods not only reduces expenses and time but also yields increased characteristics and improved performance, as evidenced by several research studies [17–19].

11.3.4 Optimizing the Efficiency of Solar Panels

Integrating AI into predictive maintenance has the capacity to reduce solar panel downtime and extend their lifespan. Furthermore, by adjusting to wind conditions and other variables, it is possible to improve the performance of photovoltaic (PV) components, including solar panels. Additionally, AI can provide accurate predictions of energy production, thereby assisting in grid management and energy distribution [7, 20].

11.4 Machine Learning Techniques in Solar Energy Conservation and Management

Physical and statistical models, which are used for solar power forecasting, are outdated and not very reliable [21]. Precise solar power prediction is crucial for efficient grid integration and optimal energy management [14, 15, 18]. The often employed ML techniques are artificial neural networks (ANNs) [7, 10, 17, 22], genetic algorithm (GA) [15–17, 19], particle swarm optimization (PSO) [20], simulated annealing (SA) [16, 20, 23], and random forest (RF) [7, 21], and hybrid algorithm [16, 20, 22], as depicted in Figure 11.2. These ML techniques are increasingly used in solar power forecasting because they can effectively analyze intricate patterns and non-linear connections in solar power generation data.

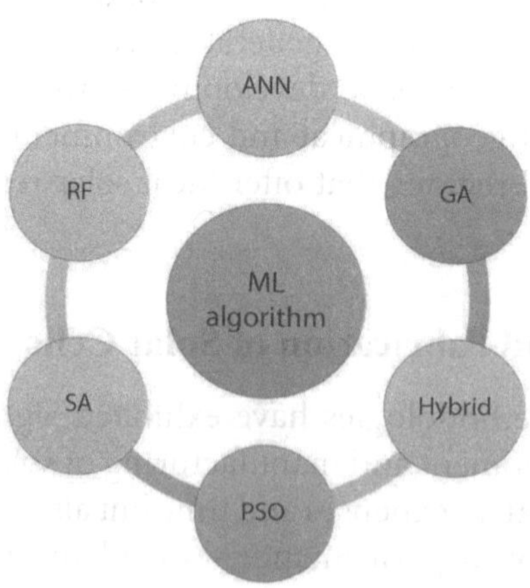

Figure 11.2 Schematic showing the most commonly used ML techniques in solar energy conservation and management [17].

11.4.1 Artificial Neural Networks (ANNs)

A node-based architecture comprises an output layer, one or more concealed layers, and an input layer, among others. Each artificial neuron, referred to as a node, is interconnected with another node and possesses a threshold and weight. Upon surpassing the predetermined threshold value, any given node's output will be initiated, thereby facilitating the transmission of data to the succeeding layer of the network. Conversely, in the absence of any data, its transmission to the succeeding layer of the network will be halted. Machine learning models rely on training data to gain knowledge and improve their accuracy over time. In a recent study, two ANN models were constructed using MATLAB NNTOOL, utilizing the multilayer perceptron methodology. These models were created to predict and evaluate the efficiency of both PV systems and PV-thermal (PV/T) systems, incorporating phase change material (PCM) integration as shown in Figure 11.3. Research indicate that training these ANN models enhances the efficiency and reliability of the prediction process, providing significant insights for future applications and enhanced management of electricity production [18].

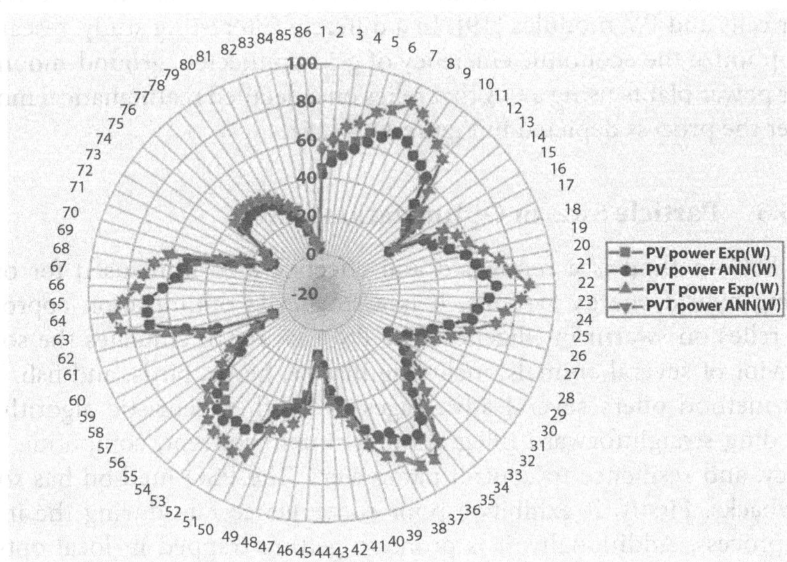

Figure 11.3 Comparison plot between power output of PV (experimental) and PV/T (experimental) vs. PV (ANN) and PV/T (ANN) [18].

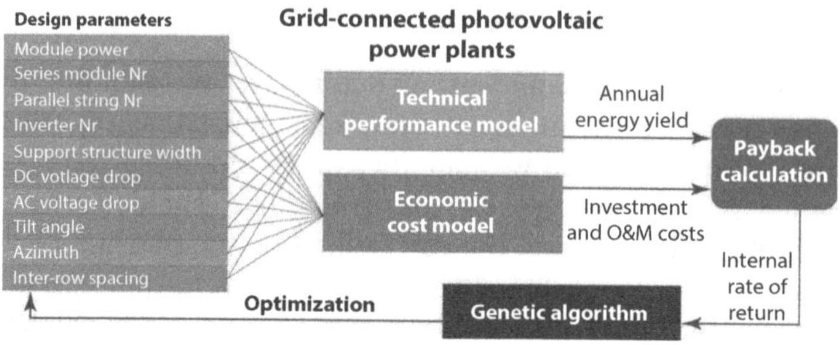

Figure 11.4 Design optimization of grid-connected solar power plants [15].

11.4.2 Genetic Algorithm

Genetic algorithm (GA) is used to solve both constrained and unconstrained optimization problems in various scientific and technical applications. The GAs are derived from the principles of genetics and selection found in nature. The GA identifies the most optimal heliostat field for a specific absorbed power. Additionally, it is utilized to determine the slope and azimuth angles for solar collectors to maximize solar radiation [19].

In a recent study, scholars devised an enhanced technique, specifically a GA utilizing non-uniform mutation, to effectively estimate the features of solar cells and PV modules [19]. In a different interesting study, researchers optimize the economic efficiency of grid-connected, ground-mounted solar power plants using a GA that relies on a detailed mathematical model as per the process depicted in Figure 11.4 [15].

11.4.3 Particle Swarm Optimization (PSO)

The PSO method is a renowned and effective algorithm used for optimizing hybrid energy systems. It is a stochastic optimization approach that relies on swarm intelligence. The PSO approach emulates the social behavior of several animals, including insects, herds, birds, and fish. The PSO method offers several advantages over other heuristic algorithms, including straightforward integration, a simple concept, computing efficiency, and resilience to control parameters. The PSO method has some drawbacks. Firstly, it exhibits a poor convergence rate during the iterative process. Additionally, it is prone to getting trapped in local optima while operating in high-dimensional spaces. Consequently, the algorithm becomes less accurate in regulating its speed and direction [20]. Recent research utilizes the PSO algorithm to ascertain the optimal PV system size for grid-connected installations. This study integrated meteorological data relevant to the locations being examined, in addition to technical and financial information concerning system devices that are commercially available [24].

11.4.4 Simulated Annealing (SA)

The SA algorithm is a computational technique that mimics the progressive cooling process, known as annealing, which is used to manufacture high-quality crystals with improved strength qualities in metals. The primary components of SA are the state transition mechanism and the cooling protocol, both aimed at minimizing the energy state. Creating an ideal crystal can be achieved by effectively managing the temperature throughout the annealing process. A recent study utilized the SA method to estimate variables for solar cell models optimally, demonstrating the significance of this technique [23].

11.4.5 Random Forest (RF)

The RF algorithm is a reliable and adaptable supervised ML technique that constructs and merges many decision trees to form a forest [7]. The findings of the research demonstrated that random forest had superior performance compared to other ML methods in forecasting solar energy generation [7]. The researchers also discovered that the precision of the AI algorithm increased by incorporating additional input parameters such as solar radiation and temperature. The algorithm generates numerous decision trees and combines them to enhance the accuracy of predictions [5, 7, 21].

11.4.6 Hybrid Algorithm

Hybrid algorithms are attracting more and more attention from researchers due to its effectiveness in integration of multiple renewal energy sources. A recent study introduces a new paradigm for identifying the most advantageous site for building solar PV farms. To identify appropriate locations for PV farms, two distinct fuzzy-based methods were employed: the analytical hierarchy process (AHP) and the Dempster-Shafer (DS) methods. According to the research, decision-makers can benefit from the DS method's increased reliability in determining the best locations for PV farms and its ability to assess the risks involved in picking a particular site for a solar farm [13].

11.5 Conclusion and Future Perspectives

AI integration in solar energy presents unprecedented opportunities for innovation, sustainability, and efficiency; it is a game-changer. This research reveals the capabilities of AI that it will inevitably have a profound impact on the trajectory of solar energy in the coming years. Adopting AI in solar energy involves more than simply technology progress; it signifies a dedication to a more intelligent and environmentally friendly future. Presently, the emphasis lies in enhancing the efficiency of the process provision of service. It is recommended to prioritize services by providing the grid network with sensors that allow for the implementation of AI approaches. As AI advances, it holds the potential to enhance solar energy optimization, thereby increasing the accessibility and efficiency of renewable energy for all individuals.

References

1. Kamran, M. and Fazal, M.R., Chapter 1 - Fundamentals of renewable energy systems, in: *Renewable Energy Conversion Systems*, pp. 1–19, Academic Press, 2021, doi: https://doi.org/10.1016/B978-0-12-823538-6.00009-9.
2. Abolhosseini, S., Heshmati, A., Altmann, J., A Review of Renewable Energy Supply and Energy Efficiency Technologies. *SSRN Electron. J.*, 8145, 1–35, 2021, doi: 10.2139/ssrn.2432429.
3. Lu, Y., Khan, Z.A., Alvarez-Alvarado, M.S., Zhang, Y., Huang, Z., Imran, M., A critical review of sustainable energy policies for the promotion of renewable energy sources. *Sustain.*, 12, 12, 1–30, 2020, doi: 10.3390/su12125078.
4. Mittal, A., Assessment of SAARC nations' solar energy potential for sustainable development. *Energy Environ.*, 34, 6, 2228–2255, 2023, doi: 10.1177/0958305X221120935.
5. Nguyen, T.H., Paramasivam, P., Dong, V.H., Le, H.C., Nguyen, D.C., Harnessing a Better Future: Exploring AI and ML Applications in Renewable Energy. *Int. J. Inf. Vis.*, 8, 1, 55–78, 2024, doi: 10.62527/JOIV.8.1.2637.
6. Sajadi, A., Strezoski, L., Strezoski, V., Prica, M., Loparo, K.A., Integration of renewable energy systems and challenges for dynamics, control, and automation of electrical power systems. *Wiley Interdiscip. Rev. Energy Environ.*, 8, 1, e321, Jan. 2019, doi: 10.1002/WENE.321.
7. Ledmaoui, Y., El Maghraoui, A., El Aroussi, M., Saadane, R., Chebak, A., Chehri, A., Forecasting solar energy production: A comparative study of machine learning algorithms. *Energy Rep.*, 10, 1004–1012, Nov. 2023, doi: 10.1016/J.EGYR.2023.07.042.

8. Paletta, Q., *et al.*, Advances in solar forecasting: Computer vision with deep learning. *Adv. Appl. Energy*, 11, 100150, Sep. 2023, doi: 10.1016/J.ADAPEN.2023.100150.
9. Raihan, A., A comprehensive review of artificial intelligence and machine learning applications in energy consumption and production. *J. Technol. Innov. Energy*, 2, 4, 1–26, Oct. 2023, doi: 10.56556/jtie.v2i4.608.
10. Deka, M.J., *et al.*, Enhancing the performance of a photovoltaic thermal system with phase change materials: Predictive modelling and evaluation using neural networks. *Renew. Energy*, 224, 120091, Apr. 2024, doi: 10.1016/J.RENENE.2024.120091.
11. Franki, V., Majnarić, D., Višković, A., A Comprehensive Review of Artificial Intelligence (AI) Companies in the Power Sector. *Energies*, 16, 3, 1077, Jan. 2023, doi: 10.3390/EN16031077.
12. Jathar, L.D., *et al.*, A comprehensive analysis of the emerging modern trends in research on photovoltaic systems and desalination in the era of artificial intelligence and machine learning. *Heliyon*, 10, 3, e25407, Feb. 2024, doi: 10.1016/J.HELIYON.2024.E25407.
13. Mokarram, M., Mokarram, M.J., Khosravi, M.R., Saber, A., Rahideh, A., Determination of the optimal location for constructing solar photovoltaic farms based on multi-criteria decision system and Dempster–Shafer theory. *Sci. Rep.*, 10, 1, 1–17, May 2020, doi: 10.1038/s41598-020-65165-z.
14. Al-Shetwi, A.Q., Sustainable development of renewable energy integrated power sector: Trends, environmental impacts, and recent challenges. *Sci. Total Environ.*, 822, 153645, May 2022, doi: 10.1016/J.SCITOTENV.2022.153645.
15. Mayer, M.J. and Gróf, G., Techno-economic optimization of grid-connected, ground-mounted photovoltaic power plants by genetic algorithm based on a comprehensive mathematical model. *Sol. Energy*, 202, 210–226, May 2020, doi: 10.1016/J.SOLENER.2020.03.109.
16. Kumar, B.A., Jyothi, B., Singh, A.R., Bajaj, M., Rathore, R.S., Tuka, M.B., Hybrid genetic algorithm-simulated annealing based electric vehicle charging station placement for optimizing distribution network resilience. *Sci. Rep.*, 14, 1, 1–28, Apr. 2024. doi: 10.1038/s41598-024-58024-8.
17. Li, F., *et al.*, Machine Learning (ML)-Assisted Design and Fabrication for Solar Cells. *Energy Environ. Mater.*, 2, 4, 280–291, Dec. 2019, doi: 10.1002/eem2.12049.
18. Bouquet, P., Jackson, I., Nick, M., Kaboli, A., AI-based forecasting for optimised solar energy management and smart grid efficiency. *Int. J. Prod. Res.*, 2024, 13, 4623–4644, Jul. 2024, doi: 10.1080/00207543.2023.2269565.
19. Saadaoui, D., Elyaqouti, M., Assalaou, K., Ben hmamou, D., Lidaighbi, S., Parameters optimization of solar PV cell/module using genetic algorithm based on non-uniform mutation. *Energy Convers. Manag. X*, 12, 100129, Dec. 2021, doi: 10.1016/J.ECMX.2021.100129.

20. Liu, H., Wu, B., Maleki, A., Pourfayaz, F., An improved particle swarm optimization for optimal configuration of standalone photovoltaic scheme components. *Energy Sci. Eng.*, 10, 3, 772–789, Mar. 2022, doi: 10.1002/ESE3.1052.
21. Shouman, E.R.M. and Shouman, E.R.M., Solar Power Prediction with Artificial Intelligence. *Adv. Sol. Photovolt. Energy Syst.*, Feb. 2024, doi: 10.5772/INTECHOPEN.1002726.
22. Jlidi, M., *et al.*, An Artificial Neural Network for Solar Energy Prediction and Control Using Jaya-SMC. *Electron*, 12, 3, 592, Jan. 2023, doi: 10.3390/ELECTRONICS12030592.
23. El-Naggar, K.M., AlRashidi, M.R., AlHajri, M.F., Al-Othman, A.K., Simulated Annealing algorithm for photovoltaic parameters identification. *Sol. Energy*, 86, 1, 266–274, Jan. 2012, doi: 10.1016/J.SOLENER.2011.09.032.
24. Kornelakis, A. and Marinakis, Y., Contribution for optimal sizing of grid-connected PV-systems using PSO. *Renew. Energy*, 35, 6, 1333–1341, Jun. 2010, doi: 10.1016/J.RENENE.2009.10.014.

12

Revolutionizing Solar PV Forecasting with Machine Learning Techniques

Supriya[1]*, Ashutosh Shukla[2], Priyanka Sharma[3] and Rupendra Kumar Pachauri[2]

[1]Department of AI & ML, School of Computing, COER University, Roorkee, India
[2]Electrical Cluster, School of Advanced Engineering, UPES, Dehradun, India
[3]Department of Mathematics, Shivalik College of Engineering, Dehradun, Uttarakhand, India

Abstract

Electricity is an essential commodity in today's world, driving both social and economic progress, a unifying catalyst serving as the cornerstone of civilization, powering and shaping life through technology. India has historically relied heavily on fossil fuels like coal and oil to power its growing economy and provide energy for its massive population. These fossil fuels are a significant source of carbon dioxide emissions, which trap heat in the atmosphere and contribute to global warming. To tackle this alarming situation, the country needs to switch to abundantly available renewable energy resources, such as solar, wind, and hydropower. This shift to clean energy sources will not only mitigate climate change but also balance biosphere. Research is being conducted on utilizing advanced technology to optimize the adoption of clean technology; one example of such technology is machine learning (ML). ML is still in its early stages of revolutionizing renewable energy but has vast potential to develop an autonomous system that is driven by sophisticated algorithms and huge datasets providing solutions to every aspect of the green energy spectrum. In light of this, a research work has been proposed with an aim to illustrate the potential of ML-based forecasting systems for solar photovoltaic (PV) power conditioning systems. This research also presents a case study of forecasting power generation by solar PV system using two prominent ML techniques, namely, support vector regression and artificial neural networks, that analyze historical data on publically available renewable energy repository. The study identifies the

**Corresponding author*: supriya@coer.ac.in

Jai Govind Singh, Rupendra Kumar Pachauri and Sasidharan Sreedharan (eds.) Forecasting Methods for Renewable Power Generation, (321–346) © 2025 Scrivener Publishing LLC

best ML approach for accurately forecasting solar power generation by comparing different performance metrics like mean square error, root mean square error, and mean absolute error.

***Keywords*:** Renewable energy, kernel tricks, support vector regression, artificial neural network, activation functions, gradient descent

12.1 Introduction

India primarily relies on fossil fuel to accomplish energy needs, fueling its rapid growth with fossil fuels; the nation faces an ever-widening gap between energy needs and supply, leaving millions in the shadows of an energy crisis. Despite being one of the world's top energy producers, nearly 53% of India's energy needs are currently met by coal. However, projections indicate that India's coal reserves are unlikely to last beyond 2050. This electricity crisis is a complex challenge, but it also presents an opportunity for transformation. The country can build a brighter future powered by clean and accessible energy. Solar, wind, and other clean energy sources offer a sustainable solution that is gaining traction due to their natural replenishment and ability to minimize environmental degradation and ecosystem disruption [1, 3].

Solar energy is the most abundant renewable energy sources; the nation can harvest vast energy due to its geographic location in a tropical region [2]; it receives about 3,000 hours of sunshine annually. The Indian government has taken initiative to harness and create a cleaner, more sustainable energy future for millions. By focusing on rooftop solar installations, smart grids, and energy efficiency measures, the plan has the potential to reduce carbon emissions, create green jobs, and improve air quality in urban centers [29].

The data shown in Figure 12.1 is taken from the Ministry of New and Renewable Energy (MNRE) of India, which evaluates the cost-competitiveness of various renewable energy options and explores avenues for further development within the Indian renewable energy sector. It can be clearly observed that solar energy and wind energy dominate the renewable energy landscape, comprising a combined 83.8 GW installed capacity and 190.6 TWh of generation, far exceeding other sources. However, hydropower also makes a significant contribution with 45.5 GW and 120.0 TWh, whereas biomass remains a smaller contributor with 10.3 GW and 55.9 TWh [4]. According to the International Renewable Energy Agency, renewable energy resources have the potential to meet 3,078 times the current global energy demand (where "X" represents the total demand), as shown in Figure 12.2. These facts highlight the immense potential of renewables to fulfill our future energy needs [5].

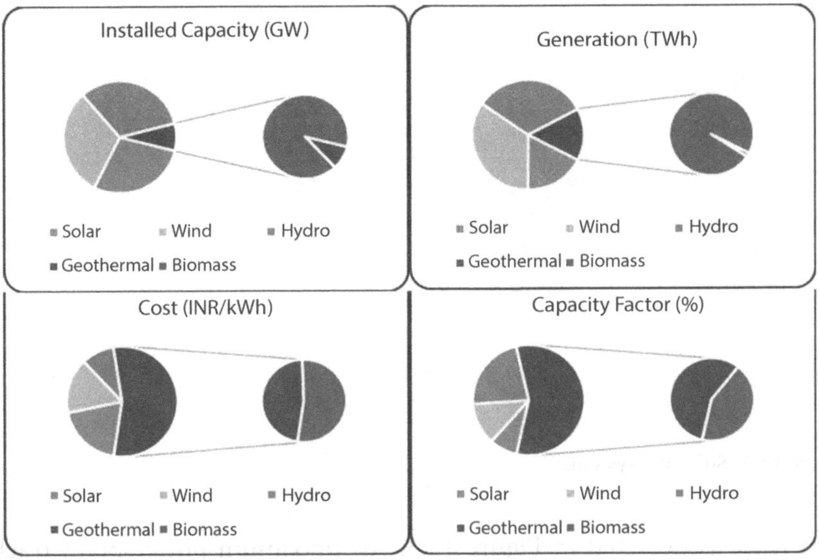

Figure 12.1 Cost-competitiveness of renewable energy resources. Data source: MNRE.

As one of the most significant applications of solar, the solar photovoltaic (PV) system is explored in this research, which consists of solar panels that harness sunlight through silicon cells, generating direct current (DC)

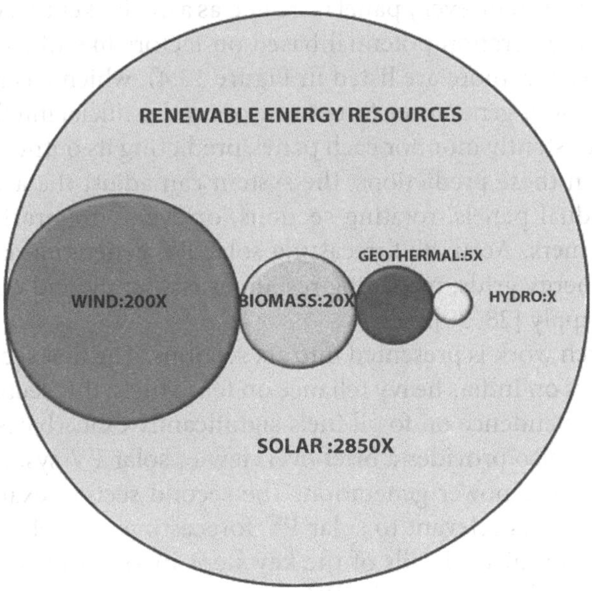

Figure 12.2 Potential of renewable energy resources [5].

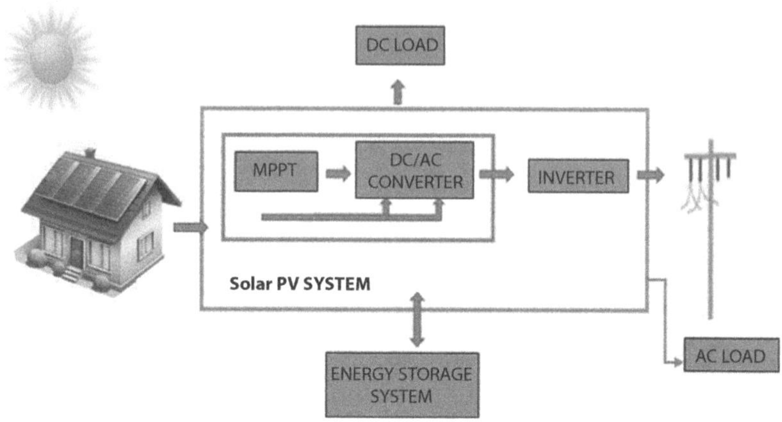

Figure 12.3 Solar PV system.

electricity, as depicted in Figure 12.3. An maximum power point tracking (MPPT) controller optimizes this output, whereas an inverter converts it to household-friendly alternating current (AC) power. A meter tracks production, and the energy then flows: either powering the building or flowing back to the grid through the electricity meter. An optional battery system stores excess solar energy for later use, providing backup during outages or smoothing out power fluctuations [27, 29].

In the above system, every panel is treated as a distinct entity, with its own unique power generation potential based on factors like tilt, shading, and temperature (a few more are listed in Figure 12.4), which has a significant effect on the power generation. Smart sensors and artificial intelligence (AI) algorithms constantly monitor each panel, predicting its output in real time [15]. Based on these predictions, the system can adjust the array's layout, tilting individual panels, rotating sections, or even temporarily disabling under performers. Accurate forecasting solar PV generation is crucial for optimizing energy grids, managing resources efficiently, and ensuring reliable power supply [28, 30].

The research work is presented into six sections. The first section, as discussed, focuses on India's heavy reliance on fossil fuels; this section explores how India's dependence on fossil fuels significantly exacerbates the greenhouse effect. It also provides a brief overview of solar PV systems and factors affecting their power generation. The second section examines some of the latest articles relevant to solar PV forecasting and related domains. The third section gives details of the key steps involved in constructing a smart system for forecasting solar PV power generation. The fourth section

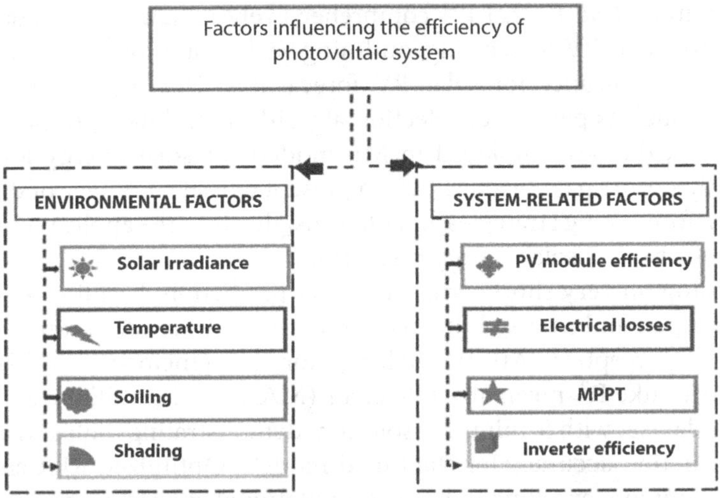

Figure 12.4 Factors influencing the efficiency of solar PV system.

discusses the importance and characteristics of two prominent machine learning (ML) algorithms, namely, support vector regression (SVR) and artificial neural network (ANN), in the context of solar PV forecasting. The fifth section presents a case study demonstrating the application of the forecasting power generation in solar PV system, and the last section summarizes the research findings and emphasizes the critical importance of transitioning from fossil fuels to renewable energy. Furthermore, it proposes potential future research directions, including exploring ensemble learning and deep learning techniques for improved accuracy in this time-series sequential data analysis.

12.2 Related Work

Connor Scott *et al.* (2023) [6] perform a comparative analysis of benchmark ML algorithms (MLAs) utilized for predicting a PV generation system within an operational University campus in Manchester, UK. The ML models implemented are ANN, support vector machine (SVM), and random forest (RF) classification algorithms. The findings discovered that the RF algorithms demonstrated the lowest average error [32 root mean square error (RMSE)] in contrast to SVM (32.3 RMSE), linear regression (LR) (36.5 RMSE), and neural network (NN) (38.9 RMSE). This study highlighted the importance of assessing the performance of diverse MLA PV forecasting models across varying dataset sizes.

S. Nikitha *et al.* (2022) [7] comprehensively review and evaluate ML models for solar PV forecasting, outlining various approaches, workflows, and ML's significance for solar PV forecasting. This paper outlines the methods, such as parameter selection algorithms and data pre-processing techniques, that are employed in ML models for solar energy forecasts. The study's principal objectives are to present a brief summary of different solar PV forecasting strategies and offer insights into the application of ML models for precise solar energy prediction.

Focusing on very short-term PV power prediction, Javadi *et al.* (2022) [8] developed and assessed the performance of LR, random forest (RFR), multilayer perceptron (MLP), and long short-term memory (LSTM) models. Metrics like R^2, mean absolute error (MAE), and RMSE were used on a 1-year dataset with 5-minute resolution, demonstrating high and acceptable prediction accuracy for the tuned models. Optimized RFR emerged as the champion, excelling in both computational efficiency and accuracy.

Abdel-Nasser *et al.* (2022) [9] present a method using a multilayer feed-forward neural network (MLFFNN) to predict the power output of a solar PV power station. This approach has the potential to optimize energy management and grid integration for solar energy. This model is trained using data from a real PV power station in Egypt. This model uses ANN model of three layers: the input layer with two inputs, the non-linear hidden layer with a hyperbolic tangent activation function, and the output layer that estimates the power of the PV power station. The estimated power by the MLFFNN is compared with the actual power obtained from a real PV power station.

Gursoy *et al.* (2022) [10] deployed an expansive dataset comprising environmental factors and generated power values from PV panels across three distinct Turkish regions. The research focused on constructing ML models (including MLP, convolutional neural network (CNN), LSTM, stacked LSTM, bidirectional LSTM, and CNN-LSTM) aimed at forecasting PV power generation through the utilization of environmental parameters within these regions. The accuracy rates achieved by these models were 98.9%, 98.6%, 95.1%, 95.0%, 95.0%, and 84.9%, respectively. The outcomes indicated the effectiveness of the proposed techniques in PV power prediction, with MLP and CNN methods exhibiting superior performance in comparison to other methodologies.

Alhmoud, Lina *et al.* (2022) [11] investigate the application of AI and ML techniques to enhance the forecasting and management of renewable PV solar supplies. The aim is to create a novel sustainable energy forecasting system. The study concludes that RF model yields the most accurate results in predicting solar power consumption, achieving a RMSE of 172.07 and a MAE of 68.7. This model predicts the consumed power output of the Yarmouk University PV solar system in Jordan.

In the article by Gabriel *et al.* (2021) [12], the power output of multiple PV technologies installed at the outdoor test facility of the Council for Scientific and Industrial Research in Pretoria was modeled using ML methods. The paper discusses the use of MLAs, specifically SVM, ANN, and extreme learning machine, for predicting daily solar irradiance in a Sao Paulo context.

Khizir, Mahmud *et al.* (2021) [13] discuss ML-based PV power generation forecasting, including the use of various regression techniques algorithms such as LSTM, RF, decision tree, polynomial, linear, and MLP. It provides a comparative analysis of their performance and suggests that RF regression performed better for the dataset.

Spyros *et al.* (2021) [14] consider the upcoming-day performance evaluation of various ML models for PV production forecasting, including Bayesian neural network, SVR, and RF models through this study. It offers useful information for creating a reliable day-ahead forecasting system. Figure 12.5 illustrate the steps of building a smart system for solar PV forecasting, it depict the overall process involved in creating a system that predicts solar power generation using ML techniques.

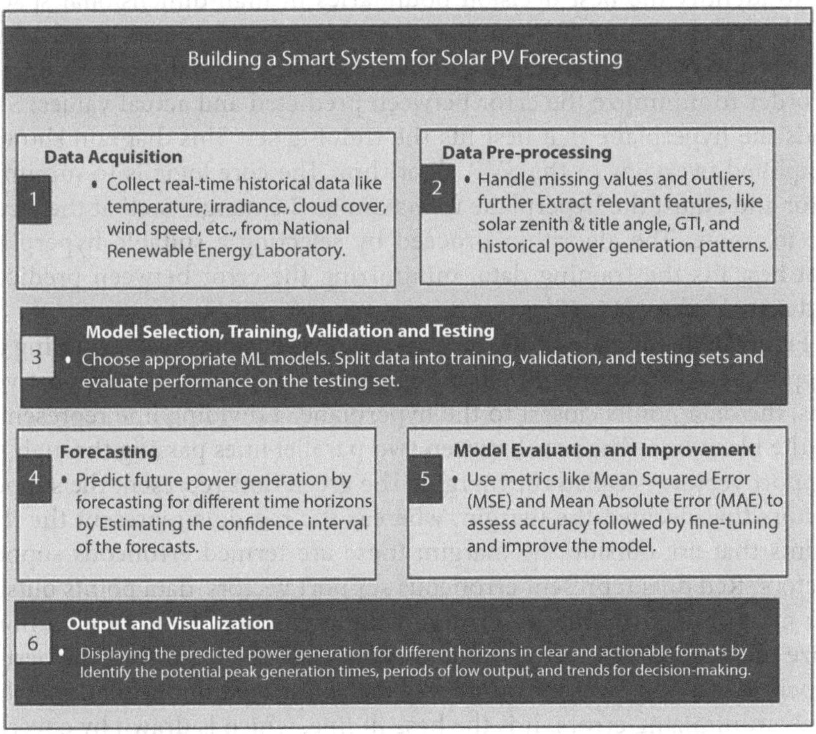

Figure 12.5 Steps of building a smart system for solar PV forecasting.

12.3 Smart System for Solar PV Forecasting

Smart system for solar PV forecasting represents a noteworthy advancement in solar energy management. By providing precise predictions, ML technology empowers researchers by optimizing energy production, which paves the way for a more efficient and sustainable future for renewable energy. This flowchart depicts all the steps involved in building a smart system for forecasting of solar PV power generation using ML techniques with historical data captured from various data repositories [19, 20].

12.4 Prominent Machine Learning Techniques for Forecasting

12.4.1 Support Vector Regression

Support vector regression (SVR) is an influential MLA for regression analysis. It is a variant of the SVM algorithm, which is prominent for its capacity to identify the best decision boundaries in high-dimensional spaces. While both SVR and SVM utilize the maximum margin principle, SVR applies it to continuous outputs, unlike SVM's focus on discrete categories in order to minimize the error between predicted and actual values; SVR finds the hyperplane that best fits the training set. This diagram shows a simplified overview of the SVR algorithm. The core logic is to minimize error and adjust the hyperplane to maximize the margin so that the errors are tolerable. The algorithm proceed by selecting a suitable hyperplane that best fits the training data, minimizing the error between predicted and actual values [16, 17].

Figure 12.6 illustrates how the hyperplane is decided by considering the support vectors and margin. These green dots represent the support vectors, the data points closest to the hyperplane, a dividing line represented by the black line. The area between two parallel lines passing through the support vectors is called the margin. The green dots represent the support vectors that decided the margin, whereas the red dots represent the data points that are outside the margin; these are termed erroneous support vectors. Red dots represent erroneous support vectors, data points outside the margin, contributing to errors in the prediction. SVR aims to minimize the sum of these errors for all data points. The black line, the best-fit hyperplane, maximizes the number of data points within the margin, effectively minimizing errors. It is the best-fit line, which is drawn by covering maximum number of data points. The goal of SVR is to find the best line

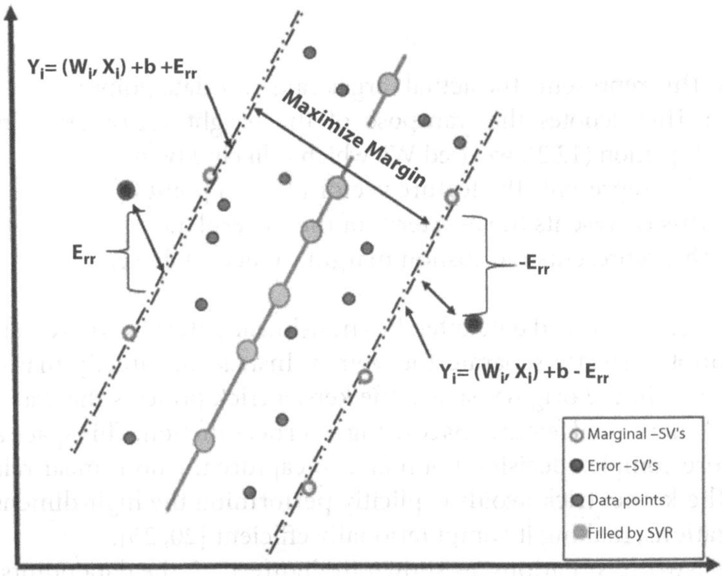

Figure 12.6 Finding a hyperplane for SVR.

that separates most of the data points with a clear gap on either side and minimize the total of these errors [18, 19, 22].

Unlike other regression algorithms, SVR offers the flexibility to specify an acceptable error margin (€) for our model. It then finds a hyperplane that minimizes this error while still fitting the data well.

$$\text{Minimize: } \frac{1}{2}\|W\|^2 \tag{12.1}$$

$$\text{Constraints: } |yi - (w^T x_i + b)| <= \varepsilon \tag{12.2}$$

Here, the equation of upper decision boundary is given by

$$W_i X_i + b_+ \varepsilon, \tag{12.3}$$

The equation of lower decision boundary is given by $WiXi + b - \varepsilon$, and the equation of the hyperplane is given by $W_i X_i + b$

where:

yi: This represents the actual target value for data point i.
w^T: This denotes the transpose of the weight vector (W). In Equation (12.2), we used Wi, which is likely a typo.
x_i: This represents the feature vector for data point i.
b: This represents the bias term of the hyperplane.
ε: This represents the epsilon margin for acceptable error.

However, real-world data often has non-linear patterns, where a straight line cannot perfectly separate the points. Instead of directly manipulating the data in the original space, the kernel trick projects the data into a higher-dimensional feature space using a kernel function. This space allows for a more complex decision boundary to capture the non-linear relationships. The kernel trick avoids explicitly performing the high-dimensional computations, making it computationally efficient [20, 25].

In real-world situations, as shown in Figure 12.7, the data points often exhibit non-linear relationships. This makes separating them using a linear hyperplane challenging, requiring us to move into a higher-dimensional space. However, kernels provide a clever solution by enabling us to find a hyperplane in this higher space without incurring the typically high computational cost of operating in such dimensions. Table 12.1 discusses three key kernels implemented in SVM along with guidance on choosing the appropriate one:

- Linear Kernel: This fundamental kernel excels for linearly separable data. It directly calculates the dot product of data points, making it efficient for large datasets.

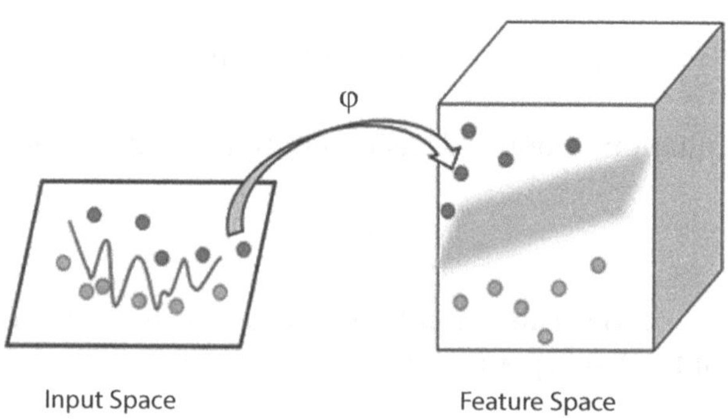

Figure 12.7 Role of kernel in SVR.

- Polynomial Kernel: When data exhibits non-linear relationships, it enhances flexibility by mapping data to a higher-dimensional space by controlling degree parameter, enabling it to capture complex patterns and feature interactions.
- Radial Basis Function (RBF) Kernel: This highly popular kernel stands out for its effectiveness with non-linear data. It maps data to an infinite-dimensional space, offering exceptional flexibility.

Table 12.1 Few well-known kernels.

Kernel type	Graph and its characteristics
1. Linear Kernel	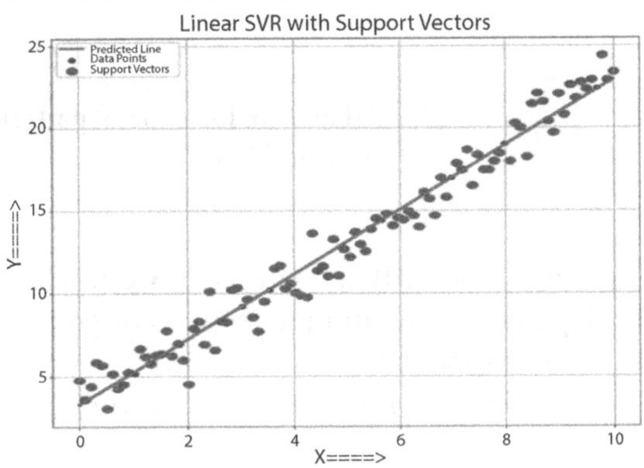 Suitable for linearly separable data, works in the original space.
2. Polynomial Kernel	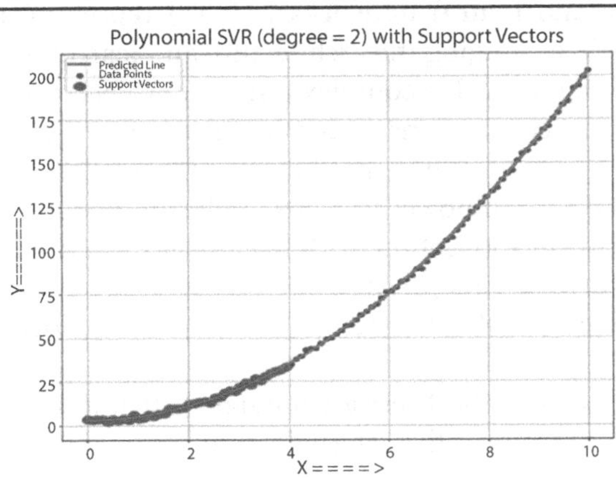 Capture interactions between feature by raising the features to a certain power.

(*Continued*)

Table 12.1 Few well-known kernels. (*Continued*)

Kernel type	Graph and its characteristics
3. Radial Basis Function (RBF) Kernel	**RBF SVR Prediction Surface** 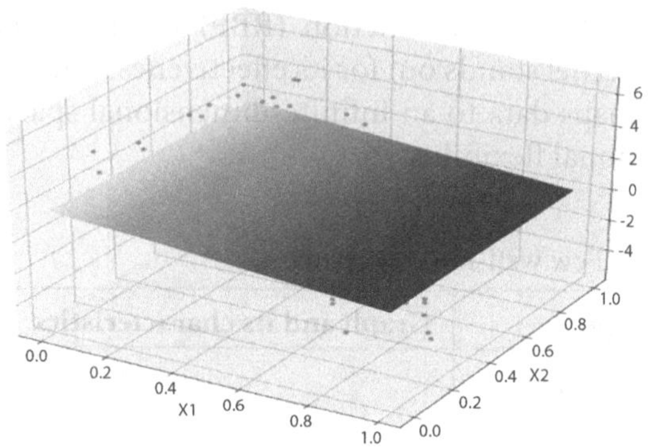 Adapts to local data patterns, powerful for complex relationships.

It is worth noting that there is "**no one-size-fits-all kernel**." The best choice depends on the unique features of your data and the desired complexity of your model. Experimentation and accurate evaluation are crucial to the optimal kernel for specific requirements.

12.4.2 Artificial Neural Network

Artificial neural networks (ANNs) represent a cutting-edge paradigm in ML, leveraging the architecture and learning algorithms of the biological brain to tackle complex tasks. As ANNs process data, they adjust their internal connections, becoming increasingly adept at recognizing patterns and making accurate predictions. The availability of vast data and powerful GPUs add power to ANN applications, revolutionizing many complex real world application fields like image detection, language translation, and self-driving cars [21, 22].

Figure 12.8 illustrates the basic building block of an ANN.

- **Inputs:** These are the signals that it receives from other neurons or the external world.

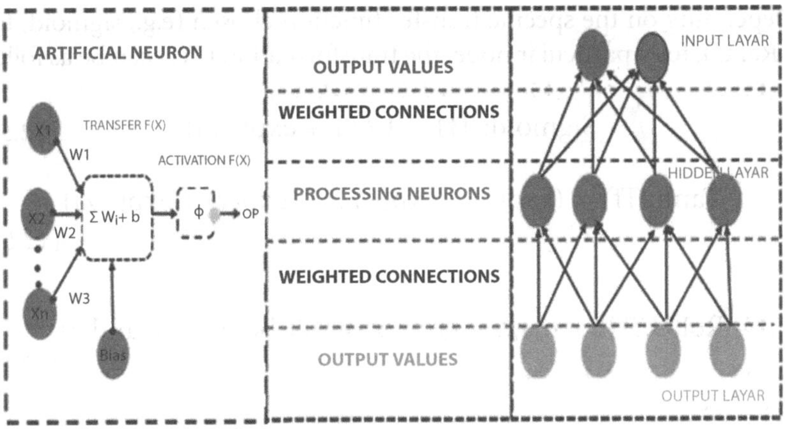

Figure 12.8 Artificial neural network architecture.

- **Weights:** Each input carries a weight, determining its impact on the neuron's decision. A high weight amplifies the influence, whereas a low weight downplays its significance.
- **Transfer Function:** Transfer functions like sigmoid, tanh, and ReLU introduce non-linearity by mapping the weighted sum of inputs to an output between specific ranges (e.g., 0 to 1 or −1 to 1). This allows the network to learn complex patterns and relationships that would not be possible with linear functions.

Computation of a transfer function is given by the following:

$$z = w_1{}^*x_1 + w_2{}^*x_2 + \ldots + w_n{}^*x_n + \text{bias} \tag{12.4}$$

where:

z: This represents the weighted sum of the inputs before applying the transfer function. It is often called the linear activation or pre-activation.

$w_1 \ldots w_n$: These represent the weights associated with each feature ($x_1 \ldots x_n$).

$x_1 \ldots x_n$: These represent the individual input features for a particular node.

Bias: This represents the bias term for the node.

Depending on the specific transfer function chosen (e.g., sigmoid, tanh, and ReLU), for a particular node, the transformation (Tf1) can be as follows:

$$\text{Sigmoid: Tf1} = 1 \,/\, (1 + \exp(-z)) \tag{12.5(a)}$$

$$\text{Tanh: Tf1} = (\exp(z) - \exp(-z)) \,/\, (\exp(z) + \exp(-z)) \tag{12.5(b)}$$

$$\text{ReLU: Tf1} = \max(0, z) \text{ (maximum between 0 and z)} \tag{12.5(c)}$$

- **Activation Function:** Activation functions introduce non-linearity by transforming the weighted sum of inputs into a non-linear output. Table 12.2 illustrates different activation functions that allow the neuron to learn complex relationships. By setting a threshold, the activation function determines if the neuron's output should be significant enough to be passed on to other neurons [23].

Assume that a sigmoid activation function on the hidden layer ***Hd1***, after implementing This function the transformation, will be represented as follows:

$$Hd1 = \text{Sigmoid}(Tf1) \tag{12.6}$$

where:

$$\text{Sigmoid}(Hd1) = e^{x}(Tf1) \,/\, (1 + e^{x}(Tf1)) \tag{12.7}$$

- **Output:** Triggered by the activation function, the neuron undergoes an activation event, sending its processed data downstream to other neurons or the output layer

The Figure 12.9 shows a block diagram of a three-layer feedforward neural network. The network has R inputs: S_1, S_2, and S_3 are neurons in three layers. The network also has a bias input that is connected to all of the neurons

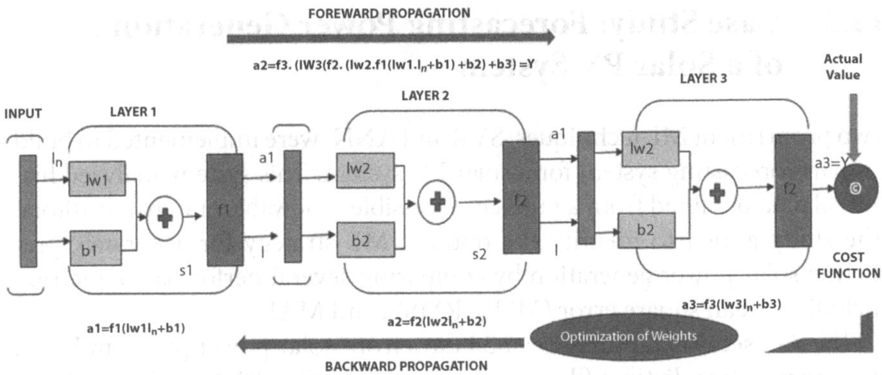

Figure 12.9 Multi-layered artificial neural networks.

in each layer. The equations shown in the figure calculate the output of each neuron:

$$a = f\,(\Sigma(wi^T * x_i) + b) \qquad (12.8)$$

where:

- a: This represents the output of the neuron.
- f: This represents the activation function applied to the weighted sum.
- wi^T: This represents the transpose of the weight vector associated with this neuron.
- x_i: This represents the input vector from neuron i^{th} in the previous layer.
- b: This represents the bias term for this neuron.

In the forward propagation phase, the network is fed input data. The input data is passed through the network's layers, and the network produces an output. During training, the network adjusts the weights based on the error between its predictions and the actual outputs. This process, called **backpropagation,** relies on the gradient of the activation function, as shown in Figure 12.9 [24]. The weights of the network are adjusted using a technique called gradient descent. **Gradient descent** works by iteratively adjusting the weights in the direction of the negative gradient of the error function. This process calculates the "slope" of the error function, i.e., how much the error changes with respect to each weight and adjusts the weights in the opposite direction of that slope [25, 26].

12.5 Case Study: Forecasting Power Generation of a Solar PV System

Two prom1inent ML techniques, SVR and ANN, were implemented to build a smart forecasting system for a solar PV system. This system analyzed historical data obtained from a publicly accessible renewable energy repository. The study aimed to identify the optimal ML strategy for accurately predicting solar power generation by comparing several performance metrics, including mean square error (MSE), RMSE, and MAE.

The dataset was captured over 34 days from solar power plants in India. It comprises two distinct files: power generation file ("Date & Time," "Plant Id," "Source Key," "DC Power," "AC Power," "Daily Yield," and "Total Yield") captures data at the inverter level, with each inverter linked to multiple lines of solar panels; and the other sensor reading file ("Date & Time," "Plant Id," "Source Key," "Ambient Temperature," "Module Temperature," and "Irradiation") gathers data at the plant level, utilizing a strategically positioned single array of sensors. Figure 12.10 illustrate the Covariance matrix of the dataset, which depicts the covariances between pairs of variables in a dataset.

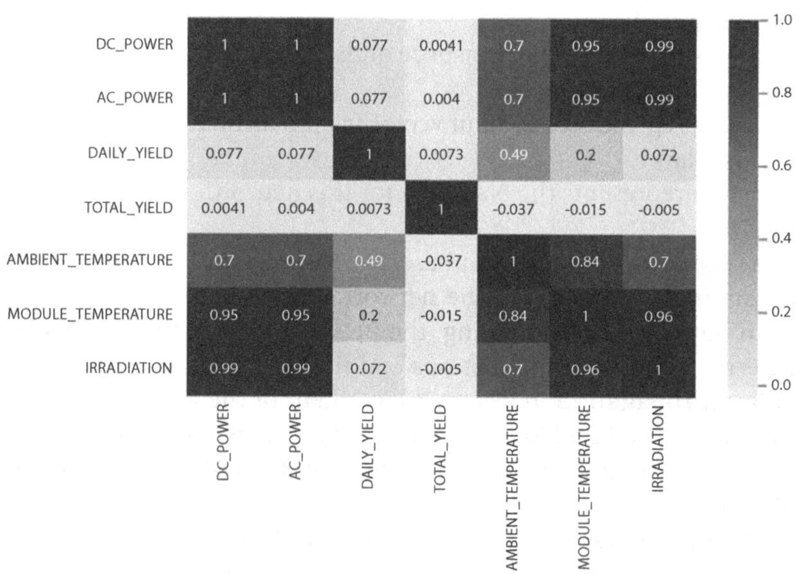

Figure 12.10 Covariance matrix of the dataset.

Table 12.2 Few well-known activation functions.

Activation function	Characteristic equation	Graphical representation	Characteristic
1. Binary Step	$f(x) = \begin{cases} 0 if\ x < 0 \\ 1 if\ x \ddagger 0 \end{cases}$	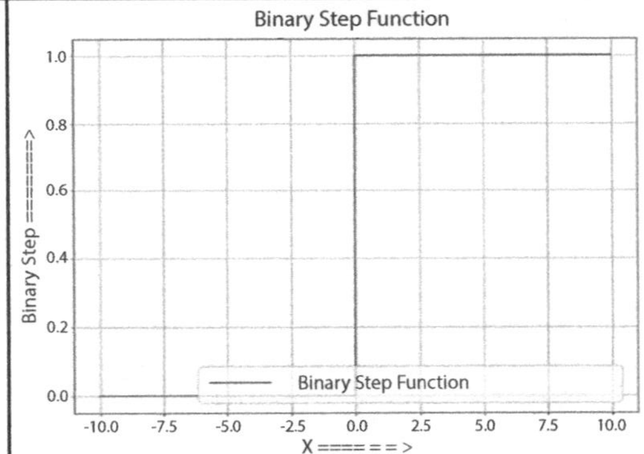	Perceptions, simple models for binary classification, rely on the binary step function. Adjusting its threshold controls neuron sensitivity, higher thresholds demanding more to activate.
2. Linear Function	$f(x) = x$	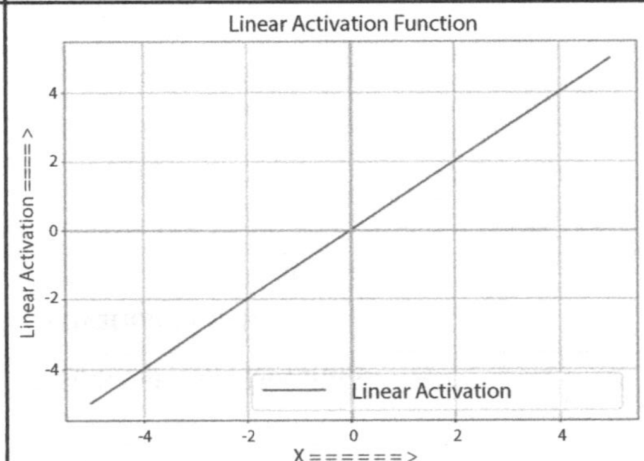	The linear function simply scales the input value by a constant factor. This means that the output is directly proportional to the input, without any distortion or compression.

(Continued)

Table 12.2 Few well-known activation functions. (*Continued*)

Activation function	Characteristic equation	Graphical representation	Characteristic
3. **Some Popular Non-Linear Neural Network Activation Functions**			
i. Sigmoid	$\sigma(x) = \frac{1}{1+e - x}$	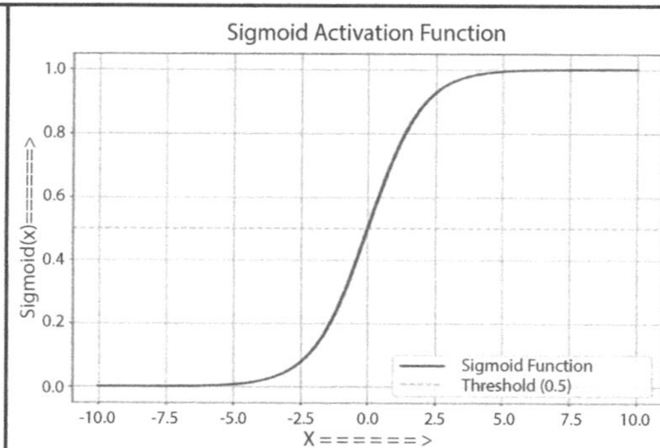	The S-shaped curve of the sigmoid function smoothly transitions between 0 and 1, making it ideal for modeling probabilities that gradually change with the input data.
ii. Hyperbolic Tangent (tanh)	$\tanh(x) = \frac{e - x - e x}{e - x + e x}$	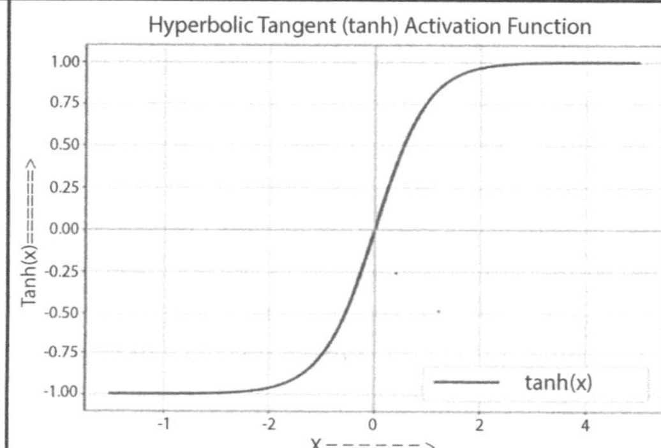	The hyperbolic Tangent activation function offers a balance between simplicity and non-linearity, making it a valuable tool in the arsenal of neural network practitioners. Its S-shaped curve, cantered output, and smooth gradient contribute to its effectiveness in various tasks.

(*Continued*)

Table 12.2 Few well-known activation functions. (*Continued*)

Activation function	Characteristic equation	Graphical representation	Characteristic
iii. Rectified Linear Unit (ReLU)	***ReLU(x)= max(0,x)***	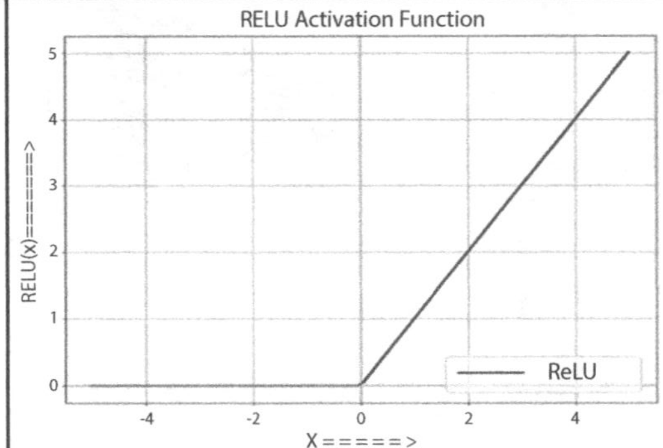	ReLU induces sparsity by nullifying negative values, thereby aiding in the minimization of overfitting.
iv. Leaky ReLU	$\textbf{\textit{Leaky ReLU}}(x)=\begin{cases}\alpha x \textit{if } x<0\\ x \textit{if } x\geq 0\end{cases}$	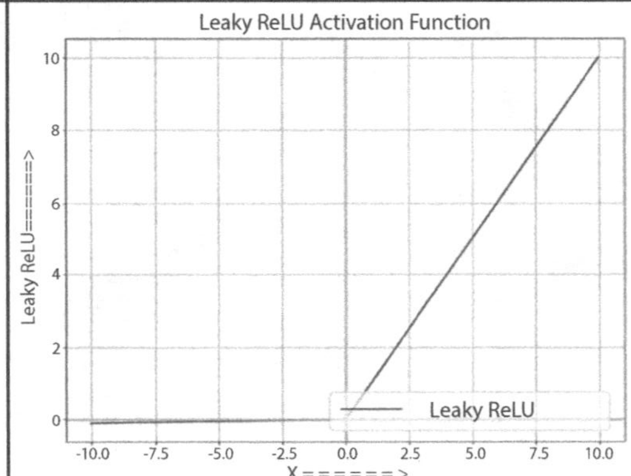	The Leaky ReLU function ensures that, for positive inputs, it behaves the same as the ReLU, whereas, for negative inputs, it allows a small, non-zero output, which is proportional to the input by a factor of α.

The above Figure 12.10 depicts the covariance matrix of key factors, including ambient and module temperature, irradiation, daily and total yield, and DC/AC power. It suggests that ambient temperature, module temperature, and irradiation are likely the most prominent predictor variables to be fed into the model for predicting DC power as their magnitude is relatively high, near to 1.

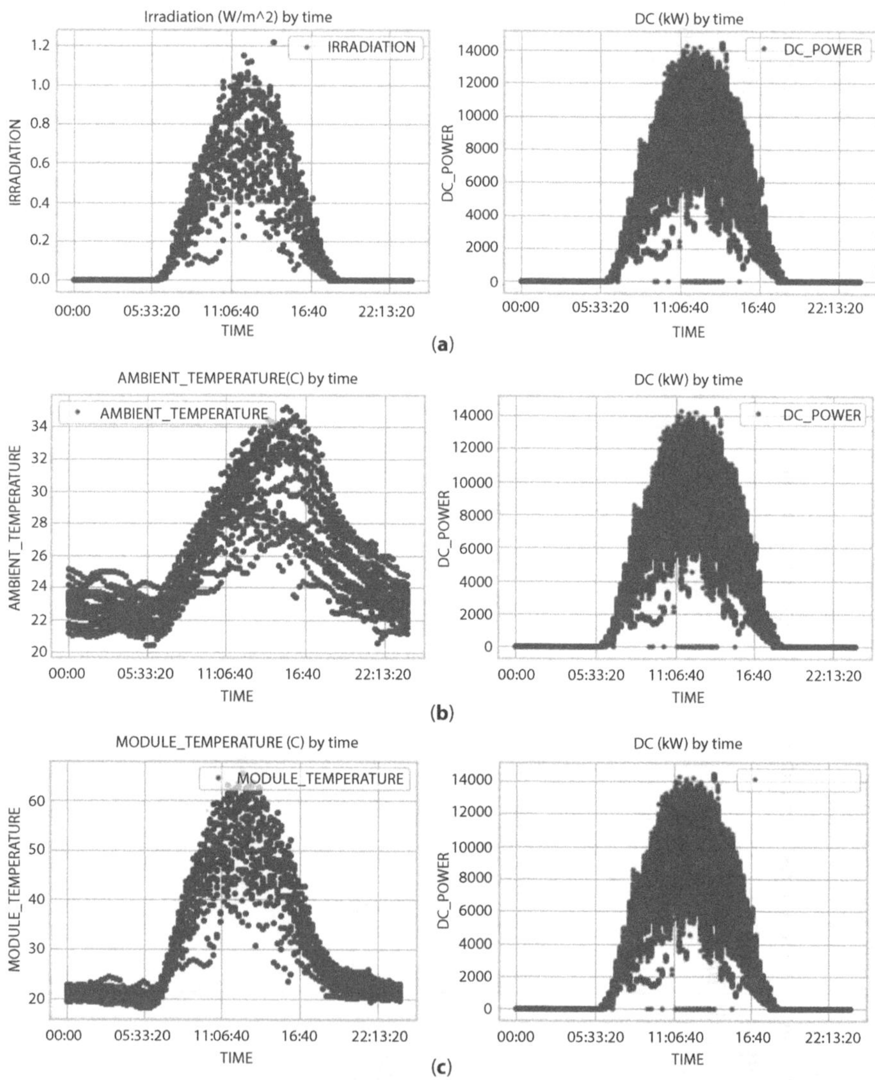

Figure 12.11 Predictor variables: (a) irradiation, (b) ambient temperature, and (c) module temperature dependencies with target variable: DC output.

Figure 12.11 depicts the dependencies of predictor variables (irradiation, ambient temperature, and module temperature) with the DC output of solar panels; these factors widely affect the power generation of solar panels.

This Table 12.3 depicts the results in terms of r2_score, MAE, MSE, and RMSE for SVM and ANN. The ANN model has the highest accuracy (97.21%), followed by the SVM model (96.98%). The ANN model also has the lowest MAE (0.23) and MSE (0.38).

While both models performed well, the ANN model is recommended based on its consistently better performance across all metrics. It is worth noting that the effectiveness of a model depends on the context, including the dataset and the target objective. Figure 12.12 depicts the performance chart of predicted and actual values of SVM & ANN.

Table 12.3 Performance comparison of the model used.

MODEL	Regression score (r2_score)	Mean absolute error (MAE)	Mean squared error (MSE)	Root mean squared error (RMSE)
ANN	0.97	0.23	0.38	0.66
SVM	0.96	0.31	0.44	0.66

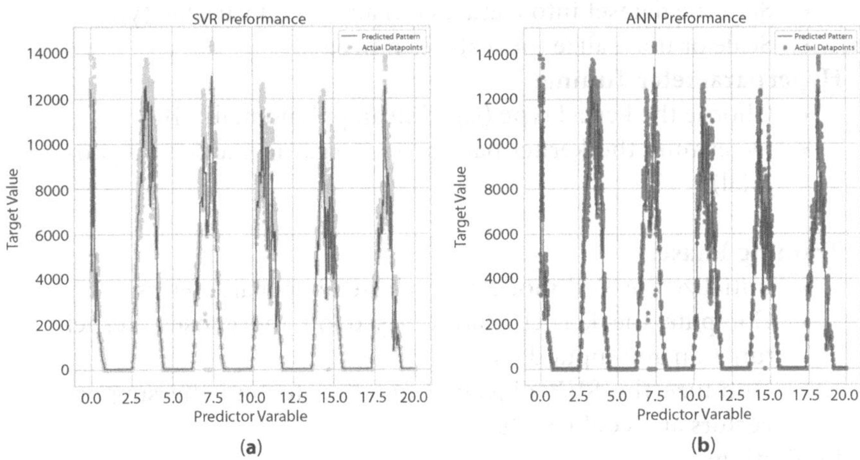

Figure 12.12 (a) SVR performance and (b) ANN performance.

12.6 Conclusion and Future Scope

The aim of this study is to explore the importance of ML techniques for forecasting solar PV generation systems. We compare the performance of two prominent MLAs, namely, SVM and ANN, using a case study. The results showed that the ANN model had the lowest average error (0.23 MAE) compared to SVM (0.31 MAE). The study highlighted the need for evaluating the performance of various ML PV forecasting models using different sized datasets and conducting critical analysis on the results. This research concludes that the evolution from fossil fuels to renewable energy is not just an environmental imperative but also an economic and health necessity. It offers a path toward a cleaner, more sustainable, and prosperous future.

Future research directions could involve implementing solar PV forecasting systems based on ensemble learning techniques like AdaBoost and XGBoost. Deep learning models, such as recurrent neural networks (RNNs) and LSTMs, are also likely to be effective, particularly owing to their capability to analyze sequential time-series data. These results can be validated using real-time data.

Appendix: Pseudo Code of Suggested Algorithms

Input: Load the Dataset
Process:

Data Preprocessing:
- Split the dataset into features (X) and target variable (y).
- Scale or normalize the features if needed.

Hyperparameter Tuning:
- Choose the kernel type (e.g., linear, polynomial, RBF).
- Determine the kernel parameters (e.g., degree for polynomial)

Training Phase:
- Initialize the SVR model with chosen hypeparameters.
- Compute the kernel matrix based on the chosen kernel function and input data.
- Optimize the SVR objective function to find the support vectors and coefficients.

Prediction:
- Compute the predicted output with the actual values.

Output: Well-trained SVR model.

Pseudo code for SVR
Input: Load the Dataset
Process:
Data Preprocessing:

- Split the dataset into features (X) and target variable (y).
- Scale or normalize the features if needed.

Initialize the neural network's parameters:

- Number of input units (features)
- Number of hidden layers and their respective units/neurons
- Number of output units (depends on the task)
- Learning rate, activation functions, and other hyperparameters.

Forward Propagation:

- Calculate the weighted sum of inputs and biases for each neuron in the hidden layers and output layer.
- Apply an activation function to the result of each neuron to introduce non-linearity.
- Pass the calculated values to the next layer for further computation until reaching the output layer.

Compute Loss: Compare the predicted output with the actual output using a loss function

Backpropagation:

- Compute the gradients of the loss function with respect to the neural network parameters.
- Update the weights and biases of the network using an optimization algorithm like Gradient Descent or its variants
- Iterate through the dataset multiple times (epochs) to improve the performance.
- **Output:** Well-trained ANN model.

References

1. Obiora, C.N., Hasan, A.N., Ali, A.A., Predicting Solar Irradiance at Several Time Horizons Using Machine Learning Algorithms. *Sustainability*, 15, 11, 8927, 2023 doi: 10.3390/su15118927.
2. Guangyin, Y., Explaining expedited energy transition toward renewables by COVID-19 in India. *Energy Policy*, 165, 112986–112986, 2022, doi: 10.1016/j.enpol.2022.112986.

3. Ellabban, O., Abu-Rub, H., Blaabjerg, F., Renewable energy resources: Current status, future prospects and their enabling technology. *Renew. Sustain. Energy Rev.*, 39, 748–764, 2014.
4. Mohit, J. and Sarah, I., Institutional Framework of Variable Renewable Energy Forecasting in India, 2023, doi: 10.2172/1985623.
5. Ismaeil, E.M., Sustainability-Based Value Engineering Management as an Integrated Approach to Construction Projects. *Building*, 14, 4, 903, 2024.
6. Scott, C., Ahsan, M., Albarbar, A., Machine learning for forecasting a photovoltaic (PV) generation system. *Energy*, 278, 127807, 2023. https://doi.org/10.1016/j.energy.2023.127807.
7. Nikitha, M.S., Nisha, K.C.R., Gowda, M.S., Aithal, P., Mudakkayil, N.M., Solar PV forecasting using machine learning models, in: *2022 Second International Conference on Artificial Intelligence and Smart Energy (ICAIS)*, Institute of Electrical and Electronics Engineers (IEEE), pp. 109–114, 2022, doi: 10.1109/ICAIS53314.2022.9742889.
8. Javadi, M., Naderi, S., Liang, X., Gong, Y., Chung, C.Y., Very Short-Term PV Power Prediction Using Machine Learning Models. *2022 IEEE Canadian Conference on Electrical and Computer Engineering (CCECE)*, pp. 55–59, 2022, doi: 10.1109/CCECE49351.2022.9918269.
9. Sharkawy, A.-N., Ali, M., Mousa, H., Ali, A., Abdel-Jaber, G., Machine Learning Method for Solar PV Output Power Prediction. *SVU-International J. Eng. Sci. Appl.*, 3, 2, 2022, https://doi.org/10.21608/svusrc.2022.157039.1066.
10. Içel, Y. and Gursoy, M., II, Machine Learning Based PV Power Prediction Using Different Environmental Parameters of Turkey, 02 August 2022, PREPRINT (Version 1) available at Research Square, https://doi.org/10.21203/rs.3.rs-1809187/v1.
11. Alhmoud, L., Al-Zoubi, A.M., Aljarah, I., Solar PV power forecasting at Yarmouk University using machine learning techniques. *Open Eng.*, 12, 1, 1078–1088, 2022. https://doi.org/10.1515/eng-2022-0386.
12. de, Freitas, Viscondi, G. and Alves-Souza, S.N., Solar Irradiance Prediction with Machine Learning Algorithms: A Brazilian Case Study on Photovoltaic Electricity Generation. *Energies*, 14, 18, 5657, 2021, doi: 10.3390/EN14185657.
13. Sushil, K., Kamlesh, K.B., Aman, S., Upcoming Power Crisis in India—Increasing Electricity Demand, 403–412, 2020, doi: 10.1007/978-981-15-2369-4_34.
14. Dubey, B., Agrawal, S., Sharma, A.K., India's Renewable Energy Portfolio: An Investigation of the Untapped Potential of RE, Policies, and Incentives Favoring Energy Security in the Country. *Energies*, 16, 14, 5491, 2023.
15. Tsai, W.C., Tu, C.S., Hong, C.M., Lin, W.M., A review of state-of-the-art and short-term forecasting models for solar pv power generation. *Energies*, 16, 14, 5436, 2023.
16. Alanqar, A.A. and Venkatratnam, C., Solar Power Monitoring and Forecasting System using Artificial Intelligence, 2023, doi: 10.1109/SCEECS57921.2023.10062972.

17. Quoc-Thang, P., Yuan-Kang, W., Quoc Dung, P., Application of a new Transformer-based model and XGBoost to improve one-day-ahead solar power forecasts, 2023, doi: 10.1109/ICPS57144.2023.10142107.
18. Altan, U., Malaquias, P., Zongjie, W., Deep Learning Architectures for Solar PV Forecasting, 2022, doi: 10.1109/NAPS56150.2022.10012176.
19. Pachauri, R.K., Mahela, O.P., Khan, B., Kumar, A., Agarwal, S., Alhelou, H.H., Bai, J., Development of arduino assisted data acquisition system for solar photovoltaic array characterization under partial shading conditions. *Comput. Electr. Eng.*, 92, 107175, 2021.
20. Younis, A. and Alhorr, Y., Modeling of dust soiling effects on solar photovoltaic performance: A review. *Sol. Energy*, 220, 1074–1088, 2021.
21. Jiao, X., Li, X., Lin, D., Xiao, W., A graph neural network based deep learning predictor for spatio-temporal group solar irradiance forecasting. *IEEE Trans. Industr. Inform.*, 18, 9, 6142–6149, 2021.
22. Padmavathi, S., and Raja, S.C., Performance Analysis of Regression Models in Solar PV Forecasting, 2023, doi: 10.1109/AISP57993.2023.10134943.
23. Khalaf, H.H,., Mohammad, A.T., Hussain, A.N., Al-Sagar, Z.S., Comparison and Assessment of PV Module Power Prediction Based on ANN for Iraq Weather, 2022, doi: 10.1109/ICOASE56293.2022.10075594.
24. Hussain, M.T., Sarwar, A., Tariq, M., Urooj, S., BaQais, A., Hossain, M.A., An evaluation of ANN algorithm performance for MPPT energy harvesting in solar PV systems. *Sustainability*, 15, 14, 11144, 2023.
25. Intra-hour PV Power Forecasting based on Multi-source Data and PSC-SVR Model, 2022, doi: 10.23919/ccc55666.2022.9901706.
26. Tsai, W.C., Tu, C.S., Hong, C.M., Lin, W.M., A review of state-of-the-art and short-term forecasting models for solar pv power generation. *Energies*, 16, 14, 5436, 2023.
27. Subramanian, E., Karthik, M.M., Krishna, G., Prasath, D., Kumar, V., Solar Power Prediction Using Machine Learning, 2023, arXiv.org, doi: 10.48550/arXiv.2303.07875.
28. Khan, A.A., Minai, A.F., Pachauri, R.K., Malik, H., Optimal Sizing, Control, and Management Strategies for Hybrid Renewable Energy Systems: A Comprehensive Review. *Energies*, 15, 17, 6249, 2022.
29. Faiz Minai, A., Khan, A.A., Pachauri, R.K., Malik, H., García Márquez, F.P., Arcos Jiménez, A., Performance evaluation of solar PV-based Z-source cascaded multilevel inverter with optimized switching scheme. *Electronics*, 11, 22, 3706, 2022.
30. Pachauri, R.K., Kansal, I., Babu, T.S., Alhelou, H.H., Power losses reduction of solar PV systems under partial shading conditions using re-allocation of PV module-fixed electrical connections. *IEEE Access*, 9, 94789–94812, 2021.

13

Machine Learning–Based Prediction of Electrical Load in the Context of Variable Weather Conditions

Ashutosh Shukla[1]*, Supriya[2] and Rupendra Kumar Pachauri[1]

[1]*Electrical Cluster, School of Advanced Engineering, UPES, Dehradun, India*
[2]*Department of AI & ML, School of Computing, COER University, Roorkee, India*

Abstract

Grid management relies heavily on accurate forecasting of seasonal power consumption, particularly in light of increasing energy demand and renewable energy reliance. Seasonal fluctuations and weather patterns significantly impact energy usage, making it challenging to predict future demand. Machine learning techniques can simulate non-linear relationships, enhancing electrical load behavior. The study employed machine learning algorithms such as random forest (RF), artificial neural networks (ANNs), and decision tree model to forecast electricity consumption using meteorological variables such as solar irradiance, humidity, and ambient temperature.

The paper explores the impact of weather elements like temperature, humidity, and solar irradiance on the accuracy of electrical load predictions in different seasons using seasonal forecasting techniques like RF and ANN, evaluated using performance metrics. The model's performance was assessed using performance metrics such as root mean square error, mean absolute error, mean square error, and R-value.

Keywords: Energy forecasting, random forest, artificial neural network, solar irradiance

**Corresponding author*: ashutiet@gmail.com

Jai Govind Singh, Rupendra Kumar Pachauri and Sasidharan Sreedharan (eds.) Forecasting Methods for Renewable Power Generation, (347–364) © 2025 Scrivener Publishing LLC

13.1 Introduction

Energy is a crucial prerequisite for the sustainable prosperity of every nation. Global energy consumption has experienced a significant surge during the past decade. Accurate energy estimates are crucial for ensuring sustainability and long-term economic success. Over the last decade, many innovative methods have been utilized to accurately predict future energy demands. Energy resources must be used and managed effectively to control energy demand. Sustainable economic growth will be ensured by controlling energy consumption to attain self-sufficiency and cost-effectiveness. Therefore, energy demand management should help with choosing ways to save energy, figuring out what we will need in the future and getting ready for it, picking the best energy sources to use first, making plans for higher energy efficiency, making policies, and coming up with strategies for lower emissions. A multitude of interrelated elements, such as the state of the economy, the calendar (both weekly and yearly periodicities), and meteorological conditions, all have an impact on electrical consumption. There are three types of power load forecasting: long-term, mid-term, and short-term, based on the forecasting step size projection. Power infrastructure construction planning often uses long-term forecasting with the goal of anticipating yearly and monthly demands, wherein the performance of the system is significantly impacted by medium-short-term forecasting, which projects weekly, daily, and hourly power loads. The quantity of electrical power utilized is known to be significantly influenced by the weather.

The meteorological variables that are generally considered include temperature, humidity, and solar radiation. Temperature and electrical power demand generally have a non-linear connection because there is a discernible rise in power consumption in response to both a drop in temperature below a threshold and an increase in temperature over a threshold. Similar relationship holds for the solar irradiance and humidity. Due to the variable in nature of solar irradiance, ambient temperature, and humidity, it is challenging for predicting the electrical load demand.

The relationship between meteorological characteristics and electrical load demand is examined in this study, and, to support the hypothesis, load forecasting models based on random forest (RF), artificial neural network (ANN), and decision tree (DT) are presented. The methods for predicting power demand that are already in use are introduced in **Section 13.2**, along with the issues and difficulties that they provide and the need to research how weather patterns affect power load demand. In **Section 13.3**, the significance of the work is discussed in detail. The suggested workflow of ANN, RF, and DT models is shown in **Section 13.4**. The model's performance on the power load dataset is displayed in **Section 13.5**.

13.2 Previous Work

Advances in machine learning research have led to the progressive integration of technologies like neural networks [1, 2] and support vector machines [3, 4] into power systems' daily load forecasting models. While workdays and holidays are added to the input matrix in some of the research, the input feature in the majority of the studies is historical data. But, weather is a significant element that impacts electricity load in addition to past trends and the nature of workdays and holidays. As previous study has seen, several researchers consider meteorological data. In [5], temperature data was used to create a load forecast model using an ANN and bagged regression trees. An analysis was conducted [6] on the relationship between temperature, solar irradiance, and load. The ANN model's accuracy was increased by using mutual information to choose the input features. A small town in Italy was given the unique characteristic of having only residential consumers in [7], when the Humidex indicator, a comprehensive indicator of atmospheric temperature and humidity, was used as an input to anticipate the regional power load of that town. Zachariadis [8] offered a projection of Cyprus' power use until 2030, derived on an econometric study of energy consumption in relation to macroeconomic factors, pricing, and meteorological conditions.

Because electrical load is highly volatile and non-linear, precise load forecasting is crucial to the efficient functioning of power systems. Appropriate prediction methods are necessary to forecast such complicated signals. The common forecasting methods fall into two categories: artificial intelligence (AI) as well as techniques based on statistics. There are three main types of energy forecasting models: black box, white box, and gray box [9]. Community microgrid load dispatch has several issues due to the inherent instability and unpredictability of renewable energy. As a result, several research studies have examined the sporadic use of renewable energy in the microgrid's load dispatch model [10, 11]. Few studies have been conducted on microgrid power demand forecasting, taking weather conditions into account. For example, Gerossier *et al.* [12] attempted to give a day-ahead load prediction by using the ambient temperature as the weather characteristic directly but did not examine the precise impact of various other meteorological features like as solar irradiance and humidity.

13.3 Significance of Work

There are benefits and drawbacks to every machine learning model. Power load forecasting, on the other hand, is primarily concerned with efficiency and accuracy [13]. Because each temporary installation might have a distinct set of characteristics and a varied length of time, it is challenging to classify and anticipate loads for smart grids [14]. Solar irradiance, wind speed and direction, humidity, and temperature are some of the most important climatic variables influencing electricity consumption patterns, especially in residential areas [13]. This research proposes a very efficient and dependable approach for load forecasting. Given the continuous growth of research on smart grids, the use of precise and accurate approaches for load forecasting may provide several benefits [15].

- Efficiently managing both the supply and demand of electricity enhances the dependability of the power system by empowering operators to make strategic choices for other participants in the market.
- By incorporating the machine learning approach with renewable energy sources, we may effectively address environmental issues such as climate change and pollution.
- Accurate load modeling enables utilities to effectively react to the demand for power supply and regulate operations, whereas load definition ensures that appropriate actions are taken.

13.4 Methodology

Load forecasting is a complex subject that involves several parameters, such as time, environmental conditions, and other variables. This technology is extensively used in all sectors of power systems, catering to diverse fundamental needs in industrial, commercial, and household environments. The fluctuation of loads across various time frames has distinct characteristics. In general, a broader time range leads to a greater impact on variables and an increase in the complexity of an issue [13]. The outcomes will be examined with different ANNs or machine learning methods. Figure 13.1 illustrates the sequence of steps in the process.

13.4.1 Input Data

Data for predicting electrical loads is obtained from publicly available databases [16], whereas meteorological data is taken from the NREL. The data for the electric power load, also known as the active power load, is generated by using the hourly voltage (V), current (I), and power factor (pf) information that is accessible at the 33/11-KV substation at Godishala, Huzurabad, Telangana state, located in India. Information on the voltage, current, and power factor was gathered on an hourly basis between 1 January 2021 and 31 December 2021. The data includes hourly load, day status (0 for weekdays and 1 for weekends), season (1 for winter, 2 for summer, and 0 for rainy), as well as hourly humidity and temperature information. Meanwhile, solar irradiance data for the same latitude and year has obtained from the NREL. In Figure 13.2, 1-year hourly data of each feature has been divided into three seasons: summer, rainy, and winter.

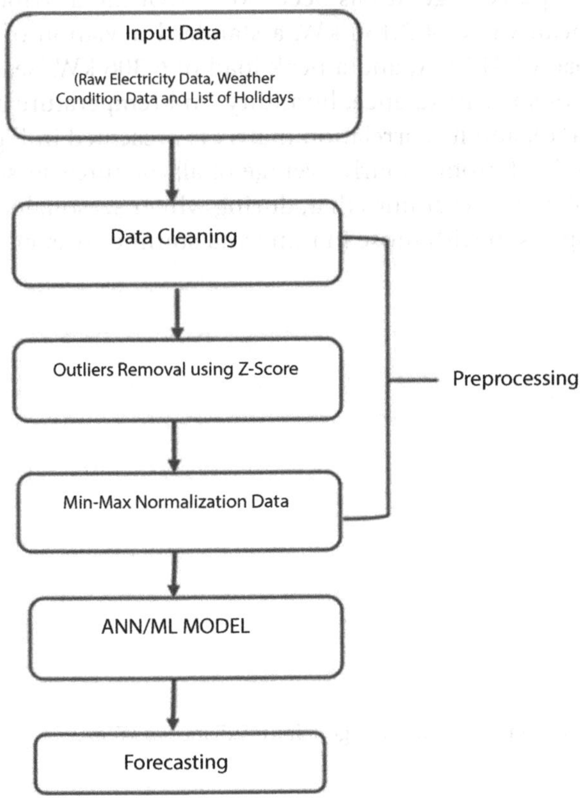

Figure 13.1 Proposed workflow.

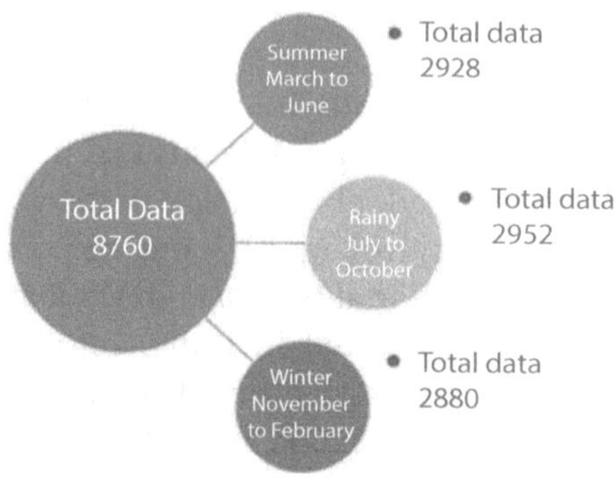

Figure 13.2 One-year data division into three seasons.

There are a total of 8,760 hourly load data values included within this collection. In this particular dataset, the load data is presented in kilowatts, the temperature is expressed in degrees Fahrenheit, and the humidity is expressed as a percentage. It has been noted that the distribution of load data has a mean value of 2,130 kW, a standard deviation of 1,302 kW, a minimum load of 412 kW, and a peak load of 6,306 kW. Seasonal hourly average data of solar irradiance, humidity, and temperature are shown in Figure 13.3(a–c), and its correlation matrix is presented in Figure 13.3(d). The electrical load profile hourly average of all the three seasons is shown in Figure 13.4, which examines that, during winter season, load demand is more in comparison with those in rainy and summer seasons.

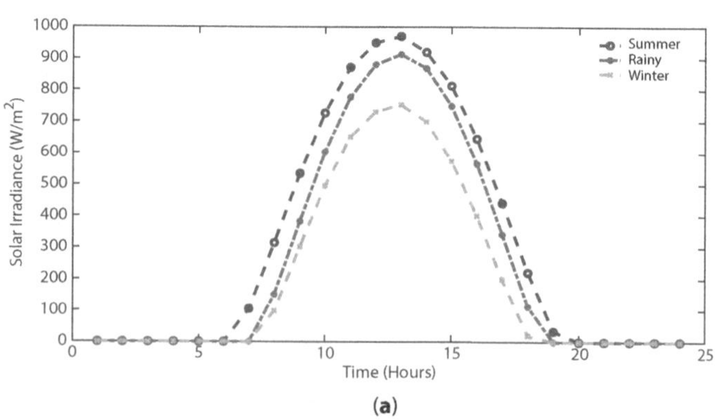

Figure 13.3(a) Seasonal 24-hour average solar irradiance in W/m^2.

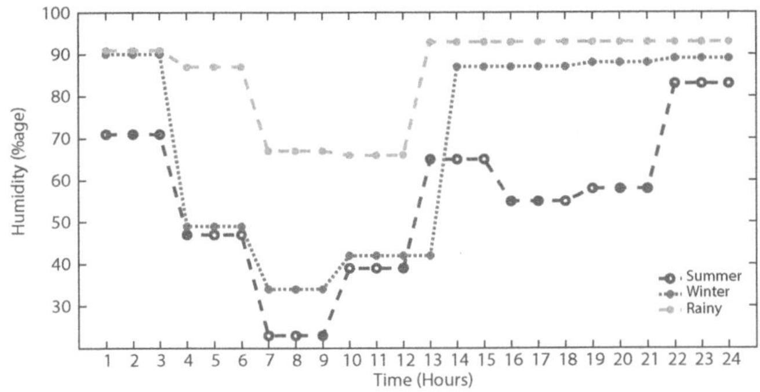

(b)

Figure 13.3(b) Seasonal 24-hour average humidity in %age.

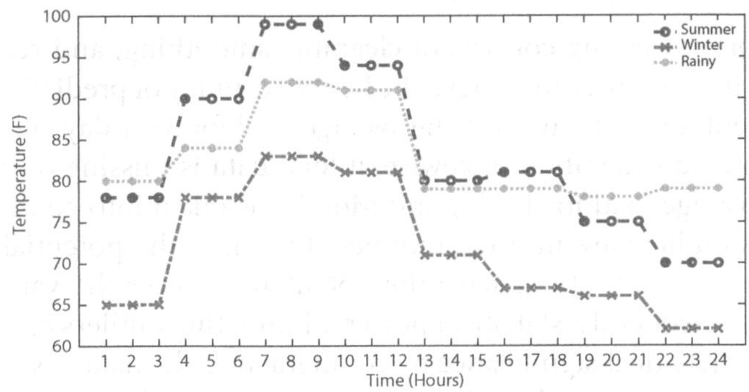

(c)

Figure 13.3(c) Seasonal 24-hour average temperature in Fahrenheit.

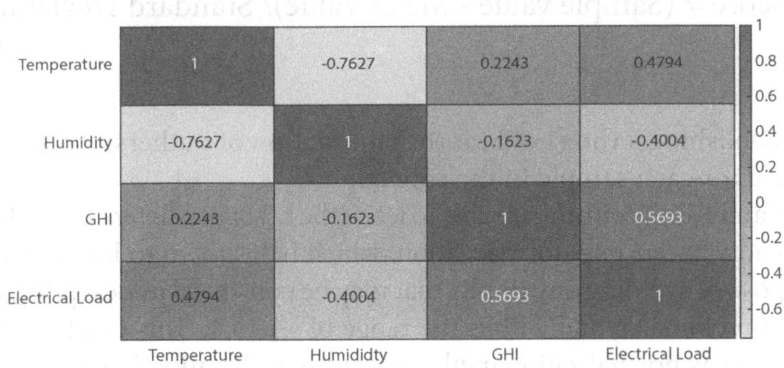

(d)

Figure 13.3(d) Correlation coefficient matrix.

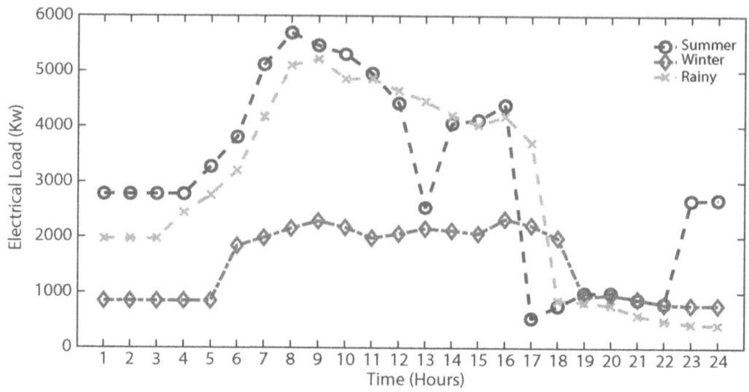

Figure 13.4 Seasonal 24-hour average electrical load in kW.

13.4.2 Data Preprocessing

The data preprocessing consists of cleaning, smoothing, and reconstruction of the data in order to achieve the better accuracy of prediction. If load data is absent for Monday, then the average load for Saturday and Tuesday is taken into consideration. Likewise, if load data is missing on Saturday, then the average load for Friday and Monday is taken into consideration. A dataset's outliers are unusual numbers that have the potential to skew statistical analysis. Outliers have the potential to lower the variability of data and, consequently, statistical power. Eliminating outliers can improve the results' statistical significance by counteracting the dataset's erroneous values. For removing such outlier data, Z-score is used, which is given by Equation (13.1):

$$\text{Z-Score} = (\text{Sample value - Mean Value}) / \text{Standard Deviation} \tag{13.1}$$

By establishing a threshold for the elimination of outliers, the Z-score is computed for each sample in the collection.

ANN and RF algorithms are able to learn the best parameters for each input more rapidly because of normalization, which helps them to learn faster. For the purpose of avoiding any results that may be construed as deceptive, inputs should approximately fall within the range of −1 to 1. This can be achieved using min-max normalization method as given in Equation (13.2).

$$S = (X - X_{MINIMUM}) / (X_{MAX} - X_{MINIMUM}) \tag{13.2}$$

13.4.3 Electrical Load Forecasting Algorithms

The load forecasting has been achieved using ANN and machine learning algorithms such as support vector machine and multiple linear regression. For the given methods, the total number of independent variables are six, which consist of day, hour, temperature (°F), humidity (%age), weekday/weekend (0/1), and solar irradiance (W/m^2), whereas the dependent variable is electrical load (kW). Mean absolute percentage error (M), root mean square error (R), and mean squared error (E) are the metrics that have been employed in order to evaluate the performance of the model shown in Equations (13.3), (13.4), and (13.5).

$$M = (1/N) \times \Sigma(|Xi_{Actual} - Xi_{Predicted}| \ / \ |Xi_{Actual}|) \times 100\% \quad (13.3)$$

$$R = sqrt((1/N) \times \Sigma((Xi_{Actual} - Xi_{Predicted})\hat{}2)) \quad (13.4)$$

$$E = (1/N) \times \Sigma(Xi_{Actual} - Xi_{Predicted})\hat{}2 \quad (13.5)$$

where

- N is the number of data points
- Xi_{Actual} and $Xi_{Predicted}$ are actual and predicted value for the i-th data point.

13.4.3.1 ANN Model for Electrical Load Forecasting

Artificial neural networks are a specialized branch within the subject of AI. These systems are engineered to replicate the functionality of the human brain through the analysis and processing of information in a manner that closely resembles human cognition. Using ANNs, the power usage pattern has been predicted in this work. In ANNs, the output is determined by the accumulation of neurons in the input and hidden layers. For the purpose of this investigation, the data has been partitioned into three distinct sets:

training (70% of data), testing (15% of data), and validation (15% of data). The mathematical model of ANN represented as Equation (13.6) as follows:

$$y_j = f\left(\sum_{i=1}^{n} W_{ij} X_i + b_j\right) \tag{13.6}$$

Here x and y are the input and output value, respectively, whereas W and b represent weight and bias. Input feature quantity dictates the input layer's dimensions. Utilizing a standard neural network with a single hidden layer, the prediction has been executed. The amount of hidden neurons is established via the use of trial and error method. The ANN architecture of all the three seasons is shown in Figure 13.5, having six input nodes and one output node. The total number of hidden layers is two, and ReLU activation function is used at each layer. The regression value for all seasons is mentioned in Figure 13.6, respectively.

The performance matrix of ANN is shown in Table 13.1 for the summer, rainy, and winter seasons. Substantial R-values and comparatively low Mean Absolute Percentage Error (MAPE) values suggest that the ANN model is functioning for summer and winter seasons with adequate overall accuracy.

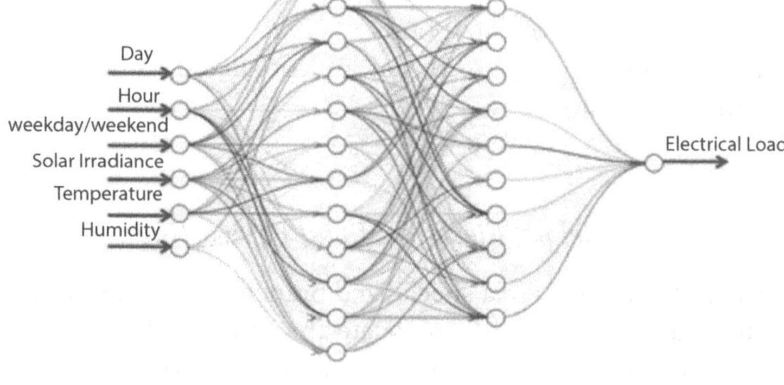

Figure 13.5 Structure of an ANN model.

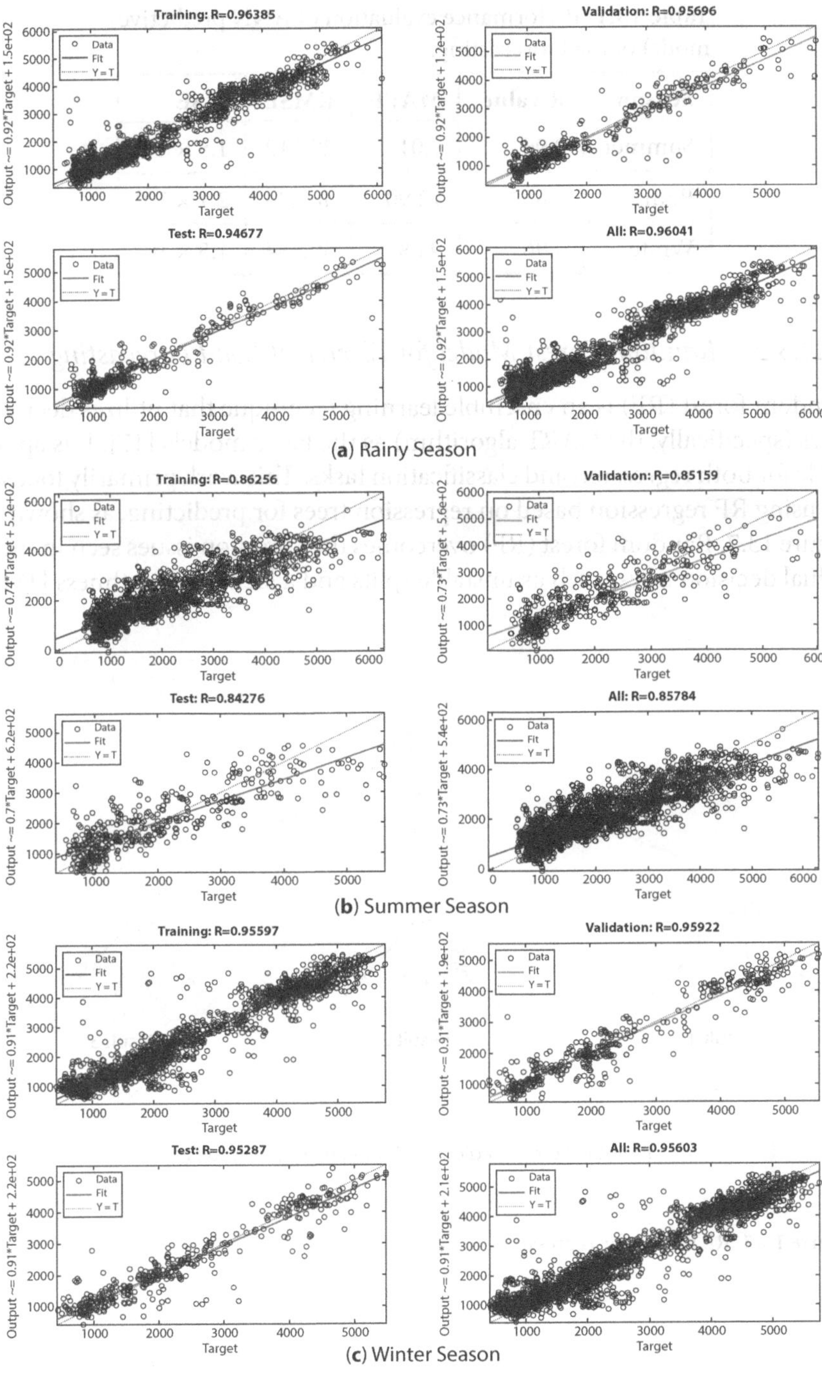

Figure 13.6 R-value of training, validation, and testing.

Table 13.1 Performance evaluation of ANN predictive model of load forecasting.

Season	R-value	MAPE	RMSE	MSE
Summer	0.96	9.01	297.12	1.5×10^5
Rainy	0.85	12.90	461.34	2.1×10^5
Winter	0.96	9.08	302.34	1.5×10^5

13.4.3.2 *Random Forest Model for Electrical Load Forecasting*

Random forest (RF) is an ensemble learning technique that utilizes decision trees (specifically, the CART algorithm) as the basic models [17]. It is applicable for both regression and classification tasks. This work primarily focuses on using RF regression based on regression trees for predicting, as shown in Figure 13.7. Random forest (RF) overcomes the common issues seen in individual decision trees, such as unstable splits and a lack of smoothness [18].

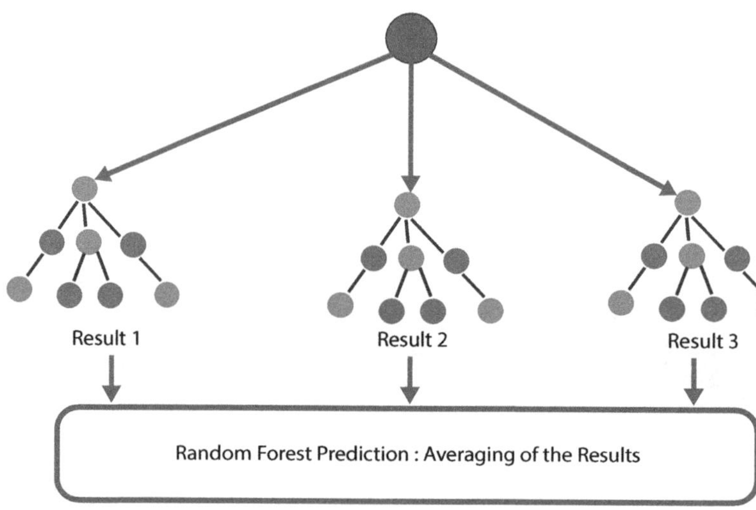

Figure 13.7 RF model for regression.

It integrates bagging with a stochastic subspace technique. The primary objective of the random subspace technique is to enhance the heterogeneity across trees by constraining them to operate on distinct random subsets of the complete predictor space. Every tree inside the forest is constructed using a bootstrap sample derived from the original dataset, so introducing an extra element of variability. The use of random predictors in the nodes of bagged trees serves to decorrelate the trees, leading to enhanced prediction accuracy and reduced model variance. The R-value for summer, rainy, and winter seasons is shown in Figure 13.8.

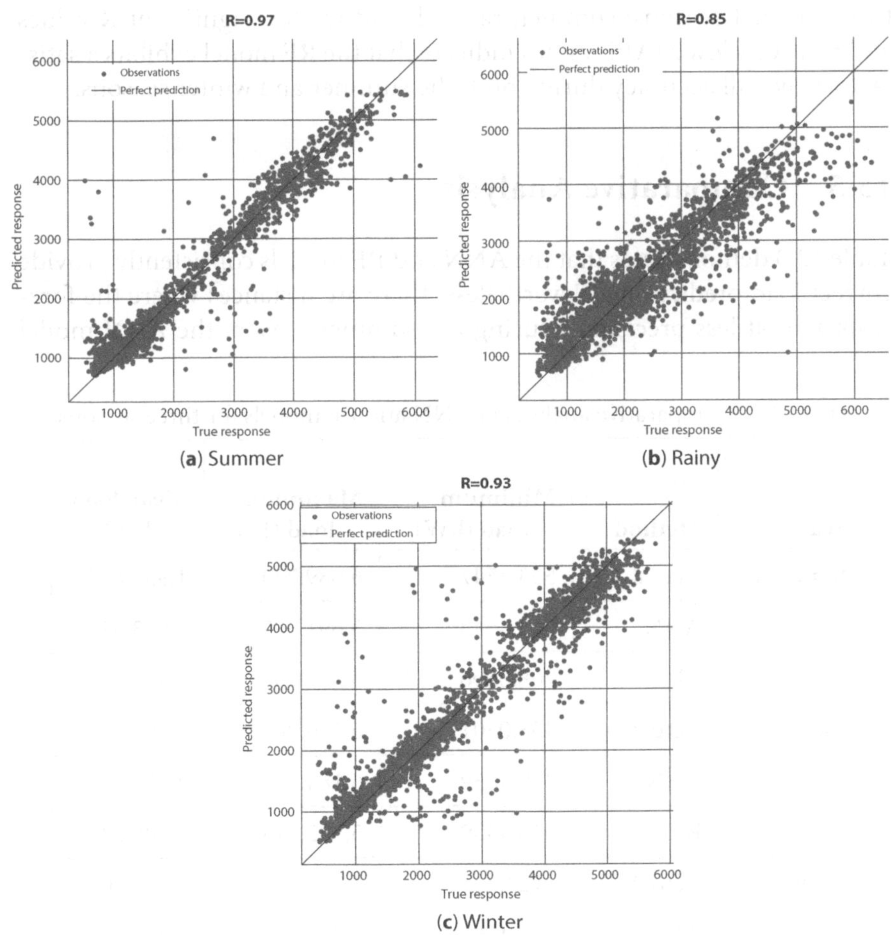

Figure 13.8 R-value for validation.

Table 13.2 Performance evaluation of random forest predictive model of load forecasting.

Season	R-value	MAPE	RMSE	MSE
Summer	0.97	7.02	210.3	4.4×10^4
Rainy	0.85	12.88	457.77	2.0×10^5
Winter	0.93	8.484	378.34	1.4×10^5

Table 13.2 presents the performance matrix of the RF model for the three distinct seasons: summer, rainfall, and winter. Significant R-values and relatively low MAPE values indicate that the RF model exhibits a satisfactory overall accuracy during both the summer and winter seasons.

13.5 Comparative Analysis

Table 13.3 demonstrates that the ANN and RF models consistently provide correct load predictions. Nevertheless, there are instances where the forecasts exhibit less precision. During the summer season, the ANN model

Table 13.3 Comparative analysis of ANN and RF models for three seasons.

Season	Method	Minimum load (kW)	Maximum load (kW)	Mean load (kW)
Summer	Actual	533.3747	6,089.724	1,932.883
	ANN	712.4548	5,691.507	1,933.317
	RF	708.0106	5,537.794	1,932.507
Rainy	Actual	458.0201	6,306.206	2,001.276
	ANN	218.9659	6,184.378	2,001.45
	RF	616.4439	5,522.614	2,000.03
Winter	Actual	412.0341	5,779.171	2,462.301
	ANN	586.4569	5,476.03	2,463.136
	RF	514.4936	5,449.614	2,462.173

tends to overestimate the demand by around 100 kilowatts (kW). The mean load prediction in all seasons is almost same for both ANN and RF models.

13.6 Conclusion

This study area has substantial potential to tackle crucial difficulties in the energy sector and provide a more sustainable and dependable electrical system. Through the integration of sophisticated machine learning algorithms and extensive analysis of meteorological data, precise and dependable predictions of seasonal electrical load may achieved, therefore facilitating effective management of power grids and fostering a more sustainable energy landscape. The ANN and RF models exhibit superior performance in forecasting load during the summer and winter seasons compared to that during the rainy season. This discrepancy may arise due to the abundance of data pertaining to the summer and winter seasons or due to the superior ability of the models to accurately represent the intricate patterns within the data for these particular seasons. Some additional details are given below:

- The peak summer demand exceeds the minimum summer load by over 10-fold. This indicates a significant surge in power usage throughout the summer months.
- The average load during the rainy season exceeds the average load during the winter season.
- This implies that individuals tend to consume a greater amount of power during the rainy season, maybe as a result of the necessity for humidity reduction or other weather-related elements.

References

1. Ping-Huan, K. and Huang, C.J.A., High Precision Artificial Neural Networks Model for Short-Term Energy Load Forecasting. *Energies*, 11, 213, 2018.
2. Wang, Y., Liu, M., Bao, Z., Zhang, S., Short-Term Load Forecasting with Multi-Source Data Using Gated Recurrent Unit Neural Networks. *Energies*, 11, 1138, 2018.
3. Hippert, H.S. and Pedreira, C.E., Neural Networks for Short-Term Load Forecasting: A Review and Evaluation. *IEEE Trans. Power Syst.*, 16, 44–45, 2001.

4. Fan, G.-F., Peng, L.-L., Hong, W.-C., Sun, F., Electric load forecasting by the SVR model with differential empirical mode decomposition and auto regression. *Neurocomputing*, 173, 958–970, 2016.
5. Dehalwar, V., Kalam, A., Kolhe, M.L., Zayegh, A., Electricity Load Forecasting for Urban Area Using Weather Forecast Information, in: *Proceedings of the 2016 IEEE International Conference on Power and Renewable Energy (ICPRE)*, Shanghai, China, 21–23 October 2016.
6. Panapongpakorn, T. and Banjerdpongchai, D., In Short-Term Load Forecast for Energy Management System Using Neural Networks with Mutual Information Method of Input Selection, in: *Proceedings of the 2019 SICE International Symposium on Control Systems (SICE ISCS)*, Kumamoto, Japan, 7–9 March 2019.
7. Beccali, M., Cellura, M., Brano, V.L., Marvuglia, A., Short-term prediction of household electricity consumption: Assessing weather sensitivity in a Mediterranean area. *Renew. Sustain. Energy Rev.*, 12, 2040–2065, 2008.
8. Zachariadis, T., Forecast of electricity consumption in Cyprus up to the year 2030: The potential impact of climate change. *Energy Policy*, 38, 744–750, 2010.
9. Amara, F. and Agbossou, K., Comparison and Simulation of Building Thermal Models for Effective Energy management. *Smart Grid Renew. Energy*, 6, 95–112, 2015.
10. Wang, L., Li, Q., Ding, R., Sun, M., Wang, G., Integrated scheduling of energy supply and demand in microgrids under uncertainty: a robust multi-objective optimization approach. *Energy*, 130, 1e14, 2017.
11. Craparo, E., Karatas, M., Singham, D.I., A robust optimization approach to hybrid microgrid operation using ensemble weather forecasts. *Appl. Energy*, 201, 135e47, 2017.
12. Gerossier, A., Girard, R., Bocquet, A., Kariniotakis, G., Robust Day-Ahead Forecasting of Household Electricity Demand and Operational Challenges. *Energies*, 11, 3503, 2018.
13. Ahmad, W., Ayub, N., Ali, T., Irfan, M., Awais, M., Shiraz, M., Glowacz, A., Towards short term electricity load forecasting using improved support vector machine and extreme learning machine. *Energies*, 13, 11, 2907, 2020. MDPI AG.
14. Hao-Tian, Z., Fang-Yuan, X., Zhou, L., Artificial neural network for load forecasting in smart grid, in: *2010 international conference on machine learning and cybernetics*, vol. 6, pp. 3200–5, IEEE, 2010.
15. Hamlich, M. and eddine Belbounaguia, N., Short-term load forecasting using machine learning and periodicity decomposition. *AIMS Energy*, 7, 3, 382–94, 2019.
16. Veeramsetty, V., Sushma Vaishnavi, G., Sai Pavan Kumar, M., Sumanth, N., Prasanna, P., Electric power load dataset. *Mendeley Data*, V1, 2022 , doi: 10.17632/tj54nv46hj.1.

17. Breiman, L., Friedman, J.H., Olshen, R.A., Stone, C.J., *Classification and Regression Trees*, Chapman and Hall, London, UK, 1984.
18. Makridakis, S., Spiliotis, E., Assimakopoulos, V., M5 accuracy competition: Results, findings, and conclusions. *Int. J. Forecast.*, 38, 1346–1364, 2022.

14

Recent Advancement in Renewable Energy with Artificial Intelligence and Machine Learning

Sakshi Chaudhary[1], Aakansha Simra[1] and Gaurav Pandey[1,2]*

[1]School of Advanced Engineering, UPES, Dehradun, India
[2]Earth Cryosphere Institute, Tyumen Scientific Center SB, RAS, Tyumen, Russia

Abstract
Developments in renewable energy have shown the importance of machine learning (ML) and artificial intelligence (AI). This chapter examines the rise and convergence of renewable energy using AI and ML, principally focusing on applications in energy production predictions, mostly including wind and solar energy systems. It also draws attention to opportunities to improve forecast accuracy and optimize energy production. Nevertheless, obstacles to its effectiveness still exist, such as the complexity of the data sparsity model. However, future technological research and development offers encouraging opportunities to overcome these limitations. The growth and prospects of AI and ML in the renewable energy sector focus on innovation and solve global energy challenges.

***Keywords*:** Artificial intelligence, machine learning, data science, renewable energy

14.1 Introduction

The use of renewable energy (RE) has become important as part of international initiatives to reduce greenhouse gas emissions and promote sustainable development. The importance of renewable energy has increased in recent decades, as evidenced by growing use and investment in the

**Corresponding author*: gaurav.pandey@ddn.upes.ac.in

Jai Govind Singh, Rupendra Kumar Pachauri and Sasidharan Sreedharan (eds.) Forecasting Methods for Renewable Power Generation, (365–386) © 2025 Scrivener Publishing LLC

sector. The world has to switch to renewable energy sources to combat climate change. The consumption patterns of various nations will be useful in assessing the degree of development. The growth of renewable energy consumption rates in a sample of nations from 2005 to 2024 is shown in Figure 14.1.

The graph depicts the trend toward higher overall RE usage. The vertical axis shows the percentage of RE usage, with higher values indicating a greater reliance on renewable energy resources. This upward trajectory indicates a joint effort toward ecological accountability and a potential decrease in dependence on non-renewable sources of energy. The graph does, however, also indicate that different nations have differing levels of development. Some countries have adopted renewable energy more quickly than others, as seen by the lines varying slopes. Additional research into nations and their policies would be required to ascertain the underlying causes of this discrepancy.

This phenomenal rise reflects growing acknowledgment of renewable energy's significance in climate change mitigation as well as its ability to boost economic prosperity. Numerous research studies have been undertaken to examine the correlation between the utilization of clean energy sources and the advancement of the economy. The majority of empirical research suggests that renewable energy consumption and real GDP are

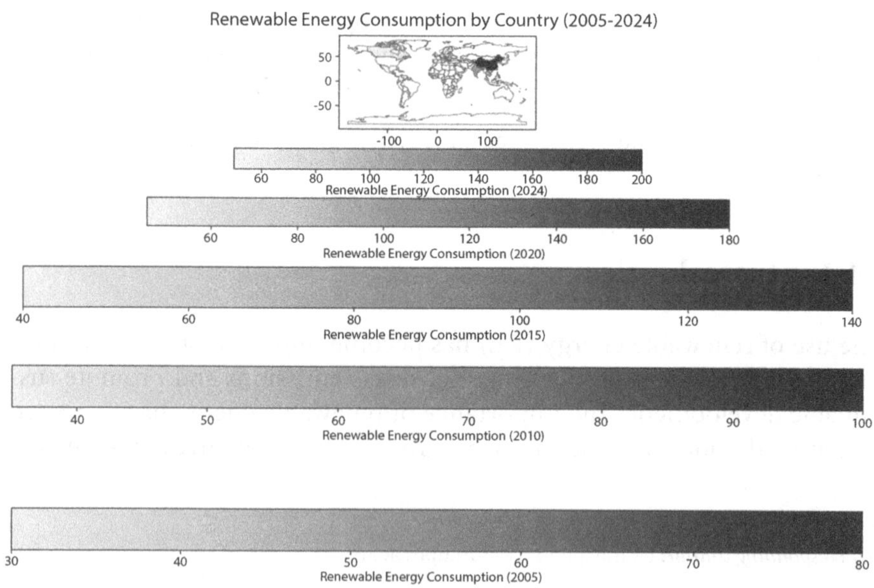

Figure 14.1 Percentage change in renewable energy consumption for five different countries (as listed in Table 14.1) between 2005 and 2024.

Table 14.1 Energy consumption across various nations from 2005 to 2024.

Years							
S. no.	**Country**	**2005**	**2010**	**2015**	**2020**	**2024**	**Energy consumption**
1.	United States	100	120	150	180	200	↓
2.	China	80	100	140	180	200	
3.	India	50	70	90	110	120	
4.	Germany	40	50	60	70	75	
5.	Canada	30	35	40	45	50	

positively correlated in both directions [1]. However, there are a few indications that the relationship is not statistically significant. This indicates that renewable energy adoption can be both a catalyst and an effect of economic expansion, highlighting the intricate relationship between renewable energy deployment and economic performance. The rise in renewable energy production, namely, from fluctuating sources such as wind and solar photovoltaics (PVs), is propelled by endeavors to decrease pollution caused by fossil fuels, regulatory and legislative structures, and incentives designed to minimize greenhouse gas emissions [2]. Renewable energy sources, which can be naturally regenerated at a rate equal to or faster than their rate of use, provide a sustainable alternative to traditional energy sources. These sources are an essential component of the worldwide plan to battle climate change and promote sustainable development because of their inherent renewability and environmental benefits, including lower air pollution and the release of gases that contribute to the greenhouse effect. The necessity of energy from green sources in the current period is, therefore, paramount. More people are becoming aware of its benefits in lowering greenhouse gas emissions, promoting sustainable development, boosting economic growth, and lowering reliance on imported fossil fuels. Energy systems that are more sustainable, dependable, and efficient may be achieved via the integration of cutting-edge technology like smart grids, which further unlocks the potential of energy from clean sources [3].

To preserve impartiality, it is crucial to emphasize the drawbacks of sources of clean energy, such as wind and solar energies. Supply interruptions are the primary obstacle for renewable energy [4]. Load demand is unpredictable, and renewable energy sources' production variation is unpredictable as well [5, 6]. Management of the supply–demand balance is thus crucial for maintaining the stability of the power system [5, 7]. For instance, clouds can cause solar panel output to drop suddenly, creating a gap between the amount of energy supplied and the amount of energy required, which can result in power disruptions. In contrast, energy waste occurs when there is an overabundance of renewable energy production [8]. Therefore, in order to ensure grid operations stability and security, it is imperative to enhance the precision of predictions for energy sources that are renewable [9]. More recently, sophisticated methods and algorithm like artificial intelligence (AI) technologies have been developed to increase the accuracy of predictions pertaining to alternative energy sources [10–13]. AI approaches utilize vast amounts of data to create sophisticated computers able to perform tasks that typically require human intelligence. Artificial intelligence techniques, which include machine learning (ML) and deep learning (DL), are being widely adopted in scientific and technical fields to address complex and non-linear data structures [14]. Methods for ML encompass several techniques such as extreme learning machines (ELMs), kernel and nearest-neighbor (k-NN), support vector machines (SVMs), and artificial neural networks (ANNs). These techniques possess a distinct advantage over statistical approaches due to their lack of reliance on internal solar system parameters. A model of a problem can be constructed utilizing AI techniques and subsequently subjected to analysis in order to accurately forecast the performance, defect diagnostics, and control capabilities and the functioning, fault diagnosis, and control capabilities of alternate energy systems. For instance, Bhuiyan and colleagues applied ML methods to forecast wind speed for the purpose of analyzing the power production of wind system.

Additional crucial uses in the renewable energy field include batteries, inverter management of PV systems, maximum power point tracking, and forecasting and planning of load demand [15]. These strategies offer possibilities to mitigate the danger of system-wide failure and ensure their reliability. The precision of renewable energy forecasts during overcast conditions (i.e., huge datasets) is one of the difficult problems that AI systems currently face, despite their promise to be more accurate, dependable, and comprehensive than previous approaches [16]. The objective is to analyze how AI is now being used for significant renewable energy applications, such as forecasting. As a result, a thorough analysis of the most recent research done in significant applications for the implementation of two separate sources of sustainable energy, such as solar and wind energy plants, is completed. Furthermore, an analysis of forthcoming challenges, strategic developments, and current events is provided.

14.2 The Growth and Intersection of AI and ML in the World of Renewable Power

The segment of energy production that depends on renewable sources is witnessing a vital shift toward more intelligent and productive energy systems through advanced technologies. The application of these technologies in the industry is extensive involving optimization, predictive analytics, and automation, all of which are vital for the efficient integration and administration of alternate sources of energy. Moghaddam *et al.* investigate how advanced technology may be used to automate energy systems in smart cities. Their research, which centers on the European Union, shows how such technologies are improving performance and productivity in energy systems, therefore changing them. Achieving sustainable future advancements with low greenhouse gas emissions requires integrating research and highlighting the capacity of modern technologies to efficiently manage and regulate power networks, especially by integrating clean energy sources like solar and wind into current supply and demand systems [15]. Kumar and Swastika conducted a comprehensive examination of the utilization of advanced technology in renewable energy; they highlighted the significant and revolutionary impact of these technologies on the domain of electrical engineering. This research showcases the significant impact of these technologies on the generation dissemination and utilization of energy through its integration, with renewable energy technology advances. This chapter offers an analysis of the effects of advanced technology on intelligent structures and services, with specific emphasis on the symbiotic relationship between these technologies and renewable energy in the development of more intelligent and efficient energy systems [17]. Focusing on emerging applications, Singh *et al.* examined AI-based renewable energy. The study examines essential methodologies that rely on ML approaches to analyze PV forecasts. The study focuses especially on resolving issues brought on by environmental factors that cause renewable energy sources' performance to fluctuate. It draws attention to AI's potential to enhance renewable energy output predictions and management. Additionally, the study highlights how important precise forecasts are to preserving grid stability and dependability [18]. Advanced algorithms and data analytics improve the efficiency of energy generation and distribution when AI and green energy are coupled. The renewable energy industry may overcome the difficulties of integrating variable energy sources, like solar and wind, into current electricity infrastructures by utilizing AI technology. The energy production and management industry can attain increased sustainability, dependability, and efficiency by utilizing AI. Therefore, the use of AI in the realm of renewable energy represents a groundbreaking advancement that is transforming the landscape of energy administration and generation [18]. For the effective integration of renewable energy sources into the global energy mix, AI is of utmost importance in automating, optimizing, and forecasting energy systems. AI is expected to become more significant in propelling innovation and efficiency in the renewable energy industry as the globe moves closer to sustainable energy alternatives. The use of data science with renewable energy, namely, in solar and wind energy systems, indicates an innovative approach that enhances the efficiency and reliability of certain energy sources. The convergence of DL, ML, and advanced data analytics to enhance energy, production, and distribution is the intersection of this field, and an intriguing depiction of this intersection is shown in Chen and Xu's analysis conducted on solar wind power data acquired from the Chinese state grid to participate in the renewable power generation forecasting competition. The study emphasizes how critical precise forecasting is to renewable energy systems, especially given how heavily power networks are reliant on renewable energy. The research used an accessible dataset including information collected from renewable energy systems located on site including windmills and solar locations over a period of 2 years. Data science provides the capacity to enhance the ability to anticipate and maintain the reliability of renewable energy sources, and this dataset plays a vital role in the creation of data-driven forecasting models [19]. A hybrid PV wind smart microgrid in Peru used a data science technique. The research centered on examining actual data of solar irradiance, wind, speed, energy consumption, and battery bank voltage to discern patterns, seasonal variation, and relationships. The insights gathered from this data-driven study were utilized to refine sizing procedures and make energy management suggestions. The smart microgrid system's performance was eventually improved by using the insights gained from this data-driven study to improve sizing practices and offer energy management suggestions. This research highlights the practical application of data science in improving the operational efficiency of renewable energy systems. Abualigah *et al.* conducted a comprehensive examination of advanced deep learning (DL) methods and machine learning (ML) applications in the context of wind, solar, and photovoltaic (PV) systems. Their analysis evaluates the capabilities of various algorithms and emphasizes the value of hybrid learning approaches for addressing energy generation challenges. The study demonstrates that hybrid methods, which combine the strengths of multiple techniques, achieve higher accuracy and efficiency compared to single-method approaches. This advantage is particularly significant when working with large datasets, where the complexity and unpredictability of renewable energy sources present sustainable challenges at the intersection of data science and renewable energy. New frameworks for data collection, processing, and analysis are being developed in addition to advanced algorithms being applied [20]. Collectively, the research shows how important data is to compare improving clean and alternate energy systems. Stakeholders in the clean energy sector can enhance the integration of alternative energy sources into the power grid by leveraging data, innovation, and efficiency. This approach enables more informed decision-making and accelerates the transition to sustainable energy systems. The convergence of data science and clean energy particularly in solar and wind energy drives significant advancements In addition to increasing the stability and predictability of renewable energy sources, data-driven approaches help create more intelligent, effective, and sustainable energy systems. As a result, data science is poised to play an increasingly critical role in the global shift toward the widespread adoption of renewable energy.

14.3 Machine Learning–Based Forecasting System for Renewable Energy Production

Renewable energy systems are being enhanced by using advanced technologies like the Internet of Things (loT) and AI to enhance their performance. This combination seeks to enhance the adaptability and flexibility of these systems. All renewable energy systems are currently implementing AI technologies, including solar PV, wind, hydro, ocean, and geothermal systems [21]. Research has demonstrated that the implementation of AI in renewable energy systems has increased the efficiency and cost-effectiveness of technologies such as PV panels, wind turbines, and solar panels. Consequently, these technologies have the potential to substitute fossil fuel–based power plants and become more widespread [22]. The ML approaches employed in renewable energy systems are illustrated in Figure 14.2 and summarized in Table 14.2. These are adaptive neuro-fuzzy inference system (ANFIS), Naïve Bayes (NB), backpropagation neural network (BPNN), genetic algorithm (GA), radial basis function (RBF), ANN, radial basis function neural network (RBFNN), SARIMA (SVM with the cyclical autoregressive integrated moving average approach), hybrid intelligent

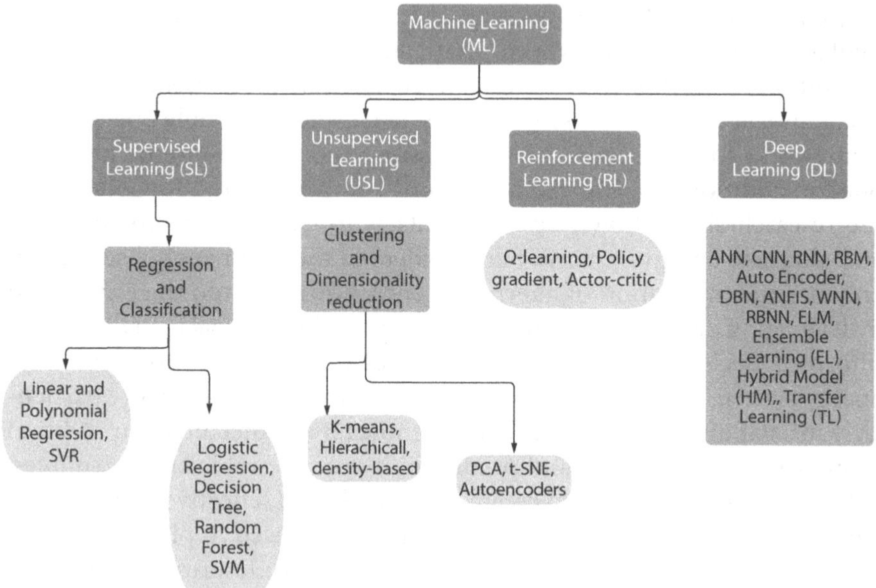

Figure 14.2 ML types and algorithms.

Table 14.2 ML and deep learning techniques including advantages and disadvantages in various applications.

Technique	Advantages	Disadvantages	Applications
Linear Regression (LR)	Simple to execute, quick training	Constrained to only involve straight-line connections	Predictive analytics
Logistic Regression	Easy to understand, performs effectively with little datasets	This method assumes linearity and is solely applicable for classification tasks.	Power outages, adverse weather, market, and healthcare predictions
Decision Trees	Can interpret categorical and continuous data	Susceptible to overfitting	Anticipatory maintenance, financial analysis

(*Continued*)

Table 14.2 ML and deep learning techniques including advantages and disadvantages in various applications. (*Continued*)

Technique	Advantages	Disadvantages	Applications
Random Forest	High precision, less susceptibility to overfitting	Computationally demanding in comparison to decision trees, challenging to interpret	Operational control strategy, picture categorization, and fraud detection
SVM	Noise-resilient linear relationships and strong connections	Calculationally costly and parameter-tunable,requires careful parameter tuning	Bioinformatics, text categorization
K-Means Clustering	Easy, rapid data exploration, and segmentation	Needs a set number of clusters and is sensitive to starting situations	Market segmentation, image segmentation
PCA	Reduces data dimensionality and noise for exploration and visualization	Can be hard to comprehend and miss important information	Recognition of images, voice, and NLP
Reinforcement Learning	Improves decision-making in dynamic contexts through trial and error	Demands a substantial amount of data and is susceptible to overfitting	Activities involving game playing and robotics
ANN	Handles big datasets, complicated interactions, and non-linear models	Data-intensive and hard to comprehend	Forecast stationary energy demand, energy resources, picture, and speech

(*Continued*)

Table 14.2 ML and deep learning techniques including advantages and disadvantages in various applications. (*Continued*)

Technique	Advantages	Disadvantages	Applications
CNN	Highly effective picture analysis; it automatically learns characteristics	Expensive, requiring large amounts of data, and not ideal for situations with limited spatial or temporal accuracy	Recognition, classification, and forecast of energy demand using satellite imagery
RNN	Supports sequential, time series, and long-term dependencies	Potential for overfitting, sluggish training, and disappearing or bursting gradients	Voice recognition, mood analysis, and time-series energy price forecasting
LSTM	Works well with long-term dependencies for time series data	May be susceptible to overfitting and need meticulous adjusting	Load forecasting, time series, voice recognition, NLP, and energy price forecasting
Autoencoders	Reduces data and noise dimensionality and for unsupervised learning	Demands substantial quantities of data and might provide challenges in terms of interpretation	Tasks including anomaly detection, image analysis, and speech recognition
ELM	Rapid training, capable of processing extensive datasets	Data that is difficult to understand may not be relevant to recent data	Renewable energy projections, visual and auditory identification, and anticipatory data analysis

(*Continued*)

Table 14.2 ML and deep learning techniques including advantages and disadvantages in various applications. (*Continued*)

Technique	Advantages	Disadvantages	Applications
GRNN	Rapid training, manage noise	Restricted to tasks using regression analysis; may not exhibit efficient scalability when used to extensive datasets	Time- series forecasting, function estimation, and energy resource forecasting
RBNN	Appropriate for non-linear classification and regression	Demands meticulous optimization of network structure and hyperparameters	Image and speech recognition, anomaly detection
WNN	Capable of processing data with varying resolutions and scales	Computationally costly, needs careful wavelet basis function selection	Signal, imagetime series prediction
ANFIS	Manage ambiguity, data exhibiting non-linearity	Requires careful selection; optimizing fuzzy rules may require significant computing resources	Systems for diagnosing faults and controlling them
DBN	Has the ability to acquire hierarchical representations of data, which proves to be very efficient for unsupervised learning.	Requires a large volume of data and is challenging to understand	Image and speech recognition, natural language processing

(*Continued*)

Table 14.2 ML and deep learning techniques including advantages and disadvantages in various applications. (*Continued*)

Technique	Advantages	Disadvantages	Applications
Ensemble Learning	Combine models to increase performance and prevent overfitting	Intensive calculation and adjustment may be needed.	Vision, speech, and NLP for renewable energy forecasting
Transfer Learning	Utilizes trained models to enhance efficiency and minimize data requirements	Generalizes poorly to fresh data, confined to comparable jobs	Energy price prediction, energy efficiency forecasting, renewable energy foresting, predictive maintenance, fault diagnosis, load forecasting image and speech recognition, and NLP

Note: NLP, Natural Language Processing.

system for modeling and integration (HISMI), SVM, generalized modular hierarchical neural network (GMHNN), and ARIMA (autoregressive integrated moving average). The ANN is a more adaptable approach among these. Compared to other modeling approaches, ANN's distinct benefits over its competitors include shorter computation times and increased in terms of data analysis parameters including accuracy, processing speed, latency, fault tolerance, volume, scalability, convergence, and performance; the same study found that the feedforward and feedback propagation ANN models suited the data the best.

14.4 AI and ML Applications for Renewable Energy

The integration of AI into renewable energy systems has led to several advancements. New algorithms have been developed to improve energy efficiency, particularly in the context of smart buildings. Bagheri *et al.* examined the application of AI algorithms across various building typologies to enhance energy efficiency. Their research focused on using AI to regulate heating systems in different types of buildings. The study showed that AI algorithms, regardless of building typologies, significantly improve energy efficiency. This optimization was achieved using a corporate server that managed heating systems through a unified energy-saving platform [23]. Abdel-Razek *et al.* investigated energy efficiency by utilizing an AI model to forecast room occupancy based on thermal comfort characteristics. Their research leveraged data on temperature, humidity, illumination, and CO_2 levels to predict room occupancy using ML methods, such as K-nearest neighbors (KNNs) and decision trees (DTs). The study revealed that KNN achieved a 99.5% accuracy rate, outperforming DT in occupancy detection by dynamically adjusting ambient settings according to occupancy. This approach demonstrated how AI can optimize energy consumption in buildings, leading to significant energy savings and improved user comfort [24]. Additionally, Abualigah *et al.* conducted an extensive analysis of advanced ML and DL methods in the context of solar, wind, and PV energy systems. Their study evaluated several learning-based modeling techniques for accurate forecasting in renewable energy sources, highlighting the potential of ML and DL algorithms to enhance system performance and reliability. The study demonstrated how well these methods work with big datasets and complicated factors, highlighting the benefits of hybrid learning approaches that blend many algorithms to provide more accurate forecasts. The research highlights the potential of new AI algorithms to optimize renewable energy production and management [20]. The development of innovative AI algorithms is a critical factor in improving energy efficiency across several industries. The development of innovative AI algorithms is a critical factor in improving energy efficiency across several industries. These developments are opening the door for more sustainable and effective energy systems, from renewable energy forecasts to smart building control. With AI's continued development, energy efficiency applications could see even bigger gains, supporting the worldwide push for sustainability and energy conservation.

14.4.1 Forecasting the Power of Photovoltaic System

Solar energy is harnessed by converting solar radiation into electrical energy via the use of solar thermal or PV systems. According to data published by the International Energy Agency, global PV capacity surpassed 179 TWh in 2021, a 22% increase over 2020. Because solar irradiance directly and substantially impacts the generation of solar power, accurate solar radiation prediction is crucial for energy output estimation. Accurate forecasting is crucial for many applications, such as microgrids, energy conservation, and management. When solar radiation data is used, AI models and algorithms perform better than traditional methods [25]. Shi *et al.* created methods to anticipate the power generation of PV systems under four typical weather conditions: overcast, foggy, rainy, and sunny. These methods make use of SVMs and weather classification. With an average mean relative error of 8.64% and a mean absolute error of 4.83%, the greatest results were achieved on sunny days. These findings demonstrate the value and promise of the proposed forecasting approach for grid-connected PV systems [26]. The proposal of Pandu *et al.* to integrate big data with AI improves solar radiation predictions [27]. Furthermore, Mandal *et al.* suggested combining wavelet transform and AI methods to estimate PV power [28], and Son *et al.* showed that a six-layer feedforward deep neural network outperformed conventional models, despite having lower accuracy in the summer and in cloudy conditions [29]. In a different study, researchers used data from previous years to assess and compare four ML methods for predicting daily solar radiation: SVM, ANN, k-NN, and DL. According to the study, all four models showed good accuracy and reliability, with ANN beating DL, SVM, and k-NN [30]. Additionally, Voyant *et al.* demonstrated that ANN outperformed other techniques in terms of prediction accuracy, processing time, and error [31]. To enhance short-term power estimates, Bouzerdoum *et al.* developed a hybrid model that blends SARIMA [32]. Their findings showed that the hybrid model outperformed the SVM and SARIMA models. Meanwhile, Lim *et al.* created a hybrid model integrating long short-term memory (LSTM) and convolutional neural network (CNN) architectures to accurately predict power production levels [33].

14.4.2 Forecasting the Power of Wind Energy System

Wind energy is increasingly gaining popularity due to its affordability, renewable nature, and environmentally friendly characteristics. Wind turbines provide a concrete illustration of sustainable energy because they promptly decrease pollution. Nevertheless, due to the erratic fluctuations and temporal unpredictability of wind energy, it remains challenging to forecast, hence rendering the continuous generation of wind power unattainable. Hence, it is essential to provide a reliable framework for forecasting wind energy [34]. Various techniques exist for forecasting wind power, such as physical, conventional statistical, and intelligent forecasting methods [35]. Wind speed data has a substantial influence on AI models and algorithms because of the association between wind direction, wind speed, and the power production of wind systems. Conventional statistical models like autoregressive moving average [36] and ARIMA [37] have shown to be more efficient in predicting short-term outcomes. However, physical methodologies are generally seen as better appropriate for making long-term wind energy estimates [36, 38, 39]. Statistical models are very accurate in producing predictions for linear data components. Nevertheless, they lack reliability in forecasting results for non-linear data [40]. Statistical techniques outperform physical and intelligent learning approaches in terms of computing efficiency. The swift progress of AI has enabled the effective creation and utilization of intelligent prediction models and algorithms, such as ANNs [41, 42] and SVMs, in the field of wind energy forecasting [43], to the quick development of AI. To estimate wind speed, researchers created and compared two models: the adaptive ANFIS and ANN. The results showed that the ANFIS algorithm outperforms ANN [44]. The authors provided convincing evidence that ANN models outperform analytical methods in terms of wind power prediction accuracy. In recent years, hybrid systems have been more often suggested for wind speed prediction because they have shown improved performance compared to single-approach forecasting [45]. Wang *et al.* devised the empirical mode decomposition–ENN (Elman neural network) technique for forecasting wind power generation [46]. To improve the accuracy of predictions, researchers have proposed hybrid SVM models [42]. Furthermore, a hybrid forecasting prototype was created to predict the transient wind power output. This prototype used a SVM and the Improved Dragonfly [47].

14.5 Approaches and Limitations in AI Application for Renewable Energy

Adenle *et al.* explore the financial challenges linked to renewable energy, specifically focusing on solar power. The authors emphasize that, although AI has promise in addressing these challenges, there are barriers that impede its adoption, such as substantial initial expenses, a scarcity of skilled workforce, and intricate technological complexities. The challenges are particularly pronounced in underdeveloped nations due to limited resources and a deficiency in sophisticated technological infrastructure [15]. Alam *et al.* specifically address the difficulties associated with incorporating innovative sources of renewable energy into the electricity grid. Wind and solar power, being intermittent alternative power sources, are known to have limitations such as restricted fault ride-through capabilities and power quality concerns. AI may be used to address these difficulties by improving predictive maintenance and optimizing grid operations. Nevertheless, the incorporation of AI presents a particular set of challenges, requiring the utilization of algorithms and data analytics capabilities [48]. Siddique *et al.* reported the effect of the Web of Things (IoT) on the renewable energy sector in Bangladesh. The proponents suggest that the IoT, a crucial component of AI applications, has the capability to facilitate the smooth integration of renewable energy sources into the power grid. Nevertheless, in order to effectively use IoT in renewable energy applications, it is essential to address challenges such as data security, network reliability, and interoperability. In order to surmount these challenges, the authors propose a collaborative strategy that integrates the efforts of both the governmental and business sectors, as well as academia [49]. While AI has immense potential to revolutionize renewable energy, there are obstacles to using AI in reality. The primary challenges are the need for sophisticated IoT infrastructure, the difficulty of integrating with the grid, and budgetary limitations. To overcome these challenges, a comprehensive strategy is needed, which involves government backing, enhancing capabilities, and advancing technology. The crucial roles that AI and IoT will play in the development of the renewable energy sector cannot be overstated. They will help overcome hurdles and unlock the whole potential of alternative energy sources.

14.6 Advances and Prospects in AI for Solar and Wind Power

Artificial intelligence is catalyzing a transformative shift in the renewable energy sector, particularly in the domains of solar and wind power. The study conducted by Abualigah *et al.* investigates the use of DL and advanced ML techniques in renewable energy systems such as solar, wind, and PV systems. They emphasize the effectiveness of these strategies in handling large datasets and complex features seen in renewable energy systems [20]. The study suggests that hybrid learning methods, which combine several AI techniques, may enhance the accuracy and efficiency of renewable energy predictions and optimizations. Dias and Morais use text mining techniques to analyze patterns in solar and wind energy, drawing on scholarly publications and patent records. Their findings indicate a significant investment in these fields for the sake of scientific and technical progress. Consequently, there may be a transition in the sector toward more AI-driven research and development [50]. The research conducted by Kumar *et al.* specifically examines the capacity for solar energy in Hungary and the Visegrád area. Their study focuses on using AI to evaluate and optimize the potential of solar energy. This is achieved via the examination of temperature distribution profiles, technical regression analysis, and geographic mapping. Research suggests that AI-powered assessments might play a crucial role in promoting the sustainable expansion of solar energy across different regions [51]. The incorporation of AI in solar and wind energy is indicated by advancements in DL and ML techniques, a growing trend in AI-driven research and development, and the use of AI in technical and geographic assessments. These trends indicate that AI will enhance the precision and effectiveness of renewable energy networks while also fostering innovation and sustainable development in the sector.

14.7 Conclusion

In conclusion, the integration of AI, ML, and renewable energy represents a significant advancement in the hunt for sustainable energy sources. This research has shown how the growth of AI and ML technologies has facilitated the creation of sophisticated forecasting systems for the production of renewable energy. These forecasting techniques improved the efficiency of solar and wind energy generation. The use of AI and ML in the renewable energy sector presents many obstacles. Using these technologies efficiently, it is crucial to tackle issues such as the computational intricacy, interpretability of models, and data integrity. To further explore and enhance AI applications in renewable energy, tactics and advancements like DL and forecasting methods provide fresh insights into improving the efficiency of renewable energy systems. By eliminating current limitations and harnessing the power of DL, AI, and ML have the potential to greatly accelerate the shift toward cleaner and more sustainable energy.

References

1. Morganti, P. and Garofalo, G., Renewable Energy and Economic Growth: An Overview of the Literature, pp. 63–78, 2020.
2. Owunna, T. and Obeagu, E., Overview of Smart Grid Technology as a Renewable Energy Source. *J. Energy Res. Rev.*, 12, 6–15, 2022. doi:10.9734/jenrr/2022/v12i3239.
3. Sarma, S., Pandey, G., Borah, U., Molokitina, N., Chauhan, G., Yadav, M., Emerging Technologies for Sustainable Energy Applications, pp. 53–86, 2024.

4. Strielkowski, W., Civín, L., Tarkhanova, E., Tvaronavičienė, M., Petrenko, Y., Renewable Energy in the Sustainable Development of Electrical Power Sector: A Review. *Energies*, 14, 24, 8240, 2021. doi:10.3390/en14248240.
5. Jie, B., Tsuji, T., Uchida, K., Impact of renewable energy balancing power in tertiary balancing market on Japanese power system based on automatic generation control standard model. *J. Eng.*, 2019, 18, 4760–4763, 2019, doi: https://doi.org/10.1049/joe.2018.9301.
6. Liu, W., Zhan, J., Chung, C.Y., Li, Y., Day-Ahead Optimal Operation for Multi-Energy Residential Systems With Renewables. *IEEE Trans. Sustain. Energy*, 10, 4, 1927–1938, 2018, doi: 10.1109/TSTE.2018.2876387.
7. Gountis, V. and Bakirtzis, A., Bidding Strategies for Electricity Producers in a Competitive Electricity Marketplace. *Power Systems IEEE Trans.*, 19, 356–365, 2004, doi: 10.1109/TPWRS.2003.821474.
8. Orkisz, M., Estimating effects of individual PV panel failures on PV array output. *IEEE Trans. Ind. Appl.*, 54, 5, 4825–4832, 2018.
9. Chen, F., Duic, N., Manuel Alves, L., da Graça Carvalho, M., Renewislands—Renewable energy solutions for islands. *Renew. Sustain. Energy Rev.*, 11, 8, 1888–1902, 2007, doi: https://doi.org/10.1016/j.rser.2005.12.009.
10. Das, U.K., Tey, K.S., Seyedmahmoudian, M., Mekhilef, S., Idris, M.Y., II, Van Deventer, W., Stojcevski, A., Forecasting of photovoltaic power generation and model optimization: A review. *Renew. Sustain. Energy Rev.*, 81, 912–928, 2018, doi: https://doi.org/10.1016/j.rser.2017.08.017.
11. de Oliveira e Lucas, P., Alves, M.A., de Lima e Silva, P.C., Guimarães, F.G., Reference evapotranspiration time series forecasting with ensemble of convolutional neural networks. *Comput. Electron. Agric.*, 177, 105700, 2020, doi: https://doi.org/10.1016/j.compag.2020.105700.
12. Mellit, A., Artificial Intelligence technique for modeling and forecasting of solar radiation data: a review. *Int. J. Artif. Intell. Soft Comput.*, 1, 52–76, 2008, doi: 10.1504/IJAISC.2008.021264.
13. Saggi, M.K. and Jain, S., Reference evapotranspiration estimation and modeling of the Punjab Northern India using deep learning. *Comput. Electron. Agric.*, 156, 387–398, 2019, doi: https://doi.org/10.1016/j.compag.2018.11.031.
14. Sobri, S., Koohi-Kamali, S., Rahim, N.A., Solar photovoltaic generation forecasting methods: A review. *Energy Convers. Manage.*, 156, 459–497, 2018, doi: https://doi.org/10.1016/j.enconman.2017.11.019.
15. Md Abul Ehsan, B., Begum, F., Ilham, S.J., Khan, R.S., Advanced wind speed prediction using convective weather variables through machine learning application. *Appl. Comput. Geosci.*, 1, 100002, 2019, doi: https://doi.org/10.1016/j.acags.2019.100002.
16. Adenle, A.A., Assessment of solar energy technologies in Africa-opportunities and challenges in meeting the 2030 agenda and sustainable development goals. *Energy Policy*, 137, 111180, 2020, doi: https://doi.org/10.1016/j.enpol.2019.111180.

17. Kumar, G.N. and Swathika, O. V. G. J. S. B. D., AI Applications to Renewable Energy - An Analysis, 2021.
18. Singh, O., Yadav, M., Yadav, P., Rishiwal, V.J. I. t. A. T. f. R. E. S, AI-Based Renewable Energy with Emerging Applications, 2021.
19. Chen, Y. and Xu, J., Solar and wind power data from the Chinese State Grid Renewable Energy Generation Forecasting Competition. *Sci. Data*, 9, 1, 577, 2022, doi: 10.1038/s41597-022-01696-6.
20. Abualigah, L., Zitar, R.A., Almotairi, K.H., Hussein, A.M., Abd Elaziz, M., Nikoo, M.R., Gandomi, A.H., Wind, Solar, and Photovoltaic Renewable Energy Systems with and without Energy Storage Optimization: A Survey of Advanced Machine Learning and Deep Learning Techniques. *Energies*, 15, 2, 578, 2022, doi: 10.3390/en15020578.
21. Jha, S.K., Bilalovic, J., Jha, A., Patel, N., Zhang, H., Renewable energy: Present research and future scope of Artificial Intelligence. *Renew. Sustain. Energy Rev.*, 77, 297–317, 2017, doi: https://doi.org/10.1016/j.rser.2017.04.018.
22. Ahmadi, M.H., Dehghani Madvar, M., Sadeghzadeh, M., Rezaei, M.H., Herrera, M., Shamshirband, S., Current Status Investigation and Predicting Carbon Dioxide Emission in Latin American Countries by Connectionist Models, vol. 12, 20191916.
23. Bagheri, A., Genikomsakis, K.N., Koutra, S., Sakellariou, V., Ioakimidis, C.S.J.B., Use of AI Algorithms in Different Building Typologies for Energy Efficiency towards Smart Buildings, 2021.
24. Abdel-Razek, S.A., Marie, H.S., Alshehri, A., Elzeki, O.M.J.S., Energy Efficiency through the Implementation of an AI Model to Predict Room Occupancy Based on Thermal Comfort Parameters, 2022.
25. Sharadga, H., Hajimirza, S., Balog, R.S., Time series forecasting of solar power generation for large-scale photovoltaic plants. *Renew. Energy*, 150, 797–807, 2020, doi: https://doi.org/10.1016/j.renene.2019.12.131.
26. Liu, L., Liu, D., Sun, Q., Li, H., Wennersten, R., Forecasting Power Output of Photovoltaic System Using A BP Network Method. *Energy Procedia*, 142, 780–786, 2017, doi: https://doi.org/10.1016/j.egypro.2017.12.126.
27. Pandu, S.-B., Britto, A.-S.-F., Sekhar, P., Vijayarajan, P., Albraikan, A.-A., Al-Wesabi, F.-N., Duhayyim, M.-A., Artificial Intelligence Based Solar Radiation Predictive Model Using Weather Forecasts, vol. 71, pp. 109–124, 2022.
28. Mandal, P., Madhira, S.T.S., haque, A.U., Meng, J., Pineda, R.L., Forecasting Power Output of Solar Photovoltaic System Using Wavelet Transform and Artificial Intelligence Techniques. *Procedia Comput. Sci.*, 12, 332–337, 2012, doi: https://doi.org/10.1016/j.procs.2012.09.080.
29. Son, J., Park, Y., Lee, J., Kim, H., Sensorless PV Power Forecasting in Grid-Connected Buildings through Deep Learning. *Sensors*, 18, 8, 2529, 2018, doi: 10.3390/s18082529.
30. Ağbulut, U., Gürel, A.E., Biçen, Y., Prediction of daily global solar radiation using different machine learning algorithms: Evaluation and comparison.

Renew. Sustain. Energy Rev., 135, 110114, 2021, doi: https://doi.org/10.1016/j.rser.2020.110114.

31. Voyant, C., Notton, G., Kalogirou, S., Nivet, M.-L., Paoli, C., Motte, F., Fouilloy, A., Machine learning methods for solar radiation forecasting: A review. *Renew. Energy*, 105, 569–582, 2017, doi: https://doi.org/10.1016/j.renene.2016.12.095.
32. Bouzerdoum, M., Mellit, A., Massi Pavan, A., A hybrid model (SARIMA–SVM) for short-term power forecasting of a small-scale grid-connected photovoltaic plant. *Sol. Energy*, 98, 226–235, 2013, doi: https://doi.org/10.1016/j.solener.2013.10.002.
33. Lim, S.-C., Huh, J.-H., Hong, S.-H., Park, C.-Y., Kim, J.-C., Solar Power Forecasting Using CNN-LSTM Hybrid Model. *Energies*, 15, 21, 8233, 2022, doi: 10.3390/en15218233.
34. Yang, W., Wang, J., Lu, H., Niu, T., Du, P., Hybrid wind energy forecasting and analysis system based on divide and conquer scheme: A case study in China. *J. Clean. Prod.*, 222, 942–959, 2019, doi: https://doi.org/10.1016/j.jclepro.2019.03.036.
35. Lin, Y., Luo, H., Wang, D., Guo, H., Zhu, K., An Ensemble Model Based on Machine Learning Methods and Data Preprocessing for Short-Term Electric Load Forecasting. *Energies*, 10, 8, 1186, 2017, doi: 10.3390/en10081186.
36. Torres, J.L., García, A., De Blas, M., De Francisco, A., Forecast of hourly average wind speed with ARMA models in Navarre (Spain). *Sol. Energy*, 79, 1, 65–77, 2005, doi: https://doi.org/10.1016/j.solener.2004.09.013.
37. Sfetsos, A., A novel approach for the forecasting of mean hourly wind speed time series. *Renew. Energy*, 27, 2, 163–174, 2002, doi: https://doi.org/10.1016/S0960-1481(01)00193-8.
38. Su, Z., Wang, J., Lu, H., Zhao, G., A new hybrid model optimized by an intelligent optimization algorithm for wind speed forecasting. *Energy Convers. Manage.*, 85, 443–452', 2014, doi: https://doi.org/10.1016/j.enconman.2014.05.058.
39. Sun, W. and Liu, M., Wind speed forecasting using FEEMD echo state networks with RELM in Hebei, China. *Energy Convers. Manage.*, 114, 197–208, 2016, doi: https://doi.org/10.1016/j.enconman.2016.02.022.
40. Gao, Y., Wang, J., Yang, H., A multi-component hybrid system based on predictability recognition and modified multi-objective optimization for ultra-short-term onshore wind speed forecasting. *Renew. Energy*, 188, 384–401, 2022, doi: https://doi.org/10.1016/j.renene.2022.02.005.
41. Tian, C. and Hao, Y., A Novel Nonlinear Combined Forecasting System for Short-Term Load Forecasting, vol. 11, p. 712, 2018.
42. Zendehboudi, A., Baseer, M.A., Saidur, R., Application of support vector machine models for forecasting solar and wind energy resources: A review. *J. Clean. Prod.*, 199, 272–285, 2018, doi: https://doi.org/10.1016/j.jclepro.2018.07.164.

43. Jiang, P. and Ma, X., A hybrid forecasting approach applied in the electrical power system based on data preprocessing, optimization and artificial intelligence algorithms. *Appl. Math. Modell.*, 40, 23, 10631–10649, 2016, doi: https://doi.org/10.1016/j.apm.2016.08.001.
44. Amoura, Y., Pereira, A., II, Lima, J., A Short Term Wind Speed Forecasting Model Using Artificial Neural Network and Adaptive Neuro-Fuzzy Inference System Models, 2022, 2022, Paper presented at the Sustainable Energy for Smart Cities, Cham.
45. Meng, A., Ge, J., Yin, H., Chen, S., Wind speed forecasting based on wavelet packet decomposition and artificial neural networks trained by crisscross optimization algorithm. *Energy Convers. Manage.*, 114, 75–88, 2016, doi: https://doi.org/10.1016/j.enconman.2016.02.013.
46. Wang, J., Zhang, W., Li, Y., Wang, J., Dang, Z., Forecasting wind speed using empirical mode decomposition and Elman neural network. *Appl. Soft Comput.*, 23, 452–459, 2014, doi: https://doi.org/10.1016/j.asoc.2014.06.027.
47. Li, L.-L., Zhao, X., Tseng, M.-L., Tan, R.R., Short-term wind power forecasting based on support vector machine with improved dragonfly algorithm. *J. Clean. Prod.*, 242, 118447, 2020, doi: https://doi.org/10.1016/j.jclepro.2019.118447.
48. Alam, M.S., Al-Ismail, F.S., Salem, A., Abido, M.A., High-Level Penetration of Renewable Energy Sources Into Grid Utility: Challenges and Solutions. *IEEE Access*, 8, 190277–190299, 2020, doi: 10.1109/ACCESS.2020.3031481.
49. Siddique, A.H., Tasnim, S., Shahriyar, F., Hasan, M., Rashid, K., Renewable Energy Sector in Bangladesh: The Current Scenario, Challenges and the Role of IoT in Building a Smart Distribution Grid, vol. 14, p. 5083, 2021.
50. Dias, C.G. and Morais, A.M., d. J. I. J. o. C. A, A Text Mining Approach in Patents and Papers: Analyzing Wind and Solar Energy Trends, vol. 175, pp. 1–12, 2020.
51. Kumar, B., Szepesi, G., Čonka, Z., Kolcun, M., Péter, Z., Berényi, L., Szamosi, Z., Trendline Assessment of Solar Energy Potential in Hungary and Current Scenario of Renewable Energy in the Visegrád Countries for Future Sustainability. *Sustainability*, 13, 10, 5462, 2021, doi: 10.3390/su13105462.

About the Editors

Dr. Jai Govind Singh

Dr. J. G. Singh received his Ph.D. (2008) in Power and Control, EED, Indian Institute of Technology, Kanpur, India; M. Tech (2003) in Power System, EED, Indian Institute of Technology, Roorkee, India; and Bachelor of Engineering (2001) in Electrical Engineering, MNIT, Allahabad, India. In addition, he completed post-doctoral research at KTH Stockholm and the University of Queensland, Brisbane, from 2008 to 2009. He works as an Associate Professor in the Department of Energy, Environment and Climate Change, School of Environment, Resources and Development (SERD) at the Asian Institute of Technology, Bangkok, Thailand. Moreover, Dr. Singh has completed 19 sponsored research projects from ERASMUS+, USAID, DST (India), BP (Thailand), IRENA, ADB, EBARA (Japan), NSTDA, IEEE PES, ADEME, PEA, ABT Associates, Tetra Tech, etc. He has published more than 120 research papers in reputed journals and conferences. He has wide research areas such as electric vehicles technologies, smart grid and micro-grid design and operation, RE generation, and integration, power system operation and control, electricity market restructuring and power trading, DG design and planning, and energy storage technologies.

Dr. Rupendra Kumar Pachauri

Dr. R. K. Pachauri is an expert in renewable energy technology and electrical hybrid power generation systems (micro-grid advancements). His major research contribution is based on hydrogen fuel cell power generation, energy management, and advanced strategies to enhance PV system performance during adverse environmental conditions (dust aerosol and partial shading conditions) and power generation forecasting using artificial intelligence and machine learning techniques. Dr. Pachauri is a distinguished academic and an associate professor holding an esteemed position within the Electrical Cluster Department at UPES in Dehradun, India since 2016. Dr. Pachauri was associated with the Department of Energy, Environment and Climate Change, School of Environment, Resources and Development, Asian Institute of Technology, Pathum Thani, Bangkok, Thailand, during his postdoctoral fellowship from June 2022 to August 2023. He completed his Ph.D. from G. B. University, Gautam Buddha Nagar, India (2016). Dr. Pachauri completed his M. Tech degree at Aligarh Muslim University, India (2009); earlier, he obtained his B. Tech degree from UPTU Lucknow, India (2006). In 2023 and 2024, Stanford University and Elsevier recognized Dr. Pachauri as one of the world's top 2% scientists. These qualifications have laid the groundwork for his illustrious academic career and expertise in the intricate electrical and electronic engineering domains. He has published more than 180 research articles in reputed journal and international conferences.

Dr. Sasidharan Sreedharan

Dr. Sasidharan Sreedharan received his Ph.D. (2010) in Sustainable Energy Transition, Asian Institute of Technology, Bangkok, Thailand, and M. Tech (1998) in Electrical Power System and B. Tech (1995) in Electrical and Electronics from Calicut University, Kerala, India. In addition, he completed his postdoctoral research from University of Hawai'i at Mānoa, USA, in 2015. After receiving his Ph.D., he was working as professor and head of the department, Calicut University and Kerala Technological University affiliated colleges, Kerala, India. Currently, he is working as assistant professor (grade C), College of Applied Sciences, Ministry of Higher Education, Sultanate of Oman. Moreover, Dr. Sreedharan has completed total 15 sponsored research projects from DOE, USAID, ADB, AICTE, DST, and KSCSTE. He has published more than 80 research papers in reputed journals and conferences. He is having wide research areas such as high performance computing, AI and ML, optimization/cyber security, smart grid operations, electrical supply restructuring, and energy storage.

Index

Also of Interest

Check out these other related titles from Scrivener Publishing

Photovoltaic Systems Technology, Edited by Mohammed Aslam Husain, MD Waseem Ahmad, Farhad Ilahi Bakhsh, Sanjeevikumar Padmanaban, and Hasmat Malik. Discover comprehensive insights into the latest advancements in solar PV technology, including power electronics, maximum power point tracking schemes, and forecasting techniques, with a focus on improving the performance of PV systems.

CLEAN AND RENEWABLE ENERGY PRODUCTION, Edited by Surajit Mondal, Adesh Kumar, Rupendra Kumar Pachauri, Amit Kumar Mondal, Vishal Kumar Singh, and Amit Kumar Sharma, ISBN: 9781394177723. This exciting new volume explores the recent advances in renewable energy production, the policies surrounding renewable energy, and practical applications.

URBAN ENERGY SYSTEMS: Modelling and Simulation for Sustainable Smart Cities, Edited by Deepak Kumar, ISBN: 9781119847441. Presenting the latest technologies and practices in this ever-changing field, this groundbreaking new volume covers the gambit for providing solutions and practical applications of smart and efficient energy systems.

ENERGY STORAGE TECHNOLOGIES IN GRID MODERNIZATION, Edited by Sandeep Dhundhara, Yajvender Pal Verma, and Ashwani Kumar, ISBN: 9781119872115. Written and edited by a team of experts, this exciting new volume discusses the various types of energy storage technologies, the applications of energy storage systems, the performance improvement of modern power systems, their role in the real-time operation of power markets, and the operational issues of modern power systems, including renewable-based generating sources.

SMART GRIDS FOR SMART CITIES VOLUME 1, Edited by O.V. Gnana Swathika, K. Karthikeyan, and Sanjeevikumar Padmanaban, ISBN: 9781119872078. Written and edited by a team of experts in the field, this first volume in a two-volume set focuses on an interdisciplinary perspective on the financial, environmental, and other benefits of smart grid technologies and solutions for smart cities.

SMART GRIDS FOR SMART CITIES VOLUME 2: Real-Time Applications in Smart Cities, Edited by O.V. Gnana Swathika, K. Karthikeyan, and Sanjeevikumar Padmanaban, ISBN: 9781394215874. Written and edited by a team of experts in the field, this second volume in a two-volume set focuses on an interdisciplinary perspective on the financial, environmental, and other benefits of smart grid technologies and solutions for smart cities.

SMART GRIDS AND INTERNET OF THINGS, Edited by Sanjeevikumar Padmanaban, Jens Bo Holm-Nielsen, Rajesh Kumar Dhanaraj, Malathy Sathyamoorthy, and Balamurugan Balusamy, ISBN: 9781119812449. Written and edited by a team of international professionals, this groundbreaking new volume covers the latest technologies in automation, tracking, energy distribution and consumption of Internet of Things (IoT) devices with smart grids.

SMART GRIDS AND GREEN ENERGY SYSTEMS, Edited by A. Chitra, V. Indragandhi and W. Razia Sultana, ISBN: 9781119872030. Presenting the concepts and advances of smart grids within the context of "green" energy systems, this volume, written and edited by a global team of experts, goes into the practical applications that can be utilized across multiple disciplines and industries, for both the engineer and the student.

SMART GRIDS AND MICROGRIDS: Concepts and Applications, Edited by P. Prajof, S. Mohan Krishna, J. L. Febin Daya, Umashankar Subramaniam, and P. V. Brijesh, ISBN: 9781119760559. Written and edited by a team of experts in the field, this is the most comprehensive and up to date study of smart grids and microgrids for engineers, scientists, students, and other professionals.

MICROGRID TECHNOLOGIES, Edited by C. Sharmeela, P. Sivaraman, P. Sanjeevikumar, and Jens Bo Holm-Nielsen, ISBN: 9781119710790. Covering the concepts and fundamentals of microgrid technologies, this volume, written and edited by a global team of experts, also goes into the practical applications that can be utilized across multiple industries, for both the engineer and the student.

INTEGRATION OF RENEWABLE ENERGY SOURCES WITH SMART GRIDS, Edited by A. Mahaboob Subahani, M. Kathiresh and G. R. Kanagachidambaresan, ISBN: 9781119750420. Provides comprehensive coverage of renewable energy and its integration with smart grid technologies.

Encyclopedia of Renewable Energy, By James G. Speight, ISBN: 9781119363675. Written by a highly respected engineer and prolific author in the energy sector, this is the single most comprehensive, thorough, and up to date reference work on renewable energy.

Green Energy: Solar Energy, Photovoltaics, and Smart Cities, Edited by Suman Lata Tripathi and Sanjeevikumar Padmanaban, ISBN: 9781119760764. Covering the concepts and fundamentals of green energy, this volume, written and edited by a global team of experts, also goes into the practical applications that can be utilized across multiple industries, for both the engineer and the student.

Energy Storage, Edited By Umakanta Sahoo, ISBN: 9781119555513. Written and edited by a team of well-known and respected experts in the field, this new volume on energy storage presents the state-of-the-art developments and challenges in the field of renewable energy systems for sustainability and scalability for engineers, researchers, academicians, industry.

Energy Storage 2nd Edition, By Ralph Zito and Haleh Ardebili, ISBN: 9781119083597. A revision of the groundbreaking study of methods for storing energy on a massive scale to be used in wind, solar, and other renewable energy systems.

Hybrid Renewable Energy Systems, Edited by Umakanta Sahoo, ISBN: 9781119555575. Edited and written by some of the world's top experts in renewable energy, this is the most comprehensive and in-depth volume on hybrid renewable energy systems available, a must-have for any engineer, scientist, or student.

Progress in Solar Energy Technology and Applications, Edited by Umakanta Sahoo, ISBN: 9781119555605. This first volume in the new groundbreaking series, Advances in Renewable Energy, covers the latest concepts, trends, techniques, processes, and materials in solar energy, focusing on the state-of-the-art for the field and written by a group of world-renowned experts.

A Polygeneration Process Concept for Hybrid Solar and Biomass Power Plants: Simulation, Modeling, and Optimization, By Umakanta Sahoo, ISBN: 9781119536093. This is the most comprehensive and in-depth study of the theory and practical applications of a new and groundbreaking method for the energy industry to "go green" with renewable and alternative energy sources.

DESIGN AND DEVELOPMENT OF EFFICIENT ENERGY SYSTEMS, Edited by Suman Lata Tripathi, Dushyant Kumar Singh, Sanjeevikumar Padmanaban, and P. Raja, ISBN: 9781119761631. Covering the concepts and fundamentals of efficient energy systems, this volume, written and edited by a global team of experts, also goes into the practical applications that can be utilized across multiple industries, for both the engineer and the student.

INTELLIGENT RENEWABLE ENERGY SYSTEMS: Integrating Artificial Intelligence Techniques and Optimization Algorithms, Edited by Neeraj Priyadarshi, Akash Kumar Bhoi, Sanjeevikumar Padmanaban, S. Balamurugan, and Jens Bo Holm-Nielsen, ISBN: 9781119786276. This collection of papers on artificial intelligence and other methods for improving renewable energy systems, written by industry experts, is a reflection of the state of the art, a must-have for engineers, maintenance personnel, students, and anyone else wanting to stay abreast with current energy systems concepts and technology.

SMART CHARGING SOLUTIONS FOR HYBRID AND ELECTRIC VEHICLES, Edited by Sulabh Sachan, Sanjeevikumar Padmanaban, and Sanchari Deb, ISBN: 9781119768951. Written and edited by a team of experts in the field, this is the most comprehensive and up to date study of smart charging solutions for hybrid and electric vehicles for engineers, scientists, students, and other professionals.

Printed in the USA/Agawam, MA
September 16, 2026

910661.126